Fundamentals of Site Remediation

Fundamentals of Site Remediation

For Metal and Hydrocarbon-Contaminated Soils

Third Edition

John Pichtel

Lanham • Boulder • New York • London

Published by Bernan Press
An imprint of The Rowman & Littlefield Publishing Group, Inc.
4501 Forbes Boulevard, Suite 200, Lanham, Maryland 20706
www.rowman.com
800-462-6420

6 Tinworth Street, London SE11 5AL, United Kingdom

Copyright © 2019 by The Rowman & Littlefield Publishing Group, Inc.

All rights reserved. No part of this book may be reproduced in any form or by any electronic or mechanical means, including information storage and retrieval systems, without written permission from the publisher, except by a reviewer who may quote passages in a review. Bernan Press does not claim copyright in U.S. government information.

ISBN: 978-1-64143-313-6
E-ISBN: 978-1-64143-314-3

∞™ The paper used in this publication meets the minimum requirements of American National Standard for Information Sciences—Permanence of Paper for Printed Library Materials, ANSI/NISO Z39.48-1992.

Contents

Preface		vii
Acknowledgments		ix
1	How Did We Get Here? A Brief History	1
2	Chemistry of Common Contaminant Elements	23
3	Hydrocarbon Chemistry and Properties	51
4	Subsurface Properties and Remediation	79
5	Environmental Site Assessments	107
6	Isolation of the Contaminant Plume	151
7	Extraction Processes	171
8	Solidification/Stabilization	199
9	Soil Vapor Extraction	221
10	Permeable Reactive Barriers	241
11	Microbial Remediation	261
12	Phytoremediation	317
13	Innovative Technologies	345
14	Nanotechnology in Site Remediation	365
15	Technology Selection	385
16	Revegetation of the Completed Site	407

Acronyms and Abbreviations	435
Glossary	439
Index	453
About the Author	475

Preface

Since earliest recorded times, humans have manipulated their surroundings in order to extract and exploit its resources. In the modern era, resource extraction, processing, and utilization have soared. There has been a concurrent production of wastes, many of which are now classified as hazardous. As an unanticipated side effect of industrial development, nations are experiencing events of soil, groundwater, and surface water contamination on an unprecedented scale; many of these contaminants pose an immediate threat to public health and the environment.

In the United States, the urgency of addressing the land contamination issue was made evident in recent legislation. Section 121(b) of CERCLA mandates the U.S. Environmental Protection Agency (EPA) to encourage the use of remedies for site cleanup that "utilize permanent solutions and alternative technologies or resource recovery technologies to the maximum extent practicable" and to encourage the use of remedial actions in which treatment "permanently and significantly reduces the volume, toxicity, or mobility of hazardous substances, pollutants and contaminants as a principal element." Additionally, EPA is directed to "avoid off-site transport and disposal of untreated hazardous substances or contaminated materials when practicable treatment technologies exist." Emphasis is placed on treatment and destruction of the contaminants affecting a site; gone are the days when contaminated soil or other media are simply abandoned, or excavated and transferred to a new location.

Unfortunately, there is no standard set of 'cookbook' instructions which can be applied toward the remediation of a contaminated site, because each situation is unique and, typically, rather complex. Each site possesses its own distinct range of soil, geologic, and groundwater conditions, which influence the migration behavior of a contaminant plume. Additionally, each contamination situation has experienced unique release characteristics (e.g., a large accidental spill vs. deliberate dumping over many years) and composition of contaminants, which vary in terms of toxicity, volatility, mobility, and so on. Based upon these considerations, a substantial body of data is required in order to fully understand contaminant behavior and how to remove the contaminant plume efficiently while minimizing damage to the site and local environs.

Few references in environmental contamination and site restoration provide a satisfactory discussion of the underlying chemical processes inherent during a contamination episode; likewise, chemical processes involved in the remedial phase of activities are often incomplete. This book is intended to serve as an introductory manual for environmental site restoration practices as mandated by CERCLA and related statutes, with emphasis on basic environmental chemistry, soil science, microbiology, and plant science. Adequate knowledge of chemistry is an essential component of designing a cleanup activity, and a certain degree of proficiency is required in order to comprehend the reactions of contaminant(s) in soil and water, and to formulate appropriate solutions for an environmental contamination episode.

The first part of this book provides background for several salient aspects of site remediation—that is, chemical properties of metal and organic environmental contaminants; properties of soils as relates to remediation; and basics of environmental site assessments. The second portion presents field remediation technologies in some detail. Included are theory of operation, practical (field) considerations, and possible environmental impacts and other consequences of the use of these technologies. Most of the technologies addressed in this book are those commonly employed in remediation of NPL (Superfund), brownfield, and UST-affected sites. However, new and promising technologies are continually coming available; hence, several are introduced herein. This book is not intended to serve as an engineering manual—the reader is referred to several excellent books already available on this subject.

This third edition of *Fundamentals of Site Remediation* has been updated where necessary, with new case studies, updated literature and resources, and new art. In addition, I have added two new chapters: one on the use of nanomaterials for remediation of contaminated soil and water, and the other on amending and revegetating barren and toxic soil.

Acknowledgments

Many individuals have contributed in many ways in bringing this work to fruition.

Sincere gratitude is expressed to Jose Ramirez, Jason Higgs, and Valerie Morris for their expert preparation of the figures, and Vaishnavi Ganesh for her editing of the manuscript.

Special thanks to all those who provided their support during the formulation of this work: my parents Rose Gonzalez, Stan Pichtel, and Ed Gonzales; my wife Theresa; and my children Yozef and Leah. I express my sincere appreciation to Werner Erhard, who supported me in discovering countless avenues for producing quality results.

Finally, I offer deep gratitude to my students: your eagerness to question, to search for underlying mechanisms, and ultimately to discover, provided me with the motivation to pursue this project.

Chapter 1

How Did We Get Here?

A Brief History

If we do not change our direction, we are likely to end up where we are headed.

—Chinese proverb

1.1 INTRODUCTION

A pollutant can be defined as any chemical or physical change to soil, water, or air that adversely affects biota and/or materials. When we use such a definition, it follows that pollutants have been released to the biosphere since the earth's earliest days. Volcanic eruptions have released tons of mercury, hydrogen sulfide, hydrocarbons, and siliceous ash; forest fires have pumped carbon monoxide and particulate matter to the atmosphere; and soil and pollen have been widely dispersed. As humans began to exercise dominion over the earth, they also began to impact local, regional, and global ecosystems. However, that impact has been relatively insignificant for most of hominids' time on the planet. For example, early *Homo sapiens* are believed to be responsible for the extinction of several species of large mammals and for local damage to ecosystems (e.g., by deliberately setting large forest fires to drive out game animals). Much later, when ancient Romans discovered the thermally insulating properties of asbestos, this mineral was mined and used extensively. Likewise, the Romans mined and smelted thousands of tons of lead for uses ranging from military supplies to linings for wine bottles. (It has been suggested that lead exposure may have contributed to the fall of the Roman Empire.)

Not until the Industrial Revolution of the eighteenth century, however, were humans responsible for environmental impacts on a massive scale. Machine labor replaced human labor (figure 1.1) and populations surged to urban centers. Global population began its exponential climb, to which there is, as yet, no leveling off. The costs of this revolution included air and water pollution, respiratory and other ailments, dehumanization, and more.

By World War I, chemical manufacturing had become sophisticated and efficient. Some of the earliest synthetic polymers were developed. Chemical weapons such as sulfur mustard and phosgene gas were used on a large scale in the trenches of Belgium and France. By World War II, plastics were used extensively in warfare (e.g., in bomber nose cones and parachutes) and to a lesser degree in the home. Chlorinated organic pesticides began to replace conventional inorganic formulations. The insecticide DDT, dubbed "the atom bomb of the insect world" by Winston Churchill, saved millions of lives by controlling disease-carrying body lice and mosquitoes.

For the first half of the twentieth century, contaminants released to soil, water, and air caused isolated damage or else were absorbed into the biosphere (at least temporarily). As the manufacture and use of anthropogenic chemicals increased, however, so did the exposure, the doses received, and, ultimately, damage to public health and the environment.

Release of mercury into Japan's Minamata Bay and the Agano River in the 1950s and 1960s resulted in the occurrence of a debilitating disorder that came to be known as Minamata disease. Nervous system dysfunction including loss

Figure 1.1   The machine works of Richard Hartmann in Chemnitz, Germany, 1868.

of motor functions and hearing was severe. Infants born to exposed mothers were found to be severely palsied, with intellectual developmental disorders. The source of the mercury was determined to be indiscriminate waste dumping by local industry.

During the postwar boom and into the 1960s, technology and manufacturing soared, with a concurrent awakening of American environmental consciousness. Photochemical smog, an artifact of industrialization and lavish lifestyles (e.g., large, inefficient automobiles possessing little or no pollution control), became a regular summer occurrence in many cities (figure 1.2). Emissions from homes heated by coal, fuel oil, or wood also contributed to local air pollution. Household wastes were incinerated in urban apartment units with little or no air pollution control. A study of sediments in Central Park Lake, New York City, correlated the accumulation of lead, tin, and zinc with use of incinerators (Chillrud et al. 1999). Congress passed the Clean Air Act (CAA) in 1963 to regulate the production of air pollutants and to set emissions standards. The CAA has since evolved from a set of principles designed to guide states in controlling sources of air pollution (the 1967 Air Quality Act), to a body of detailed control requirements (the 1970, 1977, and

Figure 1.2 Smog event over New York City skyline, 1953. *Source*: Library of Congress.

1990 amendments to the CAA) that the federal government implements and states administer. The CAA of 1970 set specific goals for emission reductions and ambient air quality improvement. The National ambient air quality standards (NAAQS), considered the centerpiece of the act, were established primarily to protect public health and secondarily to protect animals, crops, and buildings. The NAAQS were established for six pollutants: sulfur dioxide, nitrogen oxides, particulate matter, carbon monoxide, ozone, and lead.

By the late 1960s to early 1970s, concerns were heightened with regard to the management of both domestic and toxic chemical wastes. The New York City waste haulers' strikes made city dwellers painfully aware of the direct, acute hazards associated with ordinary household solid waste as mountains of refuse quickly grew on city sidewalks. Many coastal cities had disposed of municipal solid waste (MSW) and sewage sludge by dumping from barges on or beyond the continental shelf. In addition, the dumping of industrial wastes to the land was carried out legally until 1970. Hazardous and radioactive materials were routinely dumped off both the east and west coasts.

No complete records exist of the types and volumes of materials disposed in ocean waters in the United States prior to 1972; however, several reports indicate the magnitude of historic ocean dumping (U.S. EPA 2018a):

- In 1968, the National Academy of Sciences estimated annual volumes of ocean dumping by vessel or pipes:
  - hundred million tons of petroleum products;
  - two to four million tons of acid chemical wastes from pulp mills;
  - more than 1 million tons of heavy metals in industrial wastes; and
  - more than 100,000 tons of organic chemical wastes.
- A 1970 Report to the president from the council on environmental quality on ocean dumping described that in 1968 the following were dumped in the ocean in the United States:
  - Thirty-eight million tons of dredged material (34% of which was polluted),
  - 4.5 million tons of industrial wastes,
  - 4.5 million tons of sewage sludge (significantly contaminated with heavy metals), and
  - 0.5 million tons of construction and demolition debris.
- EPA records indicate that more than 55,000 containers of radioactive waste were dumped at three sites in the Pacific Ocean between 1946 and 1970. Almost 34,000 containers of radioactive wastes were dumped at three ocean sites off the East Coast from 1951 to 1962.

With the first Earth Day (1970) came a collective call for more responsible environmental management. Recycling programs became popular in many communities. President Richard Nixon signed the National Evironmental Policy Act (NEPA) into law, paving the way for the establishment of the U.S. EPA.

The Clean Water Act (CWA) had its origins in the Federal Water Pollution Control Act (FWPCA), enacted by Congress in 1972. This statute required EPA to establish nationwide effluent standards on an industry-by-industry basis. The act, now commonly known as the Clean Water Act, established the National Pollutant Discharge Elimination System (NPDES) permit program, which was administered by individual states after federal authorization. EPA, in carrying out the 1972 Act, focused on control of conventional pollutants such as biological oxygen demand and suspended solids rather than toxic pollutants. The CWA developed a uniform, nationwide approach for the protection of the country's surface water by restricting discharges of pollutants (chemical, physical, and thermal) into navigable waters. The act included a permit program, national effluent limitations, water quality standards, special provisions for oil and toxic substance spills, and grant programs for the construction of publicly owned treatment works (POTW). This framework remains in effect.

The Safe Drinking Water Act (SDWA), enacted in 1974, authorized EPA to regulate contaminants in public drinking water systems. The act greatly expanded federal authority over drinking water quality. EPA was given the responsibility of setting standards for levels of contaminants in public drinking water systems and regulating underground injection wells. The regulations established maximum contaminant levels (MCLs) and monitoring requirements for heavy metals, certain pesticides, bacteriological contaminants, fluorides, and turbidity.

In 1990, primarily in response to the *Exxon Valdez* oil spill, Congress overhauled the oil spill provisions of the CWA, thus creating the Oil Pollution Act (OPA) of 1990. Issues regarding tanker safety and oil pollution liability had been debated in Congress during the previous decades, and the act brought about rapid and sweeping changes in the oil production and transportation industry. Its stringent requirements resulted in the restructuring of the industry, created immediate demand for oil spill prevention and response technology, and catalyzed the establishment of numerous requirements at the federal, state, and local levels.

Legislation addressing the management of solid wastes dates back to the Rivers and Harbors Act of 1899. The Solid Waste Disposal Act of 1965 and the Resource Recovery Act of 1970 were designed to address management of MSW. Neither act, however, established effective regulations; rather, guidelines were offered and incentives (i.e., grant programs) were made available to municipalities. By 1976, the Resource Conservation and Recovery Act (RCRA) was enacted to address the responsible management of both solid (i.e., municipal) and hazardous wastes. The most 'muscle,' in terms of regulatory requirements, was included in the latter. Subtitle C is the centerpiece of RCRA as regards hazardous waste management. Management of hazardous wastes was to be "from cradle to grave," meaning that the waste, once produced by a generator, must be tracked throughout its entire journey to its final destination, either a designated treatment or a disposal facility. This requirement resulted in a paper trail known as the *manifest system*. RCRA requirements include the identification and quantification of hazardous wastes. Extensive requirements for generators, transporters, and treatment, storage, and disposal (TSD) facilities were established.

RCRA has been amended several times since its enactment, most importantly by the Hazardous and Solid Waste Amendments of 1984 (HSWA). The HSWA mandated far-reaching changes to the RCRA program. Requirements for waste generators varied significantly, depending on the monthly quantity of waste produced. EPA was required to regulate another 200,000 companies that produce relatively small quantities of hazardous waste. Stringent requirements were established for hazardous waste incineration systems; landfill design, construction, and operation; waste minimization; and a national land disposal ban program.

By the late 1980s underground storage tanks (USTs) were recognized as a source of soil and groundwater contamination at thousands of sites throughout the United States. USTs have contaminated the subsurface via several avenues. Some tanks and piping simply corroded or structurally failed during years of use. In other cases, spills occurred when tanks were emptied or

Figure 1.3 Leaking underground storage tanks removed from abandoned petroleum refinery.

when they overflowed during filling (figure 1.3). As of 2018, approximately 553,000 USTs were documented nationwide to store petroleum or hazardous substances (U.S. EPA 2018d). In 1988 EPA estimated that roughly 75 percent of UST systems posed a significant potential for leakage and environmental harm because the systems were constructed of steel without corrosion protection. Since then a comprehensive effort was enacted to register USTs, assess sites suspected of leaks, remove old and leaking tanks, and clean up contaminated soil and groundwater.

The Emergency Planning and Community Right-to-Know Act (EPCRA) of 1986 was initiated as a result of the Bhopal, India, incident of December 2–3, 1984. Methyl isocyanate, a potent toxin used in certain pesticides and polymers, was released from a Union Carbide facility, killing more than 3,000 local residents and injuring over 550,000 more. A year later, a similar release occurred from another Union Carbide facility in Institute, West Virginia. Casualties in the latter event were, however, low. EPCRA was a rider attached to SARA (The Superfund Amendments and Reauthorization Act) and authorized a nationwide program of emergency planning and reporting as protection against accidents involving extremely hazardous chemicals.

The Toxic Substances Control Act (TSCA) was enacted in 1976, in large part due to the discovery of extensive soil, water, and structural contamination by polychlorinated biphenyls (PCBs) and no coherent regulations to control PCB-containing material. TSCA authorizes EPA to: (1) obtain data from industry regarding the production, use, and health effects of chemical substances and mixtures, and (2) regulate the manufacture, processing, and distribution in commerce, as well as use and disposal of a chemical substance or mixture (U.S. EPA 2018b; U.S. GAO 2005). The act has been amended three times, each amendment resulting in an additional title. TSCA now contains four titles: I—Control of Toxic Substances; II—Asbestos Hazard Emergency Response Act; III—Indoor Radon Abatement Act; and IV—Lead-Based Paint Exposure Reduction Act.

By the late 1970s numerous accounts of chemical 'time bombs' were reported, revealing that toxic chemicals had been carelessly buried, often from unknown sources. One of the most notorious scenarios occurred in Niagara Falls, New York. In the 1940s and 1950s, the Hooker Chemical Company had used a number of sites within the city for disposal of more than 100,000 tons of hazardous petrochemical wastes. The wastes were placed in the abandoned Love Canal, in a large unlined pit on Hooker's property, and in other dumps throughout the city (figure 1.4). At that time, no comprehensive legislation addressing hazardous waste disposal existed in the United States. As a result, such wastes were frequently placed directly into the ground without pretreatment. An internal Hooker memorandum warned company officials: "Do not dig anywhere near this site . . . may be thousands of buried

Figure 1.4 Half-buried and leaking hazardous waste containment tanks in Love Canal neighborhood of abandoned homes. *Source*: Courtesy, University Archives, University at Buffalo, The State University of New York.

drums and tanks . . . might have fire or reaction if we dig up this junk." Portions of the affected land were later sold.

By the mid-1970s chemicals had migrated from their disposal sites. Land was subsiding in areas where containers deteriorated; noxious fumes were detected in homes; and liquids seeped into basements, surface soil, and water. The incidences of cancer, certain birth defects, and psychological problems were all well above the national average. President Jimmy Carter declared a public health emergency for the Love Canal site. Homes directly adjacent to the Love Canal were purchased by the state of New York and residents were evacuated. Numerous lawsuits were brought against Hooker Chemical, by both the U.S. government and local citizens. Issues of liability and compensation were tied up in the court system for years. Some of the difficulties related to settlement were the fact that Hooker signed a disclaimer in 1952, stating that the company assumed no liability in the event of injury to persons or loss of property resulting from the land transaction. Furthermore, there was simply no law covering the assignment of liability to responsible parties in the event of land contamination. This and similar incidents—for example, Times Beach, Missouri; Valley of the Drums, Kentucky (figure 1.5); and Kin-Buc Landfill, New Jersey (figure 1.6)—became potent catalysts for the Superfund legislation.

Figure 1.5 Valley of the Drums, Kentucky. *Source*: U.S. Environmental Protection Agency.

Figure 1.6 Kin-Buc landfill, New Jersey. *Source*: NOAA Office of Response and Restoration.

An estimated 36,000 severely contaminated sites exist in the United States (Wehr and Robbins 2011); however, certain federal agencies claim the number to be much higher (figure 1.7). With reference to current regulatory standards, an estimated 294,000 sites (range 235,000–355,000) require some form of remediation; this does not include sites with completed or ongoing remediation projects. Contaminated sites can be divided into seven groups based upon regulatory and decontamination responsibility: Superfund, Resource Conservation and Recovery Act (RCRA) Corrective Action, Underground Storage

Figure 1.7 Drums of hazardous wastes at this facility have been stockpiled improperly over long periods, resulting in severe soil contamination. *Source*: U.S. Environmental Protection Agency.

Tanks (USTs), Department of Defense (DOD), Department of Energy (DOE), Civilian Federal Agencies, and State Cleanup (U.S. EPA 2004).

The NPL or 'Superfund List' is the compendium of sites of national priority among the known releases or threatened releases of hazardous substances throughout the United States. The NPL is intended primarily to guide EPA in determining which sites warrant further investigation. As of early 2018 a total of 1,342 sites occur on the Superfund list (NPL): 1,185 nonfederal and 157 federal, with an additional 49 proposed. Sites occur on private property, company property, and many types of federal land (DOD, DOE, etc.). Beyond the federal list, states and cities possess their own lists of priority sites for cleanup.

1.2 BROWNFIELDS

EPA Region 5 defines brownfields as "abandoned, idled, or underused industrial and commercial sites where expansion or redevelopment is complicated by real or perceived environmental contamination that can add cost, time, or uncertainty to a redevelopment project." The U.S. Office of Technology Assessment definition includes a site whose redevelopment may be hindered not only by potential contamination but also by poor location, old or obsolete infrastructure, or other less tangible factors often linked to neighborhood decline (figure 1.8).

(a)

(b)

Figure 1.8 Brownfields range in size from the abandoned corner gasoline station (a) to large industrial facilities (b).

Brownfields commonly are associated with distressed urban areas, particularly central cities and inner suburbs that once were heavily industrialized but subsequently were vacated. A small percentage of brownfield sites may be contaminated to the degree that they are candidates for the NPL under CERCLA. It is estimated that there are more than 450,000 brownfields in the United States (U.S. EPA 2017a).

The stigmatic impacts of brownfields on communities are manifold. Potential investors, concerned about liability, have avoided developing abandoned industrial sites. Real estate buyers are reluctant to invest in brownfields, thus further diminishing site value. Communities ultimately lose out on property tax revenues. Many states, eager to boost local economies, are seeking to revitalize brownfields by providing economic incentives for their assessment and redevelopment.

1.3 CERCLA AND THE SUPERFUND

CERCLA was enacted to address contamination from past disposal activities. The act established a federal cleanup program (the Superfund) to finance the cleanup of contaminated properties and set guidelines for their restoration; it also established a system of assigning legal and financial liability for responsible parties. The Superfund now totals over $3.3 billion (U.S. EPA 2017b). Monies to create the fund originally were based primarily on a tax on industries ("the polluter pays" strategy), and a portion was derived from individual income taxes. Since the Superfund taxes expired in 1995, however, the burden of paying the costs shifted dramatically. Today the majority of the program's funding comes through taxpayer dollars (Anderson 2017). Regardless of the source of funds, however, cleanup costs for uncontrolled and abandoned U.S. hazardous waste sites are estimated in the hundreds of billions of dollars (U.S. GAO 2015). Because the Superfund cannot remediate all sites, EPA has formulated a system for determination of potentially responsible parties (PRPs).

1.3.1 Liability

With the enactment of CERCLA, there is no longer any question as to liability for a site that is contaminated as a result of past improper waste disposal. CERCLA establishes four classes of PRPs liable for environmental investigation and cleanup costs:

- the current owner or operator of a facility;
- person(s) who owned or operated a facility at the time of disposal of any hazardous substances;

- person(s) who arranged for disposal or treatment of hazardous substances owned by such person(s);
- person(s) who transported hazardous substances to treatment or disposal facilities.

CERCLA declares that these parties are liable for all costs of removal or remedial action incurred by the U.S. government. They are also responsible for damages for injury to, destruction of, or loss of natural resources, including the costs for assessing them. PRPs are also liable for costs of any health assessments conducted.

Liability under CERCLA is joint and several, which essentially means that all PRPs, at the outset at least, hold an equal share of liability. This relieves the government of proving the individual contributions of each PRP to the site. Courts have experienced difficulty in determining the degree of environmental harm caused by each PRP at a site where wastes having varying toxicities commingle. CERCLA allows a PRP that is liable to the government for cleanup and investigation costs to seek contribution from other PRPs. The CERCLA liability procedure was adopted by Congress to promote prompt cost recovery and equitable allocation of liability. However, the process has resulted in great complexity, which causes delay, increases transaction costs, and results in inequitable allocations of liability. The EPA policy addresses cleanup cost allocation on a case-by-case basis, thereby maintaining maximum flexibility to manage site restoration (U.S. EPA 2018c).

Throughout the history of Superfund, PRPs have searched for means to distribute the costs of cleanup as broadly as possible. Until recently, PRPs focused their efforts on other industrial PRPs and their insurance companies. They have also focused on expanding Superfund costs to U.S. taxpayers via local governments. Approximately 20 percent of sites listed on the NPL are municipal, a category that includes sites owned or operated by municipal governments, as well as privately owned properties that routinely accept MSW for disposal. The majority of those sites are landfills that have become contaminated by codisposal of industrial hazardous waste with MSW. The MSW shipped to landfills is typically of low toxicity; hazardous substances constitute less than 0.5 percent of the materials contained in ordinary MSW. However, under Superfund case law, wastes are included within the liability framework if they contain even modest quantities of hazardous materials. Industrial PRPs sued by the government have decided to sue for contribution cities whose citizens produced MSW that was brought to the site.

1.3.2 The Innocent Purchaser Defense

If a purchaser of land found subsequently to be severely contaminated can show that "due diligence" was exercised in a property assessment prior to the transaction, their liability may be reduced. Due diligence involves an appropriate investigation in previous ownership and uses of the property. Such an investigation often takes the form of an environmental site assessment; however, there is no single standardized method for conducting such an investigation. The environmental assessment process is outlined in chapter 5.

1.3.3 Removal Action versus Remediation

If a site is contaminated to the point of being placed on the NPL, a CERCLA-authorized action is conducted. Such an action is designated as either a short-term removal action or a long-term remedial response. A removal action involves cleanup or other actions taken in response to emergency conditions, an immediate threat to human health and the environment, or interim actions on a short-term basis. For the long term, remedial action is designed to stop or reduce a significant release of hazardous substances.

1.3.3.1 Removal Action

Examples of removal actions include installation of fences, waste segregation, evacuation of threatened populations, and construction of temporary containment systems (figure 1.9). The following factors are considered in determining whether to order a removal action (U.S. EPA 2009):

- exposure of populations, animals, or food chains to contaminants;
- contamination of drinking water supplies or sensitive ecosystems;
- hazardous substances in storage that pose a threat of release;
- high levels of hazardous substances at or near the soil surface that may migrate;
- weather conditions that may cause hazardous substances to be released or migrate;
- threat of fire or explosion;
- availability of other federal or state response mechanisms to respond to the release.

1.3.3.2 Remedial Action

The remedial action is a lengthy and meticulous operation whose purpose is to permanently remove, destroy, or isolate the hazard at the site. The key elements of the CERCLA remedial action process include:

- site discovery
- preliminary investigation

Figure 1.9 Workers assess the status of a cleanup at a Pennsylvania Superfund Site contaminated with heavy metals, sulfuric acid, and crude oil waste. *Source*: U.S. Army Corps of Engineers.

- site investigation
- hazard ranking system analysis
- listing on the NPL
- remedial investigation
- feasibility study
- remedy selection
- Record of Decision
- remedial design/remedial action

Remedial designs range from very simple to highly sophisticated. Because the characteristics, history, and complexity of contamination are unique at each site, designs are prepared on a case-by-case basis. Decisions on the specific remediation method are influenced by technical considerations, degree of present hazard of the site to local populations and ecosystems, and political considerations.

As noted earlier, a total of 1,342 hazardous waste sites in the United States have been assigned the highest priority for cleanup. Since the enactment of CERCLA in 1980, a total of 343 of these sites have been delisted (remediated). In some cases, this means that contaminants were neutralized, detoxified, or removed. In other cases it may simply mean that the site has been covered with an impermeable cap and is subjected to long-term monitoring of groundwater and soil.

EPA has placed great emphasis on the application of innovative technologies for site remediation. The Superfund Amendments and Reauthorization Act (SARA) requires the use of remedial technologies that permanently and significantly reduce the volume, toxicity, or mobility of contaminated materials at affected sites. In recent years the number of Superfund Records of Decision for affected sites has been greatest for novel techniques as compared with conventional technologies.

1.4 "IT AIN'T OVER 'TIL IT'S OVER"—YOGI BERRA

Since Earth Day 1970, the U.S. government and numerous state governments have enacted stringent, comprehensive regulations with far-reaching benefits in terms of protection of public health, soil, water, air, and ecosystems. Some states have made much greater strides than others.

Contaminated sites continue to be discovered throughout the United States. Additionally, illegal and improper waste disposal continues as unscrupulous owners, operators, and employees seek the lowest-cost methods for waste disposal. The following abbreviated list presents some recent news reports:

"Guilty pleas in illegal shipping of hazardous waste," Associated Press, June 19, 2018.

Three people have pleaded guilty in federal court to illegally transporting 9 million pounds of hazardous waste from Mississippi to Missouri without acquiring the proper permits. Two individuals also pleaded guilty to placing a person in imminent danger by releasing hazardous waste. The company made plastic pellets that are used to remove paint from tanks and other equipment on military bases. A second company was supposed to recycle the powdery waste, which is often contaminated with cadmium, chromium, and lead; however, the waste was not recycled. The owner of the recycling company later pleaded guilty to dumping the waste.

"Plan for 600 homes near Erie stalled when lost chemical waste dump was found nearby," *The Denver Post*, **February 12, 2018.**

A real estate developer had planned to clean up an old landfill and build 600 single-family homes. However, an environmental site assessment of the 300 acres revealed 84,120 gallons of chemical waste, legally buried, but "lost" over time. More than a thousand drums are thought to have been entombed in the landscape. Back in the 1960s, the waste—most of it the by-product of magnetic tape and developer produced by IBM—was legally dumped, burned, or buried at the landfill, according to a 1990 assessment of the area by EPA. But the disposal was not executed adequately. The landfill's permit was revoked by Weld County in 1969 because of poor operating practices. The waste being unearthed consists mainly of industrial solvents. Workers have found trichloroethylene, a known carcinogen, and 2-butanone, which has been reported as causing neuropsychological effects. Drums containing liquid waste will be taken to an incinerator in Kimball, Nebraska.

"US military base in Incheon found to have severe soil contamination," *The Hankyoreh*, **October 29, 2017.**

In South Korea, a U.S. military base, Camp Market, is set to be returned to the South Korean government due to extremely high levels of soil contamination. The article states that the soil at the base is contaminated with dioxins, heavy metals, and PCBs. Groundwater at the base is also contaminated with trichloroethylene and total petroleum hydrocarbons. The elevated levels of soil contaminants at the base are most likely linked to its past as a military vehicle reutilization center and as a waste handling facility. The U.S. military says it does not want to conduct a remediation effort at the former base due to potential health risks to human life. The South Korean government is requesting that the U.S. government be held liable for the cleanup of the former base even after it is formerly transferred to the South Korean government. The U.S. government has withheld information of contamination in the past with other bases that have been transferred to the South Korean government.

"EPA cleanup of Pillsbury site nearly complete; lawsuits pending." *The State Journal-Register*, **October 31, 2017.**

The EPA ordered remediation of Pillsbury Mills, a former food processing facility, due to the presence of asbestos-containing material (ACM), laboratory chemicals, and universal wastes. The cleanup occurred over nine months and cost $1.8 million. The ACMs were construction debris and pipe-wrap and boiler insulation materials. Approximately 2,200 tons of asbestos-contaminated debris and 1,160 cubic yards of bulk asbestos were transported to hazardous material landfills. One of the associates of Pillsbury Mills is facing federal charges for improperly removing the asbestos when originally ordered to.

"Group Wants Radiation Testing at Pruitt-Igoe Site," CBS St. Louis, **November 13, 2017.**

In St. Louis, Missouri, an environmental group is asking that clearing and demolition work be halted at an old housing complex that was potentially contaminated with radiation. The group believes that the housing development was contaminated with radioactive cadmium sulfide during a Cold War experiment more than forty years ago. They warn that unknown amounts of radioactive material could be released to surrounding residential areas during the site cleanup. The group's director is requesting that a Health Hazard Evaluation be carried out before work continues. St Louis city officials state that the EPA has already carried out an evaluation, and found the site to be within the safe limits for clearing and demolition work.

"**Colorado landfills are illegally burying low-level radioactive waste from oil and gas industry, *Denver Post* learns,**" *Denver Post*, **September 22, 2017.**

A loophole in Colorado state environmental laws had allowed landfills within the state to accept and dispose of low-level radioactive materials for years. This practice was discovered and put to an end in September of this year. There are only two landfills within the state that are properly equipped and designed to accept radioactive materials. The Colorado Public Health Department is working to close the loophole in the legislation in order to protect the health of residents. The health department has no estimate as to how much radioactive waste has been improperly disposed over the years.

"**EPA Inspection Reveals Hazardous Waste Violations at Diamond-Vogel Paint Company in Burlington, Iowa,**" **U.S. EPA News Releases, February 1, 2016.**

An EPA compliance evaluation inspection revealed violations of the Resource Conservation and Recovery Act (RCRA) related to the storage and handling of hazardous waste. In a settlement filed by EPA in Lenexa, Kansas, the company will pay a $21,700 civil penalty to the United States. One of Diamond's facilities was improperly storing approximately 9,625 gal of waste paint materials. The hazardous waste containers were left open and many were not labeled. According to EPA, weekly inspections were not carried out in Diamond's hazardous waste storage area. Diamond additionally was cited for failing to provide a safe work environment for its workers by mishandling these materials. Diamond Vogel has been ordered to send the EPA photo evidence that they are now storing the hazardous waste chemicals properly. The company was also ordered to submit an updated contingency plan identifying the secondary emergency coordinators, and documentation showing regular inspections of the hazardous waste areas.

"**State: Exide mishandled hazardous waste in Frisco.**" *The Dallas Morning News*, **October 1, 2013.**

State regulators allege that Exide Technologies mishandled and improperly treated hazardous waste for years at its now-closed plant in Frisco, Texas. The ten violations cited in a notice of enforcement are based on an investigation by the Texas Commission on environmental quality. A key finding was that waste buried in the closed portions of Exide's landfill contains hazardous levels of lead. The TCEQ officials said that the alleged violations date to 1998. The company also failed to adequately test for cadmium before sending about 3,388 tons of treated waste from the Frisco plant to a disposal facility between May 2012 and January 2013. Other violations involve hazardous waste piles stored alongside its landfill and numerous two-ton sacks of waste stored in areas of the plant not permitted for hazardous waste.

"Wal-Mart pleads guilty to dumping hazardous waste." *USA Today*, **May 28, 2013.**

In May of 2013, Walmart agreed to pay over $110 million for violating the CWA by illegally dumping hazardous materials into local trash bins and sewer systems in California and improperly handling pesticides in Missouri. The company also violated a number of state environmental laws. Walmart did not have a training program regarding how to handle and dispose hazardous waste. A similar incident occurred in 2010 in which the company had to pay over $25 million. In both instances, the materials were either dumped in sewage systems or transported to facilities improperly. A training program is now in place for all employees.

QUESTIONS

1. Which act promulgated by the U.S. Congress created the "innocent landowner defense" for a property buyer who is concerned about environmental issues, as a means of protection? Under this act the purchaser may attempt to show "due diligence" in conducting a property investigation.
2. Which of the following is correct? PRPs under a CERCLA situation are as follows: (a) owner or operator of the facility; (b) owner of the hazardous storage facility; (c) the hazardous waste generator; (d) transporter of the hazardous substance; (e) all of the above.
3. In assessing cleanup costs among multiple PRPs at an EPA-funded cleanup, how does the EPA apportion costs?
4. A primary goal of CERCLA is to establish a mechanism to respond to releases of hazardous substances at uncontrolled waste sites. True or false?
5. CERCLA liability does not necessarily involve proving negligence. True or false? Explain.
6. Which of the following can be PRPs at a CERCLA-contaminated site? (a) operator of the disposal facility; (b) generator of the hazardous substances; (c) the local regulatory agency; (d) owner of the disposal facility; (e) transporter of the hazardous substances.
7. Search the brownfields database in your community. How many sites are listed? What are the primary industry types (e.g., heavy industry, petrochemical, manufacturing, gasoline stations)? What types of contaminants occur?
8. Do NPL sites exist in your county or state? Check the EPA website for NPL sites and determine: (1) site history, including production activities; (2) waste types generated; and (3) Record of Decision. Have the sites been delisted?

9. CERCLA liability is strict, joint and several, and retroactive. Define and/or explain these three components of liability. You may want to check the EPA website, www.epa.gov/superfund/ or some of the references listed in this chapter.

REFERENCES

Anderson, B. 2017. Taxpayer dollars fund most oversight and cleanup costs at Superfund sites. *The Washington Post*. https://www.washingtonpost.com/national/taxpayer-dollars-fund-most-oversight-and-cleanup-costs-at-superfund-sites/2017/09/20/aedcd426-8209-11e7-902a-2a9f2d808496_story.html?utm_term=.00304b3e1b53 (Accessed January 24, 2018).
Butler, J. C. III, M. W. Schneider, G. R. Hall, and M. E. Burton. 1993. Allocating Superfund costs: Cleaning up the controversy. *Environmental Law Reporter* 23:10133–44.
Chillrud, S. N., R. F. Bopp, H. J. Simpson, J. M. Ross, E. L. Shuster, D. A. Chaky, D. C. Walsh, C. C. Choy, L. Tolley, and A. Yarme. 1999. Twentieth-century atmospheric metal fluxes into Central Park Lake, New York City. *Environmental Science & Technology* 33:657–61.
Hess, K. 2007. *Environmental Site Assessments, Phase 1: A Basic Guide*, 3rd ed. Boca Raton, FL: CRC Press.
Nardi, K. J. 2016. Underground storage tanks. In *Environmental Law Handbook*, 23rd ed., edited by Christopher Bell, et al., pp. 201–250. Lanham, MD: Bernan Press.
Rockwood, L. L., and J. L. Harrison. 1993. The Alcan decisions: Causation through the back door. *Environmental Law Reporter* 23:10542–45.
Steinzor, R. I., and M. F. Lintner. 1992. Should taxpayers pay the cost of Superfund? *Environmental Law Reporter* 22:10089–90.
U.S. EPA (Environmental Protection Agency). 2018a. *Learn about Ocean Dumping*. https://www.epa.gov/ocean-dumping/learn-about-ocean-dumping#Before.
———. 2018b. *Summary of the Toxic Substances Control Act*. https://www.epa.gov/laws-regulations/summary-toxic-substances-control-act.
———. 2018c. *Superfund Enforcement: 35 Years of Protecting Communities and the Environment*. https://www.epa.gov/enforcement/superfund-enforcement-35-years-protecting-communities-and-environment#pay (Accessed January 24, 2018).
———. 2018d. *Underground Storage Tanks (USTs)*. https://www.epa.gov/ust (Accessed October 21, 2018).
———. 2018e. *Emergency Planning and Community Right-to-Know Act (EPCRA)*. https://www.epa.gov/epcra (Accessed April 25, 2019).
———. 2017a. *Overview of the Brownfields Program*. https://www.epa.gov/brownfields/overview-brownfields-program (Accessed November 20, 2017).
———. 2017b. *Superfund Special Accounts*. https://www.epa.gov/enforcement/superfund-special-accounts (Accessed November 20, 2017).

———. 2004. *Cleaning Up the Nations Waste Sites: Markets and Technology Trends*. EPA 542-R-04-015. Technology Innovation and Field Services Division. Office of Solid Waste and Emergency Response. http://www.epa.gov/tio/download/market/2004market.pdf.

———. 2009. *Superfund Removal Guidance for Preparing Action Memoranda*. Washington, DC: Office of Emergency Management.

U.S. GAO (Government Accountability Office). 2015. *Hazardous Waste Cleanup. Numbers of Contaminated Federal Sites, Estimated Costs, and EPA's Oversight Role*. Testimony Before the Subcommittee on Environment and the Economy, Committee on Energy and Commerce, House of Representatives. GAO-15-830T. Washington, D.C.

———. 2005. *Chemical Regulation: Approaches in the United States, Canada, and the European Union*. GAO-06-217R, November 4, 2005. www.gao.gov/docsearch/abstract.php?rptno=GAO-06-217R.

Wehr, K., and P. Robbins. 2011. *Green Culture*. Los Angeles: Sage Publishing.

Chapter 2

Chemistry of Common Contaminant Elements

Life should be like the precious metals, weigh much in little bulk.

—Seneca

2.1 INTRODUCTION

Thousands of sites in industrialized nations are contaminated with toxic metals such as cadmium, chromium, mercury, and lead. Certain metals, when concentrated in soil, pose a threat to public health and the environment. Metals enter the biosphere from disposal of wastes in landfills or uncontrolled dumps; from fugitive dusts and atmospheric fallout; and from spills and leaks in factories, mines, distribution facilities, or during transportation. Metals can become attached to particulates (e.g., soil, dust, fly ash) and carried in air currents; leached through the soil profile; carried away in surface runoff; and accumulate in plants to the point at which they become phytotoxic and/or are translocated through the food chain. Metal contamination of soils by anthropogenic inputs has been discovered in industrial areas, mine sites, and urban areas, and near metal smelters in waste disposal areas and along roadsides (figure 2.1).

As part of a comprehensive environmental assessment and remediation program for a metal-affected site, soil testing must be conducted in order to determine total concentrations of metals and, ideally, the chemical forms (i.e., species) present. It is also important to measure background metal levels for comparison and to assess the degree of contamination. Soils are composed of mineral and organic constituents that vary in composition across the land

Figure 2.1 Abandoned metal plating facility. Soil at the site is heavily contaminated with chromium.

surface as well as downward through the profile. It is, therefore, a challenge to establish a minimum concentration below which a soil may be considered uncontaminated by a particular pollutant. Average concentrations of trace metallic contaminants in soils are given in table 2.1.

The mobility of trace elements in soils is controlled by a range of chemical and physical phenomena. For example, finer-sized soil fractions (e.g., clays, silts, Fe oxides, organic matter) bind metals by cation exchange and specific adsorption. For contamination situations in which metal contamination is very high—that is, thousands of mg/kg—the sorption capacity of soils is exceeded, and the contamination will be present as discrete metal-mineral phases (Kabata-Pendias and Szteke 2015; Davis and Singh 1995). Metal ions are immobilized in these soils by the formation of insoluble precipitates, incorporation into the crystalline structure of clays and metal oxides, and/or by physical entrapment in the immobile water within soil pores. Conversely, metals can become mobilized under acidic pH regimes, under reducing conditions and in high salt content soils. The significance of these properties is discussed below and in subsequent chapters.

In this chapter, the chemical properties and reactions of several common metallic and semimetallic contaminants are presented, with emphasis on mobility/immobility, toxicity, transformations in the biosphere, and remediation practices.

Table 2.1 Distribution of trace elements in world soils

Element	Common Range (mg/kg)	Average
Arsenic (As)	1–65	5
Cadmium (Cd)	0.01–0.70	0.06
Chromium (Cr)	1–1,100	100
Lead (Pb)	2–200	10
Mercury (Hg)	0.01–0.3	0.03
Barium (Ba)	100–3,000	430
Boron (B)	2–100	10
Copper (Cu)	2–140	30
Manganese (Mn)	20–3,000	600
Nickel (Ni)	5–500	40
Selenium (Se)	0.1–2	0.3
Silver (Ag)	0.01–5	0.05
Tin (Sn)	2–200	10
Zinc (Zn)	10–770	50

Source: Hooda 2010; Kabata-Pendias 2010; U.S. Environmental Protection Agency 1995.

2.2 THE ELEMENTS

2.2.1 Arsenic

Based on its position in the Periodic Table, arsenic (As) is a semimetal, or metalloid, and possesses attributes of both metals and nonmetals. Arsenic is widely distributed throughout the biosphere, occurring in over 200 mineral species including native As, sulfides, sulfosalts, and oxides. As with many of the minerals discussed in this chapter, arsenic is highly chalcophilic, which means that it can crystallize in a reducing environment to form sulfide species. Significant anthropogenic sources of As to the biosphere include metal processing, including smelting of copper, lead, and zinc ores; processing of sulfur and phosphorus minerals; coal combustion; land application of contaminated biosolids; and spraying of As-containing pesticides (Wenzel 2013; Kabata-Pendias 2010). The worldwide range of As in natural soils is < 1 to 95 mg/kg; however, the As content of contaminated soils has been reported to be as high as 2,000 mg/kg (Kabata-Pendias and Szteke 2015).

Arsenic is present in contaminated sites as As_2O_3 or its derivative compounds. Arsenic also may occur as methylarsinic acid, $H_2AsO_3CH_3$, an ingredient in many pesticides, as well as arsine, AsH_3, and its methyl derivatives including dimethylarsine $As(CH_3)_2$. The chemistry of As in soil, sediments, and water is somewhat complex. Its oxidation states include −3, 0, +3, and +5. The As(V) compounds predominate in aerobic soils and sediments; As(III) compounds occur in slightly reduced soils; and arsine, methylated arsines, and elemental As predominate under very reduced conditions. Under most conditions As(V) is present as the $H_2AsO_4^-$ species, while As(III) as the H_3AsO_3 species is dominant in low pH and reducing environments (Wenzel 2013; Crecelius et al. 1986). The anions AsO_2^-, AsO_4^{3-}, $HAsO_4^{2-}$, and $H_2AsO_3^-$ are the most common mobile forms of As. Arsine and methylated derivatives are highly volatile and will vaporize after formation.

Arsenic compounds in soils, sediments, and water undergo complex reactions including oxidation-reduction, ligand exchange, and biotransformation. These processes are affected by pH, salinity, oxidation-reduction status, types and amounts of anions present, clay and hydrous oxide content, sulfide concentration, and populations of microorganisms. The various biological and chemical transformations that control the mobility and availability of As are shown in figure 2.2.

Many As compounds adsorb strongly to soils and sediments, specifically to clays, oxides, hydroxides, and organic matter; therefore, leaching is minimal (Wenzel 2013). In soil, arsenates (AsO_4^{3-}) behave similarly to phosphates; that is, they are sorbed by Fe and Al oxides, aluminosilicates, and layer silicate clays. Arsenates are also readily fixed by phosphate gels, Ca, and humus. The most strongly sorbing solids for As, however, are Fe and Al oxides.

Several plant species, including those established on mine wastes and other As-contaminated media, have been found to tolerate high levels of tissue As. Many plants take up soil As passively—a fairly linear relationship between soil As concentration and plant uptake of As has been reported. Plants are also capable of translocating root As to aboveground biomass. Toxicity of As to plants depends on the concentration of soluble As in soil; however, plant response may also be related to total soil As and additionally to soil properties. For example, significant growth reduction may occur with 1,000 mg/kg As in clayey soil; however, in a sandy soil 100 mg/kg As is equally toxic (Kabata-Pendias 2010). Arsenic toxicity has been noted in plants growing on mine spoil, on soils treated with arsenical pesticides, and on soils amended

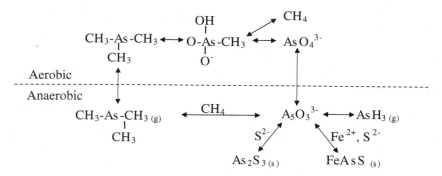

Figure 2.2 Transformations of arsenic in soil and sediment.

with biosolids (i.e., sewage sludge). Symptoms of As toxicity include leaf wilting, root discoloration, and cell plasmolysis. The most common symptom, however, is growth reduction.

Depending on soil properties and pollution source, plants may accumulate extremely large quantities of As. Arsenic concentrations are also dependent upon species/cultivar/ecotype (Fitz and Wenzel 2006). On polluted sites plants may accumulate several thousand mg/kg As in shoot dry matter and more than 10,000 mg/kg in roots (Fitz and Wenzel 2002). Kabata-Pendias (2010) reported tissue As levels in some plants as high as 6,000 mg/kg dry weight. Arsenic poisoning from plants to animals is relatively uncommon; however, negative health effects of high As concentration in crop and forage plants are possible.

The toxicity of As in soil may be overcome by several approaches, depending on the source of As and soil properties. In flooded soils, for example rice paddies, increasing the oxidation state (by draining) limits As bioavailability (Kabata-Pendias 2010; U.S. HHS 1992; Hanada et al. 1975). Application of amendments that precipitate As in soil, for example, ferrous sulfate and calcium carbonate, can also be employed effectively.

2.2.2 Barium

Barium (Ba) and its compounds are used in several industrial and other applications including oil and gas drilling muds; insecticides; plastic stabilizers; paint, glass, and rubber manufacture; lubricating oils; and jet fuels. Barium compounds continue to have important medical uses.

Barium sulfate (barite, $BaSO_4$) and barium carbonate (witherite, $BaCO_3$) occur in ore deposits and comprise approximately 0.05 percent of the earth's crust. The average Ba concentration in soils is higher than that of most of the trace elements, with a worldwide range of 19 to over 2,300 mg/kg (Kabata-Pendias 2010). Barium occurs naturally in drinking water and foods as a result of mineral dissolution and normal cation exchange reactions at the soil-root interface. Barium occurs only in the +2 oxidation state. In soils it associates geochemically with feldspars and biotites. The Ba^{2+} ion substitutes readily for K^+ in these structures as their ionic radii are similar (1.35 and 1.38 A° for Ba^{2+} and K^+, respectively). As soils weather, released Ba^{2+} can occur as the dissolved cation; however, it can also be immobilized by precipitation with sulfate or carbonate; concentration into Mn and P concretions; specific adsorption onto oxides and hydroxides; or by fixation on high-charge layer silicate clays such as smectites and vermiculites (Kabata-Pendias and Szteke 2015; Madejon 2013; Smith et al. 1995).

Barium mobility in soils is a function of soil characteristics such as CEC and $CaCO_3$ content. In soils with high CEC values, Ba mobility is limited due to adsorption. Clay and humus exchange sites have a preferential cation exchange selectivity for Ba^{2+} over Ca^{2+} and Mg^{2+}. High $CaCO_3$ content limits mobility by precipitating Ba as $BaCO_3$. In the presence of sulfate ions, Ba precipitates as $BaSO_4$. Barium also forms salts of low solubility with arsenate, chromate, hydroxide, and phosphate ions. As a result, Ba is relatively immobile in soil (Madejon 2013). Barium may be mobilized under certain conditions, however. Its concentration in soil solutions show substantial variation, for example, depending on soil texture (Kabata-Pendias 2010). In the presence of Cl^- ions, Ba is mobile and is more likely to be leached into groundwater because of the relative solubility of $BaCl_2$. In soil water, Ba can form complexes with natural organic compounds.

Barium is a common component of plant tissue; however, it is not considered to be an essential element for plant growth. Relative to the quantities present in soils, little Ba is concentrated by plants. Tissue Ba content ranges from 1 to 198 mg/kg (dry weight), the highest in leaves of cereals and legumes and the lowest in grains and fruits. The highest concentrations of Ba, up to 10,000 mg/kg (dw), are reported for various trees and shrubs (Chaudry et al. 1977; Shacklette et al. 1978). Raghu (2001) measured a concentration of 3,500 mg/kg Ba in shoots of indigo (*Indigofera cordifolia*). There are limited reports of toxic Ba concentrations in plants. Concentrations considered toxic have ranged from 220 mg/kg to as much as from 1 percent to 3.5 percent (dw) (Madejon 2013; Uminska 1993). Plants may take up Ba readily from acid soils. Toxicity effects of Ba to plants may be reduced by addition of Ca, Mg, and S salts to the growth medium, as antagonistic interactions between these elements and Ba occur in both plant tissue and soil (Kabata-Pendias 2010).

2.2.3 Cadmium

Cadmium (Cd) occurs in the earth's crust as sulfide ores of lead, zinc, and copper. Significant Cd releases to the biosphere have resulted from the smelting of metallic sulfide ores; burning of fossil fuels; incineration of MSW; metal plating operations; and disposal of Cd-containing wastes such as biosolids, Ni-Cd batteries, and Cd-plated steel. Cadmium rarely occurs in significant levels in water, although it can leach into groundwater from waste disposal sites.

In soil, Cd concentrations in excess of about 1 mg/kg are considered evidence of anthropogenic pollution (Uminska 1993; Chaudry et al. 1977). The range of Cd in natural soils is 0.06 to 1.1 mg/kg (McBride 1994), and levels in contaminated sites have reached as high as 0.15 percent (Kabata-Pendias 2010).

Cadmium is rarely found in nature as the pure metal. At contaminated sites, Cd exists primarily as the Cd(II) ion (the predominant oxidation state in nature), Cd-CN⁻ complexes, or $Cd(OH)_2$, depending on pH and waste processing prior to disposal. During weathering Cd readily solubilizes. The most significant environmental factors that control Cd ion mobility are pH and oxidation-reduction potential. Cadmium is most mobile in soils at pH 5 or below, whereas in neutral and alkaline soil Cd is rather immobile. At pH values < 8 it exists as the Cd^{2+} ion; in addition, it may form complex ions such as $CdCl^+$, $CdHCO_3^+$, $CdCl_4^{2-}$, $Cd(OH)_3^-$, and $Cd(OH)_4^{2-}$ and organic chelates. Monovalent hydroxy ion species (e.g., $CdOH^+$) may occur that do not readily occupy cation exchange sites. In the alkaline pH range, Cd^{2+} precipitates as $Cd(OH)_2$ and the less soluble $CdCO_3$. The Cd^{2+} ion can also coprecipitate with $CaCO_3$. In strongly oxidizing environments Cd is likely to form minerals such as CdO and $CdCO_3$ and may also accumulate in phosphate and other deposits (Kabata-Pendias 2010). Under reducing conditions and in the presence of sulfides, CdS forms. Cadmium will also precipitate with arsenate, chromate, phosphate, selenate, and selenite, with solubilities being a function of pH and geochemistry.

Cadmium adsorption often correlates with the CEC of clay minerals, carbonate minerals, oxides, and organic matter in soils. Various anions such as chloride and sulfate complex with Cd, maintaining it in solution. Conversely, the presence of other soil cations (Ca^{2+}, Mg^{2+}, Fe^{2+}, etc.) reduces Cd sorption due to competitive adsorption.

Cadmium is associated with Zn in its geochemistry. The Cd^{2+} cation is more mobile than Zn^{2+} in acidic oxidizing conditions. This mobility is attributed to weak adsorption of Cd to organic matter, silicate clays, and oxides. At high concentrations, Cd complexes with humic substances or other organic molecules. Soil microbial activity is believed to influence the transformations of Cd in soils. Cadmium does not form volatile compounds.

Because Cd is readily available to plants from both atmospheric and soil sources, its concentration rapidly increases in plants grown in polluted areas. Furthermore, the origin of soil Cd is an important factor controlling its solubility and phytoavailability. Oats absorbed a much higher proportion of anthropogenic Cd (added as CdO) than of lithogenic origin (Grupe and Kuntze 1987). A substantial proportion of soil Cd accumulates in root tissues (Saengwilai et al. 2017; Phusantisampan et al. 2016; Pichtel et al. 2000; Kabata-Pendias 1979), although it is known to occur in substantial amounts in shoot tissues of many leafy vegetables as well as in grains (Smolders and Mertens 2013). The permissible limit of cadmium in plants recommended by The World Health Organization (1996) is 0.02 mg/kg.

Cadmium is considered a toxic element to plants, with the disruption of enzyme activities being the main cause of damage. The greatest concern regarding plant uptake, however, is as a Cd reservoir and a pathway to humans and animals. Thus, tolerance of some plant species to higher Cd levels poses a public health risk. In human and animal nutrition, Cd is a cumulative poison. Cadmium accumulates in food crops and grasses, and in domestic livestock and wildlife. Application of sewage sludge to soil may increase soil Cd levels; this process can result in elevated Cd levels in crops. EPA has set guidelines for the application of sewage sludges and other materials to agricultural soils in efforts to limit Cd to food and feed crops (U.S. EPA 1983).

Liming a soil beyond pH 7.0 will result in the formation of relatively immobile Cd species (e.g., hydroxides, oxides, and carbonates). This reaction is relevant for the development of reclamation techniques for the management of Cd-enriched soils.

2.2.4 Chromium

Significant sources of chromium (Cr) to the biosphere include the chemical manufacturing industry and combustion of fossil fuels (e.g., natural gas, oil, and coal). Other sources include wastewaters from electroplating, leather tanning, and textile industries; incineration of MSW and sewage sludge; cement manufacture; and emissions from air conditioning cooling towers that use Cr compounds as rust inhibitors.

The variability in the oxidation states of Cr in soils is significant in terms of public health and environmental hazard. Chromium exhibits variable oxidation states (from 0 to +6); however, it exists primarily in three forms: metallic chromium (Cr[0]), trivalent (Cr[III]), and hexavalent (Cr[VI]). Chromium occurs in natural soils primarily as the trivalent and hexavalent species, in concentrations ranging from 0.5 to 250 mg/kg, with average values between approximately 40 and 70 mg/kg (Richard and Bourg 1991; Coleman 1988).

The Cr(VI) ion, common in industrial uses, forms anions such as chromate (CrO_4^{2-}), bichromate ($HCrO_4^-$), or dichromate ($Cr_2O_7^{2-}$). These remain soluble in soils and sediments; thus, the risk of groundwater contamination is significant. Hexavalent Cr readily passes through cell membranes and can damage DNA; hence, it is carcinogenic. Most soil Cr, however, exists as Cr(III) (chromic) and occurs within mineral structures or as mixed Cr(III) and Fe(III) oxides. In general, Cr(III) resembles Fe(III) and Al(III) in ionic size and in geochemical properties; for example, Cr(III) readily substitutes for Fe(III) in mineral structures (Kabata-Pendias 2010). The compounds of Cr(III) are generally stable in soils because they are only slightly mobile in acid media, and have precipitated by pH 5.5 and above. Aqueous concentrations are generally below water quality standards. The Cr(III) species is therefore less of a public health and environmental concern. In contrast, Cr(VI) is unstable in soil and is readily mobilized in both acid and alkaline conditions (Kabata-Pendias 2010).

The environmental behavior of Cr depends on various factors including soil pH, oxidation state, mineralogical properties, and presence of organic matter. Chromium behavior is governed strongly by both soil pH and redox potential (Gonnelli and Renella 2013; James and Bartlett 1983a, 1983b). The adsorption of chromate by soils and subsurface materials is rather poorly understood. The CrO_4^{2-} anion may be adsorbed to Fe and Al oxides and with other positively charged colloids. Chromate may also be adsorbed by ligand exchange, and $HCrO_4^-$ may behave similarly to $H_2PO_4^-$. Reduction of Cr(VI) to Cr(III) with subsequent precipitation or adsorption of the trivalent species may occur in the presence of reductive solids, Fe(II), and organic material (Bartlett and Kimble 1976b; James and Bartlett 1983c). Soil pH also affects the rate of reduction of Cr(VI) to Cr(III). Conversely, some Cr(III) can be oxidized to chromate, CrO_4^{2-}, at high pH (Bartlett and James 1979). Soil organic matter is known to mobilize a portion of soil Cr.

Chromium is a required element in human and animal nutrition. Stimulating effects of Cr on plants have been observed by several researchers; however, phytotoxicity of Cr is a common phenomenon.

In most plants, the range of critical leaf Cr concentrations is 1–10 mg/g dry matter (Zayed and Terry 2003). Stunted growth, a poorly developed root system, curled and discolored leaves, leaf chlorosis, narrow leaves, chlorotic bands, and yield reduction are common visual symptoms of Cr toxicity in plants (Pratt 1966; Hunter and Vergnano 1953; Kabata-Pendias 2010; Hara and Sonoda 1979). Under field conditions, Cr is rarely toxic to plants.

The form most available to plants is Cr(VI), which is also toxic to plants and other organisms. There is evidence that Cr(VI) can be transformed in plant cells into Cr(III), which react with DNA and protein compounds. Cr(VI) compounds exhibit a mutagenic effect on bacterial cells while Cr(III) compounds result in no or slight mutagenic activity (Mamyrbaev et al. 2015). Organic exudates from the plant root (e.g., acids) may reduce Cr(VI) in the rhizosphere. The Cr(III) species is generally unavailable to plants, and, due to low solubility, Cr(III) compounds are not translocated through cell membranes (Singh 2005). The minimal mobility of soil Cr(III) may be responsible for an inadequate Cr supply to plants.

Remediation of Cr-contaminated soil apparently is closely linked to oxidation state. For soils enriched in Cr(VI), washing with hot water has been effective (Ososkov and Bozzelli 1994). In the case of Cr(III) contamination, however, removal processes become much more complex due to the formation of insoluble complex ions and precipitates. Some researchers have experienced modest success with simple dissolution of Cr minerals using chelating agents (e.g., EDTA, NTA) and acids (HCl) (Pichtel and Pichtel 1997).

Others (Thirumalai Nivas et al. 1996) have applied anionic surfactants to soil to remove Cr as colloidal micelles.

2.2.5 Lead

Lead (Pb) has highly chalcophilic properties; its primary form in nature is galena, PbS. Lead is released to the biosphere primarily from metal smelting and processing, secondary metals production, Pb battery manufacturing, pigment and chemical manufacturing, and disposal of Pb-containing waste. Specific Pb sources include industrial waste buried in landfills, high-explosive burn sites, firing ranges, and storage and disposal of lead acid batteries.

The fate of anthropogenic Pb in soil is of significant concern because this metal is hazardous to humans and animals from two sources: the food chain and soil dust inhalation and/or ingestion.

Lead enters soil in various and complex compounds. During weathering of natural Pb sulfides, slow oxidation processes tend to form carbonates, oxides, hydroxides, and other minerals; Pb is also incorporated into clay minerals. Contaminated soil at battery processing sites likely contains $PbSO_4$ with various metallic Pb forms. Lead 'pigs' (90%–95% Pb) have been recovered (figure 2.3), and metallic Pb battery components such as electrode screens, terminals, plates, or chips are commonly found. Lead concentrations of up to 30 percent have been reported for soil intermingled with battery casing scrap and associated Pb compounds (Royer et al. 1992). The main Pb pollutants emitted from smelters occur in mineral forms (e.g., PbS, PbO, $PbSO_4$,

Figure 2.3 Lead pigs unearthed at an automotive battery recycling facility. *Source:* Mark Steele.

and PbO·PbSO$_4$) (see figure 2.4), while Pb in auto exhausts is mainly in the form of halide salts (e.g., PbBr, PbBrCl, Pb(OH)Br, and (PbO)$_2$PbBr$_2$). Tetramethyl lead, a relatively volatile organolead compound, may form as a result of biological alkylation of organic and inorganic lead by microorganisms in anaerobic sediments (Smith et al. 1995).

An upper limit for the Pb content of a normal soil is approximately 70 mg/kg. However, due to widespread Pb pollution, most soils are likely to be enriched in this metal, especially in the surface horizon (Kabata-Pendias 2010). It is therefore difficult to locate a genuinely non-contaminated soil.

Lead occurs mainly as Pb(0), Pb(II), and, to a lesser degree, the Pb(IV) species. Lead is reported to be the least mobile of the heavy metals. Contamination of soils with Pb is mainly an irreversible and, therefore, a cumulative process in surface soils. Several observations of the Pb balance in various ecosystems show that the input of this metal greatly exceeds its output (Kabata-Pendias 2010).

Following release at a site, most Pb is retained strongly by soil; very little is transported to surface water or groundwater. The fate of Pb is affected by adsorption, ion exchange, precipitation, and complexation to organic matter. In soil, Pb forms stable complexes with both inorganic (e.g., Cl^-, CO_3^{2-}) and organic (e.g., humic and fulvic acid) ligands (Alloway 2013; Bodek et al. 1988). Soluble Pb may react with carbonates, sulfides, sulfates, and phosphates to form low-solubility compounds. Farrah and Pickering (1977) suggest that Pb sorption on montmorillonite is a cation exchange process,

Figure 2.4 Scanning electron micrograph of soil at a battery recycling facility. The mixed aggregates contain both anglesite (PbSO$_4$) and metallic Pb.

while on kaolinite and illite Pb is competitively adsorbed. Abd-Elfattah and Wada (1981) found a higher selective adsorption of Pb by Fe oxides, halloysite, and imogolite than by humus, kaolinite, and montmorillonite. In other studies, the greatest affinity to sorb Pb was reported for Mn oxides. Lead is also associated with Al hydroxides. In some soils Pb may be highly concentrated in calcium carbonate particles or in phosphate concretions. In soil with high organic matter content and a pH ranging from 6.0 to 8.0, Pb may form insoluble organic complexes; if the soil has less organic matter at the same pH, hydrous lead oxide complexes or lead carbonate or lead phosphate precipitates may form.

Increased concentrations of Pb in surface soils have been reported for various terrestrial ecosystems. Although soil Pb is relatively insoluble and often occurs in adsorbed and precipitated forms, some Pb is known to be available for plant uptake (Kabata-Pendias and Szteke 2015; Steinnes 2013; Pichtel et al. 2000). Furthermore, certain soil and plant factors (e.g., low pH, low P content of soil, organic ligands) promote both Pb uptake by roots and subsequent translocation to aboveground shoots. The levels of Pb in soil that are toxic to plants are a function of plant species, Pb concentration, and Pb species; thus, threshold levels vary over a wide range. Several authors have reported concentrations ranging from 100 to several thousand mg/kg (Kabata-Pendias 2010; Pichtel et al. 2000; Pichtel and Salt 1998).

Several plant species and genotypes are adapted to growth in high Pb concentrations in soils. A number of species and ecotypes (as well as bacterial species) are known to possess Pb-tolerance mechanisms (Fahr et al. 2013; Yangh et al. 2005; Thurman 1991; Verkleij and Schat 1990). This is reflected by variable quantities in plants, as well as a varied distribution of Pb between roots and aboveground biomass (Kabata-Pendias 2010). The highest bioaccumulation of Pb generally is reported for leafy vegetables (mainly lettuce) grown in proximity to nonferrous metal smelters where plants are exposed to Pb sources in both soil and air. In these locations, highly contaminated lettuce may contain up to 0.15 percent Pb (dw) (Smith et al. 1995; Roberts et al. 1974). Lead has also been found to accumulate in certain crops grown on biosolids-amended soils (Carbonell et al. 2016; Pichtel and Anderson 1997).

The mode of Pb uptake into plants is passive (Adriano 1986). Although not typically mobile in soil, Pb can become slightly soluble in the rhizospere, absorbed by root hairs and stored in cell walls (Kabata-Pendias 2010). When Pb is present in soluble forms in nutrient solution, roots may take up substantial quantities, the rate increasing with increasing concentration and time. The translocation of Pb from roots to tops is typically limited, however. Airborne Pb, a major source of Pb pollution, is also taken up by plants through foliage (Kabata-Pendias 2010).

Adequate soil P content is known to reduce the effects of Pb toxicity. This interference is due to the ability of Pb to form insoluble phosphates in both plant tissue and in soil. Sulfur is known to inhibit the transport of Pb from root to shoot. Sulfur deficiency markedly increases Pb movement into shoots (Smith et al. 1995). Huang et al. (2005) demonstrated that arbuscular mycorrhizal fungi could protect host plants from the phytotoxicity of excessive lead by changing the speciation from bioavailable to nonbioavailable forms. This suggests that mycorrhiza offer a unique mechanism for protecting their host from excessive lead phytotoxicity via chemical transformation in the soil.

The solubility of Pb can be greatly decreased by liming. A high soil pH may precipitate Pb as hydroxide, phosphate, or carbonate, as well as promote the formation of Pb-organic complexes that are rather stable.

2.2.6 Mercury

Mercury (Hg) is a chalcophile, and in unweathered rocks is commonly found as cinnabar, HgS. Sources of soil contamination with Hg include metal-processing industries, certain chemical works (especially chlor-alkali), and use of fungicides containing Hg. Sewage sludges and other wastes may also be sources of Hg contamination. Mercury has been used in the gold amalgamation process in gold mining regions in the Brazilian Amazon basin (Grimaldi et al. 2015). Common secondary Hg sources are spent batteries, fluorescent lamps, switches, dental amalgams, measuring devices, and laboratory and electrolytic refining wastes.

Mercury is a persistent environmental pollutant and is notorious for bioaccumulation ability in fish, animals, and humans (Chang et al. 2009). The mechanism and extent of toxicity depend strongly on the type of Hg compound and its redox state (Wagner-Döbler n.d.).

Background levels of Hg in soils are difficult to estimate due to widespread Hg pollution. The background contents of soil Hg are estimated to range between 50 and 300 µg/kg, depending on soil type. Thus, Hg contents in excess of these values should be considered an indicator of anthropogenic contamination (Kabata-Pendias 2010).

At contaminated sites, Hg exists in the mercuric (Hg[II]), mercurous (Hg_2[II]), elemental (Hg[0]), or alkylated form (e.g., methyl and ethyl mercury). The behavior of Hg in contaminated soil is of concern because its bioavailability poses a significant health hazard. In soil and surface water, solid forms partition to colloidal particles whereas volatile forms (e.g., metallic mercury and dimethylmercury) are released to the atmosphere. Mercury exists primarily in the mercuric and mercurous forms as complexes having varying solubilities. In soil and sediments, sorption is one of the most important controlling pathways for removal of Hg from solution.

Mercury is strongly sorbed to humic materials and is retained by soils mainly as slightly mobile organocomplexes. In acid conditions, Hg is leached from soil profiles as organic complexes. Other removal mechanisms include flocculation and coprecipitation with sulfides. $Hg(OH)_2$ is likely to predominate over other aqueous species at pH values near or above neutrality. Some investigators believe that $Hg(OH)_2$ is the preferred sorbed species (Kabata-Pendias 2010).

The most important transformation process in the environmental fate of Hg is biotransformation. Any Hg form entering sediments, groundwater, or surface water under the appropriate conditions can be microbially converted to the methylmercuric ion. Sulfur-reducing bacteria are responsible for most Hg methylation in the environment, with anaerobic conditions favoring their activity. The methylation of elemental Hg plays a key role in environmental cycling of Hg. Methylated Hg is mobile and readily taken up by organisms including some higher plants. Humic substances are known to mediate the chemical methylation of inorganic Hg by releasing labile methyl groups. Methylation of Hg can also occur abiotically. Mercury methylation processes have apparently been involved in earlier events of catastrophic Hg poisoning. Methylmercury, being soluble and mobile, rapidly enters the aquatic food chain (Cristol et al. 2008; Baralkiewicz et al. 2006; Webb et al. 2006). Concentrations of methylmercury in carnivorous fish can be 10,000 to 100,000 times the concentrations found in ambient waters (Callahan et al. 1979).

Volatile elemental Hg may be formed through the demethylation of methylmercury or the reduction of inorganic Hg, with anaerobic conditions favoring the reactions (figure 2.5). Rapid conversion of organic Hg and Hg^{2+} ions to the elemental state (Hg°) in contact with humic substances may occur. Several types of bacteria and yeasts have been shown to promote the reduction of cationic Hg^{2+} to the elemental state (Hg°).

Plants take up Hg from solution culture. There is evidence that high soil Hg concentrations increase Hg contents of certain plants (Moreno-Jimenez et al. 2006). The rate is highest for roots, but leaves and grains also accumulate substantial Hg. Various organic Hg compounds (e.g., methyl, ethyl, phenyl) added to soil are partly decomposed or adsorbed by soil constituents. However, all these compounds are taken up by plants. Methyl-Hg is the most available, while phenyl- and sulfide-Hg are the least available Hg forms to plants. Plants are known to directly absorb Hg vapor (Ericksen and Gustin 2004; Ericksen et al. 2003).

The distribution of Hg in plants has been studied because of the potential for entry into the food chain. Therefore, substantial information is available related to the Hg content of plant foods. The background levels of Hg in

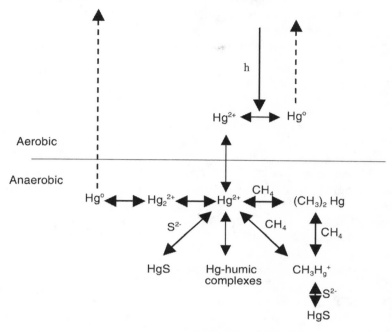

Figure 2.5 Transformations of mercury in soil and sediment.

vegetables and fruits vary from 2.6 to 86 μg/kg (dw). Plants grown on contaminated sites may accumulate much higher than normal amounts of Hg, however. Certain plant species, lichens, carrots, lettuce, and mushrooms, in particular, are likely to take up more Hg than other plants grown at the same location. Also, some parts of plants have a greater ability to adsorb Hg, as is the case of apple flesh (Kabata-Pendias 2010).

Plants differ in their ability to take up Hg and can also develop a tolerance to high tissue Hg concentrations when grown in soils overlying Hg deposits. The Hg tolerance mechanism or the physiological barrier is not known; however, it is probably related to the inactivation of Hg at membrane sites (Kabata-Pendias 2010).

Regarding plant growth on contaminated soils, there is the additional concern for generation of methyl-Hg from microbial methylation of Hg by bacteria and fungi under both aerobic and anaerobic soil conditions. Even simple Hg salts or metallic Hg create a hazard to plants and soil biota from the Hg vapor (Kabata-Pendias 2010). The symptoms of Hg toxicity are, most commonly, stunting of seedling growth, restricted root development, and an inhibition of photosynthesis and, as a consequence, yield reduction. Mercury accumulated in root tissue inhibits K^+ uptake by plants.

Mercury uptake by plant roots may be minimized by neutralizing the soil pH with liming materials. Also, sulfur-containing compounds and rock phosphates may inactivate mercurial fungicides or elemental Hg in soils.

2.2.7 Nickel

Anthropogenic sources of nickel (Ni), from industrial activity in particular, have resulted in a significant increase in the Ni content of soils in numerous locations. Nickel is released in emissions from mining and metal-processing operations, municipal waste incineration, and combustion of coal and oil. Land application of biosolids and certain phosphate fertilizers may also contribute significant Ni to soil. Sources to water include stormwater runoff, wastewater from municipal sewage treatment plants, and leachate from landfills. Worldwide soil Ni concentrations range from 0.2 to 450 mg/kg. Highest Ni contents occur in clay and loamy soils. Nickel is also elevated in soils formed over basic (i.e., relatively low in silica) and volcanic rocks, and in some organic-rich soils (Kabata-Pendias 2010).

The +2 oxidation state is the only stable Ni form in the environment. In surface soil horizons Ni occurs mainly in organically bound forms, part of which may be soluble chelates. Nickel species such as Ni^{2+}, $NiOH^+$, $HNiO_2^-$, and $Ni(OH)_3^-$ also occur.

Generally, the solubility of soil Ni is inversely related to soil pH. Nickel sorption on Fe and Mn oxides is especially pH dependent, probably because $NiOH^+$ is preferentially sorbed (Kabata-Pendias 2010).

Environmental Ni pollution greatly influences its concentrations in plants. In ecosystems where Ni is an airborne pollutant, plant tops are likely to concentrate the majority of Ni, most of which can be washed from leaf surfaces. Biosolids have also been shown to be a source of Ni to plants. Nickel in sludge that is present mainly in organic chelated forms may be readily available (Alloway 2013) and therefore phytotoxic. As Ni is readily mobile in plants, fruit and grains are reported to contain elevated Ni concentrations (Kabata-Pendias 2010).

The soluble, adsorbed (exchangeable), and organically bound fractions of soil Ni are the forms most available to plants (Pichtel and Anderson 1997). Nickel is readily taken up from soil, and until certain concentrations in plant tissues are reached, uptake is positively correlated with soil Ni concentration. Both plant and pedological factors affect Ni uptake by plants, but the most significant factor is soil pH (Kabata-Pendias 1995).

Nickel is a rather toxic element to plants. The mechanism of Ni toxicity is not well understood, although impaired growth of plants and injuries caused by excess concentrations have been observed. Elevated concentrations in plant tissue may inhibit photosynthesis and transpiration. The most common symptom of Ni phytotoxicity is chlorosis. With plants under Ni stress, the absorption of nutrients, root development, and metabolism are restricted. In addition, low N_2 fixation by soybean plants may be caused by Ni excesses (Kabata-Pendias 2010).

Generally the range of excessive or toxic quantities of Ni in most plant species varies from 10 to 100 mg/kg (dw). Most sensitive species are affected by much lower Ni concentrations, ranging from 10 to 30 mg/kg (dw). Several species are known for their tolerance and hyperaccumulation of Ni; about 300 Ni hyperaccumulating plants have been identified (Prasad 2005; Nkoane et al. 2005; Massoura et al. 2004). *Thlaspi oxyceras* (Boiss) Hedge was found to contain 35,600 mg/kg (dw) Ni, and *Thlaspi cypricum* Bornm contained 52,120 mg/kg dw (Prasad 2005).

Soil treatments such as additions of lime, phosphate, or humified organic matter, decrease Ni availability to plants (Kabata-Pendias 2010).

2.2.8 Selenium

Excessive levels of selenium (Se) in soil have been determined both in naturally Se-enriched areas (typically semiarid and arid environs) and in industrial areas where Se is released to the atmosphere. Examples of the former include various coal deposits and clay-rich sediments. Examples of the latter include certain metal-processing operations, combustion of coal, and MSW incineration. Several emissions, including Se-enriched fly ash, may be a significant source of Se, which is relatively available to plants (Kabata-Pendias 2010). Soil Se ranges worldwide from 0.01 to 2 mg/kg. Elevated concentrations (up to 1,200 mg/kg) have been reported for some organic-rich soils derived from black shales (Sherwood Lollar 2005).

Selenium, a chalcophile, tends to be associated with sulfide minerals in rocks. Natural weathering results in the oxidation of insoluble reduced forms, including elemental Se (Se°), selenides (Se_2^-), and selenium sulfides, to the more soluble selenites (SeO_3^{2-}) and selenates (SeO_4^{2-}).

Several oxidation states are possible for Se in soil, and the soil oxidation-reduction potential strongly affects Se behavior. In alkaline, oxidized soils, the selenates predominate. They bond weakly to oxides and other minerals resulting in high mobility in neutral to alkaline soils. In slightly acid oxidized soils, selenites are common; these are of lower mobility than selenates. This is because of the ability of SeO_3^{2-} to chemisorb strongly onto oxides and aluminosilicates, particularly Fe oxides and montmorillonite, respectively, and to precipitate as the insoluble ferric selenite. In reducing environments, that is, wet or organic-rich soils, the insoluble reduced forms (selenides and Se-sulfides) predominate so that Se mobility and bioavailability are low. Under reducing conditions biological methylation of Se may occur; this process may form volatile compounds (e.g., dimethyl selenides).

The solubility of Se in most soils is rather low; therefore, many agricultural areas produce crop plants and forage having low Se content. However, in naturally Se-enriched soils, in calcareous soils, in arid zones, and in contaminated soils in industrial areas, Se as selenate may be accumulated by plants in concentrations high enough to be toxic to grazing livestock. Some legumes (e.g., sweet clover) grown on coal ash are known to contain as much as 200 mg/kg (dw) Se. Some hyperaccumulating plants accumulate > 1000 mg Se/kg d.w. and thrive in Se-rich regions of the world; these include Stanleya, Astragalus species, Conopsis, Neptunia, and Xylorhiza. The so-called secondary accumulators show no signs of toxicity up to 100–1000 mg Se/kg d.w.—for example, *Brassica juncea*, *Brassica napus*, Broccoli, Helianthus, Aster, Camelina, and *Medicago sativa* (Gupta and Gupta 2016).

Application of P, S, and N may detoxify Se, which may be a result either of depressing Se uptake by roots or of establishing a more favorable ratio of Se to these elements. The application of S is an important remedial treatment on Se-enriched soils (Kabata-Pendias 2010; Johnson 1975).

2.2.9 Zinc

Zinc ores include zinc sphalerite or zinc blende (ZnS and marmatite, ZnS with some Fe). Zinc is extracted from ores by roasting to form zinc oxide (ZnO). The main use of Zn is as a corrosion-resistant coating on iron or steel. Zinc is used to make the casings of zinc-carbon dry cell batteries, sacrificial anodes, and die-cast parts. Brass alloy is used in shell casings, tubing, and electrical equipment. Zinc dust is used in priming compounds, paints, alkaline dry cell batteries, and metals processing.

Soil contamination with Zn has reached high concentrations in certain areas. Zinc is released to the atmosphere as dust and fumes from zinc production facilities, automobile emissions, and fuel combustion. Additional sources include MSW incineration, coal combustion, smelter operations, and some metalworking industries. Municipal biosolids applied to cropland can also be enriched with trace metals including Zn. Hazardous waste sites are additional sources: Zn was found in over 700 of the 1,300 NPL hazardous waste sites (U.S. EPA 1994). These sources, along with releases of Zn through metal corrosion and tire wear, contribute to urban runoff contamination.

Zinc occurs naturally in soils ranging from 17 to 125 mg/kg. The most common and mobile Zn form in soil is Zn^{2+}, but other ionic species may occur. Factors controlling Zn mobility in soil are similar to those for Cu, but Zn appears to occur in more readily soluble forms. Clay minerals, hydrous oxides, and pH are among the key factors controlling Zn solubility in soils. Of lesser importance are precipitation as hydroxide, carbonate, and sulfide compounds; and complexation by organic matter. Zinc can enter some layer silicate clays (e.g., montmorillonite) and become immobilized. Soil organic matter is capable of bonding Zn in stable forms (Kabata-Pendias 2010). Because Zn is adsorbed by mineral and organic moieties, its accumulation in surface horizons is fairly common.

Zinc is considered to be readily soluble relative to the other heavy metals in soil. Zinc is most readily mobile and available in acid light (sandy, loamy) mineral soils. Solubility and availability of Zn are inversely correlated with Ca and P compounds present in soil. This relationship presumably is a result of both adsorption and precipitation processes (Kabata-Pendias 2010).

Zinc is an essential trace element in both plant and animal nutrition. Toxicity and tolerance in plants have been of interest because prolonged use of Zn fertilizers, as well as inputs from industrial pollution have resulted in increased Zn content of surface soil. Several plant species and genotypes are known to possess a significant tolerance to soil Zn. Some genotypes grown in Zn-rich soil or areas of heavy atmospheric Zn deposition may accumulate extremely large quantities without exhibiting symptoms of toxicity (Thurman 1991; Verkleij and Schat 1990). Zinc is not considered to be highly phytotoxic. Zinc-Cd interactions are common in the literature, with reports both of antagonism and synergism between the two elements in uptake-transport processes (Sarwar et al. 2015; He et al. 2004; Smith et al. 1995).

Additional concerns related to Zn pollution are changes in metal speciation. For example, in soil (pH 6.8) amended with Zn-enriched compost (composted biosolids, composted MSW), an increase of readily available Zn species, and weakly bound or exchangeable Zn forms, was observed (Alloway 2013; Pichtel and Anderson 1997). Soluble Zn-organic complexes that occur in municipal biosolids are mobile in soil and therefore available to plants (Lavado et al. 2005). Amelioration of Zn-contaminated soil is commonly based on controlling its availability by addition of lime or organic matter or both.

QUESTIONS

1. The toxicity of heavy metals depends in large part on its form, for example, ionic versus organically bound. Explain.
2. Explain the factors affecting Cd mobility in the environment. Include soil pH, oxidation-reduction status, clay mineralogy, and other factors.
3. Which chromium species, for example, +3 versus +6, is a suspected mutagen? An essential trace element? Discuss the factors that result in transformation of one species to the other.

4. Which heavy metals/metalloids are known to form methylated compounds? What is the environmental significance of such methylation reactions?
5. Volcanic eruptions are a significant source of which metal(s) to the biosphere?
6. Which metal is more mobile in soils, all other factors being equal: Cd or Pb? Explain your reasoning.
7. Due to widespread Pb pollution, most soils may be somewhat enriched in this metal. True or false? Discuss.
8. What is (are) the common form(s) of lead in soils?
9. Pb may be present in soils and aquatic ecosystems in the Pb^{2+} form. True or false?
10. Lead will not readily leach from soil; that is, rainfall and other natural conditions will not promote the downward movement of Pb out of the soil profile. Explain the processes responsible for this phenomenon.
11. Mercury is the most volatile of all the heavy metals. True or false?
12. Which form(s) of mercury is (are) readily mobile and easily taken up by organisms?
13. How do Hg species react with humic materials—is sorption possible? Discuss.

REFERENCES

Abd-Elfattah, A., and K. Wada. 1981. Adsorption of lead, copper, zinc, cobalt and cadmium by soils that differ in cation-exchange materials. *Journal of Soil Science* 32(2):271.

Adriano, D. C. 1986. *Trace Elements in the Terrestrial Environment.* Berlin: Springer-Verlag.

Alloway, J. 2013. *Heavy Metals in Soils: Trace Metals and Metalloids in Soils and Their Bioavailability.* New York: Springer.

Baralkiewicz, D., H. Gramowska, and R. Golstrodyn. 2006. Distribution of total mercury and methyl mercury in water, sediment and fish from Swarzedzkie lake. *Chemistry and Ecology* 22(1):59–64.

Bartlett, R. J., and B. James. 1979. Behavior chromium in soils: III. Oxidation. *Journal of Environmental Quality* 8:31–5.

Bartlett, R. J., and J. M. Kimble. 1976a. Behavior of chromium in soils: I. Trivalent forms. *Journal of Environmental Quality* 5(4):379–83.

———. 1976b. Behavior of chromium in soils: II. Hexavalent forms. *Journal of Environmental Quality* 5(4):383–86.

Bodek, I., W. J. Lyman, W. F. Reehl, and D. H. Rosenblatt. 1988. *Environmental Inorganic Chemistry: Properties, Processes, and Estimation Methods.* Elmsford, NY: Pergamon.

Brearley, F. Q., and A. D. Thomas. 2015. *Land-Use Change Impacts on Soil Processes: Tropical and Savannah Ecosystems*. CAB International.

Brooks, R. R., J. Lee, and T. Jaffre. 1974. Some New Zealand and New Caledonian plant accumulators of nickel. *Journal of Ecology* 62(2):493–98.

Callahan, M. A., M. W. Slimak, and N. W. Gabel. 1979. *Water-Related Environmental Fate of 129 Priority Pollutants. Vol. 1, Introduction and Technical Background, Metals, and Inorganics, Pesticides and PCBs*. EPA-440/4-79-029a, Report to U.S. Environmental Protection Agency, Office of Water Planning and Standards, Washington, DC. Springfield, VA: Versar Incorporated.

Carbonell, G., M. Torrijos, J. A. Rodríguez, and Á. Porcel M. 2016. Uptake and metal transfer from biosolid-amended soil to tomato (*Solanum lycopersicum* Mill L.) plants. *Journal of Plant Chemistry and Ecophysiology* 1(1):1002.

Chang, T. C., S. J. You, B. S. Yu, C. M. Chen, and Y. C. Chiu. 2009. Treating high-mercury-containing lamps using fullscale thermal desorption technology. *Journal of Hazardous Materials* 162(2–3):967–72.

Chaudry, F. M., A. Wallace, and R. T. Mueller. 1977. Barium toxicity in plants. *Communications in Soil Science and Plant Analysis* 8(9):795–97.

Cline, S. R., B. E. Reed, and M. Matsumoto. 1993. Hazardous and industrial wastes. In: *Proceedings of the 25th Mid-Atlantic Industrial Waste Conference*, edited by A. Davis. Lancaster, PA: Technomic.

Coleman, R. N. 1988. Chromium toxicity: Effects on microorganisms with special reference to the soil matrix. In: *Chromium in the Natural and Human Environments*, edited by J. O. Nriagu and E. Nieboer. New York: Wiley Interscience.

Crecelius, E. A., N. S. Bloom, C. E. Cowan, and E. A. Jenne. 1986. *Speciation of Selenium and Arsenic in Natural Waters and Sediments. Arsenic Speciation*. EPRI project 2020-2. Battelle Pacific North Lab, Sequim, WA, 32.

Cristol, D. A., R. L. Brasso, A. M. Condon, R. E. Fovargue, S. L. Friedman, K. K. Hallinger, A. P. Monroe, and A. E. White. 2008. The movement of aquatic mercury through terrestrial food webs. *Science* 320(5874):335.

Davies, B. E. 1977. Heavy metal pollution of British agricultural soils with special reference to the role of lead and copper mining. In: *Proceedings of the International Seminar on Soil Environment and Fertility Management in Intensive Agriculture*, Tokyo, 394.

Davis, A., and I. Singh. 1995. Washing of zinc(II) from a contaminated soil column. *Journal of Environmental Engineering* 121(2):174–85.

Dixon, J. B., and S. B. Weed. 1977. *Minerals in Soil Environments*. Madison, WI: Soil Science Society.

Ericksen, J. A., and M. S. Gustin. 2004. Foliar exchange of mercury as a function of soil and air mercury concentrations. *Science of the Total Environment* 324(1–3):271–79.

Ericksen, J. A., S. E. Lindberg, J. S. Coleman, M. S. Gustin, D. E. Schorran, and D. W. Johnson. 2003. Accumulation of atmospheric mercury in forest foliage. *Atmospheric Environment* 37(12):1613–22.

Fahr, M., L. Laplaze, N. Bendaou, V. Hocher, M. El Mzibri, D. Bogusz, and A. Smouni. 2013. Effect of lead on root growth. *Frontiers in Plant Science*. doi: 10.3389/fpls.2013.00175.

Farrah, H., and W. F. Pickering. 1977. The sorption of lead and cadmium species by clay minerals. *Australian Journal of Chemistry* 30(7):1417.

Fitz, W., and W. Wenzel. 2002. Arsenic transformations in the soil-rhizosphere-plant system: Fundamentals and potential application to phytoremediation. *Journal of Biotechnology* 99(3):259–78.

Fitz, W., and W. Wenzel. 2006. Sequestration of arsenic by plants. In: *Managing Arsenic in the Environment - From Soil to Human Health,* edited by R. Naidu, E. Smith, G. Owens, P. Bhattacharya, and P. Nadebaum. Collingwood: CSIRO Publishing.

Gonnelli, C., and G. Renella. 2013. Chromium and nickel. In: *Heavy Metals in Soils: Trace Metals and Metalloids in Soils and Their Bioavailability,* edited by J. Alloway. New York: Springer.

Gough, L. P., and R. C. Severson. 1979. Impact of point source emission from phosphate processing on the element content of plants and soils. In: *Trace Substances in Environmental Health,* edited by D. D. Hemphill, vol. 10, 225. University of Missouri, Columbia: Soda Spring, Idaho.

Grimaldi, M., S. Guédron, and C. Grimaldi. 2015. Impact of gold mining on mercury contamination and soil degradation in Amazonian ecosystems of French Guiana. Chapter 9. In: *Land-Use Change Impacts on Soil Processes: Tropical and Savannah Ecosystems,* edited by F. Q. Brearley, and A. D. Thomas. CAB International.

Grupe, M., and H. Kuntze. 1987. Zur Ni-mobilitat einer geogen belasteten Braunerde. *Mitteiliunger Deutsch Bodenkundl Gesellschl* 55:333.

Gupta, M., and S. Gupta. 2016. An overview of selenium uptake, metabolism, and toxicity in plants. *Frontiers in Plant Science.* doi: 10.3389/fpls.2016.02074.

Gutenmann, W. H., C. A. Bache, W. D. Youngs, and D. J. Lisk. 1976. Selenium in fly ash. *Science* 191(4230):966.

Hanada, S., M. Nakano, H. Saitoh, and T. Mochizuki. 1975. Studies on the pollution of apple orchard surface soils and its improvement in relation to inorganic spray residues. *Bulletin of the Faculty of Agriculture* (Hirosal University, Japan) 25:13.

Hara, T., and Y. Sonoda. 1979. Comparison of the toxicity of heavy metals to cabbage growth. *Plant and Soil* 51(1):127–33.

He, P. P., X. Y. Lv, and G. Y. Wang. 2004. Effects of Se and Zn supplementation on the antagonism against Pb and Cd in vegetables. *Environment International* 30(2):167–72.

Hooda, P. S. 2010. *Trace Elements in Soils.* New York: Wiley.

Huang, P. M. 1975. Retention of arsenic by hydroxy-aluminum on surface of micaceous mineral colloids. *Soil Science Society of America Proceedings* 39(2):271.

Huang, Y., S. Tao, and Y.-J. Chen. 2005. The role of arbuscular mycorrhiza on change of heavy metal speciation in rhizosphere of maize in wastewater irrigated agriculture soil. *Journal of Environmental Sciences* 17(2):276–80.

Hughes, M. K., N. W. Lepp, and D. A. Phipps. 1980. Aerial heavy metal pollution and terrestrial ecosystems. *Advances in Ecological Research* 11:217.

Hunter, J. G., and O. Vergnano. 1953. Trace-element toxicities in oat plants. In: *Annals of Applied Biology*, edited by R. W. Marsh and I. Thomas. Cambridge, UK: University Press.

Hutchinson, T. C., and L. M. Whitby. 1974. Heavy-metal pollution in the Sudbury mining and smelting region of Canada. I. Soil and vegetation contamination by nickel, copper, and other metals. *Environmental Conservation* 1:123–32.

James, B. R., and R. J. Bartlett. 1983a. Behavior chromium in soils: VI. Interactions between oxidation-reduction and organic complexation. *Journal of Environmental Quality* 12:173–76.

———. 1983b. Behavior of chromium in soils: VII. Adsorption and reduction of hexavalent forms. *Journal of Environmental Quality* 12(2):177–81.

———. 1983c. Behavior of chromium in soils: V. Fate of organically complexed Cr(III) added to soil. *Journal of Environmental Quality* 12(2):169–72.

Johnson, C. M. 1975. Selenium in soils and plants: Contrasts in conditions providing safe but adequate amounts of selenium in the food chain. In: *Trace Elements in Soil-Plant-Animal Systems*, edited by D. J. D. Nicholas and A. R. Egan. New York: Academic Press.

Jones, L. H. P., S. C. Jarvis, and D. W. Cowling. 1973. Lead uptake from soils by perennial ryegrass and its relation to the supply of an essential element (sulphur). *Plant and Soil* 38(3):605.

Kabata-Pendias, A. 1979. *Effects of Inorganic Air Pollutants on the Chemical Balance of Agricultural Ecosystems*. United Nations ECE Symposium on Effects of Air-Borne Pollution on Vegetation, Warsaw, August 20, 134.

———. 2010. *Trace Elements in Soil and Plants*. 4th ed. Boca Raton, FL: CRC Press.

Kabata-Pendias, A., and B. Szteke. 2015. *Trace Elements in Abiotic and Biotic Environments*. Boca Raton, FL: CRC Press.

Kitigashi, K., and I. Yamane, eds. 1981. *Heavy Metal Pollution in Soils of Japan*. Tokyo: Japan Science Society Press.

Lavado, R. S., M. B. Rodriguez, and M. A. Taboada. 2005. Treatment with biosolids affects soil availability and plant uptake of potentially toxic elements. *Agriculture, Ecosystems and Environment* 109(3–4):360–64.

Laxen, D. P. H. 1985. Trace metal adsorption/coprecipitation on hydrous ferric oxide under realistic conditions. *Water Research* 19(10):1229–32.

Lindsay, W. L. 1972. Zinc in soils and plant nutrition. *Advances in Agronomy* 24:147.

Madejon, P. 2013. Barium. In: *Heavy Metals in Soils: Trace Metals and Metalloids in Soils and their Bioavailability*, edited by J. Alloway. New York: Springer.

Malm, O., W. C. Pfeiffer, C. M. M. Souza, and R. Reuther. 1990. Mercury pollution due to gold mining in the Madeira River Basin, Brazil. *Ambio* 19:11–5.

Mamyrbaev, A. A., T. A. Dzharkenov, Z. A. Imangazina, and U. A. Satybaldieva. 2015. Mutagenic and carcinogenic actions of chromium and its compounds. *Environmental Health and Preventive Medicine* 20(3):159–67.

Massoura, S. T., G. Exhevarria, E. Leclerc-Cessac, and J. L. Morel. 2004. Response of excluder, indicator, and hyperaccumulator plants to nickel availability in soils. *Australian Journal of Soil Research* 42(8):933–38.

McBride, M. B. 1994. *Environmental Chemistry of Soils*. New York: Oxford University Press.

Miller, W. P., and W. W. McFee. 1983. Distribution of cadmium, zinc, copper, and lead in soils of industrial northwestern Indiana. *Journal of Evironmental Quality* 12(1):29–33.

Moore, J. N., W. H. Ficklin, and C. Johns. 1988. Partioning of arsenic and metals in reducing sulfidic sediments. *Environmental Science and Technology* 22(4):432–37.

Moore, R. E., B. E. Reed, and M. Matsumoto. 1993. Hazardous and industrial wastes. In: *Proceedings of the 25th Mid-Atlantic Industrial Waste Conference*, edited by A. Davis. Lancaster, PA: Technomic.

Moreno-Jimenez, E., R. Millan, J. M. Penalosa, E. Esteban, R. Gamarra, and R. O. Carpena-Ruiz. 2006. Mercury bioaccumulation and phytotoxicity in two wild plant species of Almaden area. *Chemosphere* 63(11):1969–73.

Nivas, B., D. A. Sabatini, B. Shiau, and J. H. Harwell. 1996. Surfactant enhanced remediation of subsurface chromium contamination. *Water Research* 30(3):511–20.

Nkoane, B. B. M., W. Lund, G. M. Sawula, and G. Wibetoe. 2005. Identification of Cu and Ni indicator plants form mineralised locations in Botswana. *Journal of Geochemical Exploration* 86(3):130–42.

Nriagu, J. O. 1988. Production and uses of chromium. In: *Chromium in the Natural and Human Environments, Advances in Environmental Science and Technology*, edited by J. O. Nriagu and E. Nieboer, vol. 20. New York: Wiley Interscience.

Olivera, L. J., L. D. Hylander, and S. E. De Castro. Mercury behavior in a tropical environment: The case of small-scale gold mining in Pocone, Brazil. *Environmental Practice* 6(2):121–34.

Ososkov, V., and J. W. Bozzelli. 1994. Removal of Cr(VI) from chromium contaminated sites by washing with hot water. *Hazardous Waste and Hazardous Materials* 11(4):511–17.

Pacha, J., and R. Galimska-Stypa. 1988. Mutagenic properties of selected tri- and hexavalent chromium compounds. *Acta Biologen Sil Katowice, Poland* 9:30.

Peterson, P. J. 1975. Element accumulation by plants and their tolerance of toxic mineral roles. In: *International Conference on Heavy Metals in the Environment, Pathways and Cycling*, edited by T. C. Hutchinson, vol. 2. Toronto: Institute for Environmental Studies, University of Toronto.

Phusantisampan, T., W. Meeinkuirt, P. Saengwilai, J. Pichtel, and R. Chaiyarat. 2016. Phytostabilization potential of two ecotypes of *Vetiveria zizanioides* in cadmium-contaminated soils: Greenhouse and field experiments. *Environmental Science and Pollution Research International* 23(19):20027–38.

Pichtel, J. 2005. Phytoextraction of lead-contaminated soils: Current experience. In: *Heavy Metal Contamination of Soils: Problems and Remedies*, edited by I. Ahmad, S. Hayat, and J. Pichtel. Enfield, NH: Science Publishers.

Pichtel, J., and M. Anderson. 1997. Trace metal bioavailability in municipal solid waste and sewage sludge composts. *Bioresource Technology* 60(3):223–29.

Pichtel, J., and T. M. Pichtel. 1997. Comparison of solvents for ex-situ removal of Cr and Pb from contaminated soil. *Environmental Engineering Science* 14(2):97–103.

Pichtel, J., and C. A. Salt. 1998. Vegetative growth and trace metal accumulation on metalliferous wastes. *Journal of Environmental Quality* 27(3):618–24.

Pichtel, J., K. Kuroiwa, and H. T. Sawyerr. 2000. Distribution of Pb, Cd and Ba in soils and plants of two contaminated sites. *Environmental Pollution* 110:171–78.

Prasad, M. N. V. 2005. Nickelophilous plants and their significance in phytotechnologies. *Brazilian Journal of Plant Physiology* 17(1). doi: 10.1590/S1677-04202005000100010.

Pratt, P. F. 1966. Chromium. In: *Diagnostic Criteria for Plants and Soils*, edited by H. D. Chapman, 136–41. Riverside: University of California.

Raghu, V. 2001. Accumulation of elements in plant and soils in and around Mangampeta and Vemula barite mining areas, Cuddapah District, Andra Pradesh, India. *Environmental Geology* 40:1265–77.

Richard, F., and A. C. M. Bourg. 1991. Aqueous geochemistry of chromium: A review. *Water Research* 25(7):807–16.

Roberts, T. M., W. Gizyn, and T. C. Hutchinson. 1974. Lead contamination of air, soil, vegetation, and people in the vicinity of secondary lead smelters. In: *Trace Substances in Environmental Health*, edited by D. D. Henry, vol. 8, 155. Columbia: University of Missouri Press.

Royer, M. D., A. Selvakumar, and R. Gaire. 1992. Control technologies for remediation of contaminated soil and waste deposits at Superfund lead battery recyling sites. *Journal of the Air and Waste Management Association* 42(7):970–80.

Saengwilai, P., W. Meeinkuirt, J. Pichtel, and P. Koedrith. 2017. Influence of amendments on Cd and Zn uptake and accumulation in rice (*Oryza sativa* L.) in contaminated soil. *Environmental Science and Pollution Research International* 24(18):15756–67.

Sarwar, N., W. Ishaq, G. Farid, M. R. Shaheen, M. Imran, M. Geng, and S. Hussain. 2015. Zinc-cadmium interactions: Impact on wheat physiology and mineral acquisition. *Ecotoxicology and Environmental Safety* 122:528–36.

Shacklette, H. T., J. A. Erdman, and T. F. Harms. 1978. Trace elements in plant foodstuffs. In: *Toxicity of Heavy Metals in the Environment, Part 1*, edited by F. W. Oehme. New York: Marcel Dekker.

Sherwood Lollar, B. 2005. *Environmental Geochemistry. Treatise on Geochemistry*, vol. 9. New York: Elsevier.

Singh, V. P. 2005. *Toxic Metals and Environmental Issues*. New Delhi: Sarup and Sons.

Smith, L. A., J. L. Means, A. Chen, B. Alleman, C. C. Chapman, J. S. Tixier, S. E. Brauning, A. R. Gavaskar, and M. D. Royer. 1995. *Remedial Options for Metals-Contaminated Soils*. Boca Raton, FL: CRC Press.

Smolders, E., and J. Mertens. 2013. Cadmium. In: *Heavy Metals in Soils: Trace Metals and Metalloids in Soils and Their Bioavailability,* edited by J. Alloway. New York: Springer.

Steinnes, E. 2013. Lead. In: *Heavy Metals in Soils: Trace Metals and Metalloids in Soils and their Bioavailability*, edited by J. Alloway. New York: Springer.

Thirumalai Nivas, B., D. A. Sabatini, B. Shiau, and J. H. Harwell. 1996. Surfactant enhanced remediation of subsurface chromium contamination. *Water Research* 30(3):511.

Thurman, D. A. 1991. Mechanism of metal tolerance in higher plants. In: *Effect of Heavy Metal Pollution on Plants, Metals in the Environment*, edited by N. W. Lepp, vol. 2, 239–49. London: Applied Science Publishers.

Tobin, J. M., D. G. Cooper, and R. J. Neufeld. 1984. Uptake of metal ions by *Rhizopus arrhizus* Biomass. *Applied and Environmental Microbiology* 47(4):821.

Uminska, R. 1993. Cadmium contents of cultivated soils exposed to contamination in Poland. *Environmental Geochemistry and Health* 15(1):15–9.

U.S. Department of Health and Human Services. 1992. *Toxicological Profile for Barium and Compounds*. Clement International Corporation. Washington, DC: Agency for Toxic Substances and Disease Registry.

U.S. Environmental Protection Agency. 1983. *Process Design Manual. Land Application of Municipal Sludge*. EPA-625/1-83-016. Cincinnati, OH: Municipal Environmental Research Laboratory.

———. 1994. *Common Chemicals Found at Superfund Sites*. EPA 540/R-94/044. Washington, DC: Office of Solid Waste and Emergency Response.

———. 1995. *Contaminants and Remedial Options at Selected Metal-Contaminated Sites*. EPA/540/R-95/512. Cincinnati: National Risk Management Research Laboratory.

Verkleij, J. A. C., and H. Schat. 1990. Mechanisms of metal tolerance in higher plants. In: *Heavy Metal Tolerance in Plants: Evolutionary Aspects*, edited by J. A. Shaw, 179–93. Boca Raton, FL: CRC Press.

Wagner-Döbler. No date. *Microbiological Treatment of Mercury Loaded Waste Water—The Biological Mercury-decontamination-system*. German Research Centre for Biotechnology. http://www.gbf.de/mercury remediation1/pdf-documents/Information%20leaflet.PDF.

Webb, M. A. H., C. B. Schreck, M. Plume, C. Wong, D. T. Gunderson, G. W. Feist, M. S. Fitzpatrick, and E. P. Forester. 2006. Mercury concentrations in gonad, liver, and muscle of white sturgeon *Acipenser transmontanus* in the lower Columbia River. *Archives of Environmental Contamination and Toxicology* 50(3):443–51.

Welch, A. H., M. S. Lico, and J. L. Hughes. 1988. Arsenic in groundwater of the western United States. *Groundwater* 26(3):333–47.

Wenzel, W. W. 2013. Arsenic. In: *Heavy Metals in Soils: Trace Metals and Metalloids in Soils and Their Bioavailability*, edited by J. Alloway. New York: Springer.

World Health Organization. 1996. *Permissible Limits of Heavy Metals in Soil and Plants*. Geneva, Switzerland.

Woolson, E. A. 1992. Introduction to arsenic chemistry and analysis. In: *Arsenic and Mercury. Workshop on Removal, Recovery, Treatment, and Disposal*. EPA/600/R-92/105.

Woolson, E. A., J. H. Axley, and P. C. Kearney. 1973. The chemistry and phytotoxicity of arsenic in soils II. Effects of time and phosphorus. *Soil Science Society of America Proceedings* 37(2):254.

Yang, G., J. Wu, and Y. Tang. 2005. Research advances in plant resistance mechanisms under lead stress. *Chinese Journal of Ecology* 24(12):1507–12.

Zayed, A. M., and N. Terry. 2003. Chromium in the environment: Factors affecting biological remediation. *Plant and Soil* 249(1):139–56.

Zimdahl, R. L. 1975. *Entry and movement in vegetation of lead derived from air and soil sources.* Paper presented at the 68th annual meeting of the Air Pollution Control Association, Boston, June 15.

Chapter 3

Hydrocarbon Chemistry and Properties

The healthful balm, from Nature's secret spring,
The bloom of health, and life, to man will bring;
As from her depths the magic liquid flows,
To calm our sufferings, and assuage our woes.

—Seneca Oil advertisement, circa 1850

3.1 INTRODUCTION

Hydrocarbon-based chemicals are the foundation of numerous industries including fuel refining; plastics manufacture; paint, solvent, and pesticide manufacture; dry cleaning; asphalt production; detergents; electronics; and many others. The use of hydrocarbons worldwide has soared since World War II.

Release of hydrocarbons to the biosphere are of increasing concern. Decades following passage of the Oil Pollution Act (1990), massive oil spills to the world's oceans continue to make headlines (Madrigal 2018). Contamination of soil has occurred following accidents in pipeline and truck transport of petroleum products. Fuels such as gasoline, diesel jet fuel have contaminated soil and water from improper use or storage. Also common is spillage of motor oil onto soil due to improper handling. Immense quantities of pesticides, solvents, and other petrochemicals are used on the commercial and individual scale in the United States, and transport and application of these chemicals is of public health and environmental concern.

Certain petroleum hydrocarbons are acutely hazardous to public health or the biosphere, either via explosion hazard, direct toxicity, defatting of biological tissue, or simple asphyxiation. Long-term exposure to hydrocarbon compounds such as benzene (a common gasoline additive) or benzo[a]pyrene (which occurs in diesel exhaust fumes, tobacco smoke, and charbroiled food) is known to increase cancer risk.

Hazardous waste-contaminated sites are often heterogeneous mixtures of hydrocarbon (organic) compounds that may be at least partially water soluble, resistant to bacterial degradation, and/or toxic in small doses. In order to successfully implement a remediation program for a hydrocarbon-affected site, adequate knowledge of the chemical and physical properties of those hydrocarbons is essential. Understanding the chemical nature of organic contaminants is important in assessing routes of exposure to organisms and paths of migration in the biosphere. Operators need to know which compounds are expected to be toxic, which will float upon or sink within an aquifer, which will volatilize, and so on.

Petroleum products are mixtures of hydrocarbon compounds; that is, they are a class of chemical compounds that contain the elements hydrogen and carbon. Although a broad range of compounds exists, the general categories of hydrocarbons encountered in nature as well as in commercial use are rather predictable.

3.2 STRUCTURE AND NOMENCLATURE

Organic chemistry is the study of carbon-containing compounds including hydrocarbons that are bound by a covalent bond. In a covalent bond, electrons are shared between similar or chemically similar atoms (e.g., C—C; C=O; C—N) as opposed to bonds such as ionic (electrostatic), which involve the attraction of oppositely charged ions (e.g., K^+Cl^-).

Hydrocarbon molecules are composed of carbon and hydrogen in myriad combinations. An important rule of thumb for the correct representation of organic structures is that, in a stable hydrocarbon molecule, carbon always, *ALWAYS*, has four bonds!

3.2.1 Aliphatic Hydrocarbons

The aliphatics are a category of straight-chain or branched hydrocarbons that can be divided into three groups depending on the number of bonds between adjacent carbon atoms. Alkanes contain only single bonds between carbon atoms, alkenes possess a double bond, and alkynes contain a triple bond. In those molecules having a combination of bonds, the compound is classified based on the highest number of bonds.

3.2.2 Alkanes

Alkanes possess a single bond between all carbon atoms. They may occur as straight chains, branched chains, or cyclic alkanes. The general formula for straight-chain alkanes is C_nH_{2n+2}. The simplest alkane is methane, CH_4.

$$\begin{array}{c} H \\ | \\ H-C-H \\ | \\ H \end{array}$$

We can build on this simple foundation and create a two-carbon chain:

$$\begin{array}{cc} H & H \\ | & | \\ H-C-C-H, \text{ also represented as } CH_3CH_3 \text{ or } C_2H_6, \text{ ethane.} \\ | & | \\ H & H \end{array}$$

Adding a single C to the chain (and observing the "four bonds to carbon" rule), we continue:

$$\begin{array}{ccc} H & H & H \\ | & | & | \\ H-C-C-C-H \text{ or } CH_3(CH_2)CH_3 \text{ or } C_3H_8 \text{ (propane)} \\ | & | & | \\ H & H & H \end{array}$$

$CH_3(CH_2)_2CH_3$	butane
$CH_3(CH_2)_3CH_3$	pentane
$CH_3(CH_2)_4CH_3$	hexane
$CH_3(CH_2)_5CH_3$	heptane
$CH_3(CH_2)_6CH_3$	octane
$CH_3(CH_2)_7CH_3$	nonane
$CH_3(CH_2)_8CH_3$	decane
.	
.	
.	
$CH_3(CH_2)_{14}CH_3$	hexadecane, etc.

Methane, ethane, and propane possess low molecular weights, have very low boiling points, and are the primary components of natural gas, a common fuel.

The above alkanes are all known as normal, or *n*-alkanes, which indicate that the molecule exists as a straight chain without branches. Branched alkanes may be represented by the prefix *iso-*, for example, isobutane,

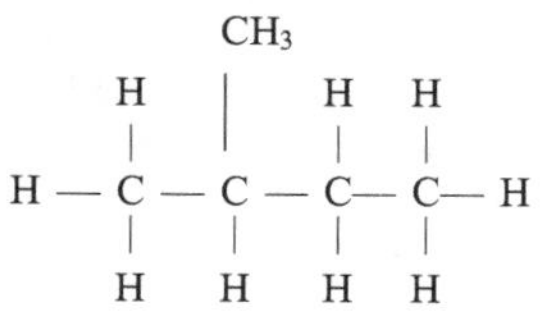

$$\begin{array}{c} \quad\quad\quad CH_3 \\ \quad H \quad\; | \quad\; H \;\; H \\ \quad\;\; | \quad\;\; | \quad\;\; | \;\; | \\ H-C-C-C-C-H \\ \quad\;\; | \quad\;\; | \quad\;\; | \;\; | \\ \quad H \;\; H \;\; H \;\; H \end{array}$$

Branched isomers of hydrocarbons are important in gasoline formulations because they burn at a higher rate compared to the corresponding *n*-alkane—that is, they have a higher *octane number*. This number is essentially a rating of a fuel sample, used to indicate antiknock performance in motor vehicle engines. The higher the octane number, the greater the resistance to engine knock and the better the acceleration. Octane number is discussed further in section 3.3.3, 'Gasoline.'

Alkanes also occur in cyclical structures, for example, cyclohexane, C_6H_{12}:

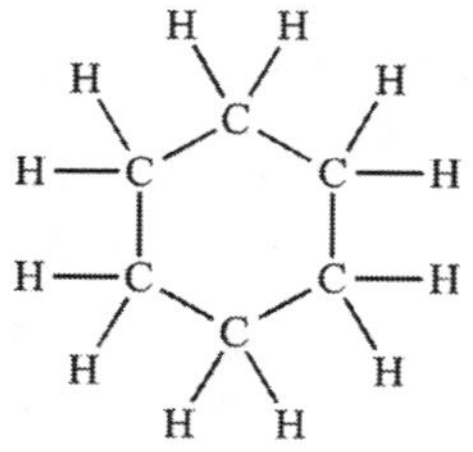

or abbreviated as:

Where each apex of the hexagon is a carbon atom. Two H atoms are understood at each apex in order to form a total of four bonds to each C. Cyclohexane is a common component of gasoline.

3.2.3 Substitution in Aliphatics

In industrial, petrochemical, and other applications, aliphatics may contain substituents bound to the base molecule. For example, alkyl halides contain a halogen atom in place of a hydrogen:

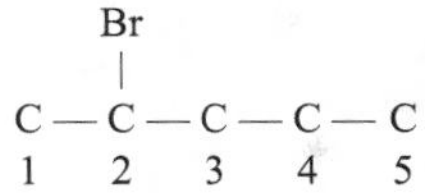

$$\begin{array}{c} \text{Br} \\ | \\ \text{C}-\text{C}-\text{C}-\text{C}-\text{C} \\ 1 \quad 2 \quad 3 \quad 4 \quad 5 \end{array}$$

which is 2-bromopentane. The halogens comprise Group VIIB of the Periodic Table and include F, Cl, Br, I, and At. Note that, in the structure above, the '2-' indicates the carbon position to which the substituent (the Br atom) is attached. Numbering is such that we arrive at the smallest possible number for a substituent location. The molecule:

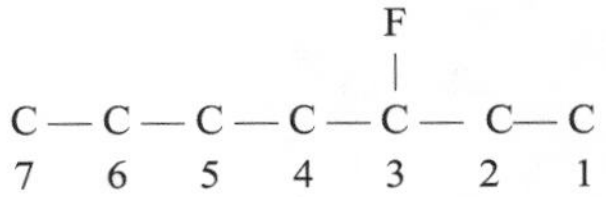

$$\begin{array}{c} \text{F} \\ | \\ \text{C}-\text{C}-\text{C}-\text{C}-\text{C}-\text{C}-\text{C} \\ 7 \quad 6 \quad 5 \quad 4 \quad 3 \quad 2 \quad 1 \end{array}$$

is 3-fluoroheptane, *not* 5-fluoroheptane.

Another example is carbon tetrachloride, a common industrial solvent, which is essentially a substituted methane:

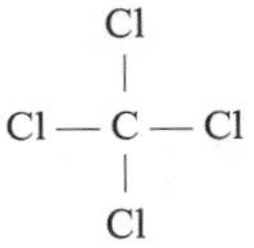

$$\begin{array}{c} \text{Cl} \\ | \\ \text{Cl}-\text{C}-\text{Cl} \\ | \\ \text{Cl} \end{array}$$

It should be noted that the simple addition of a halide to a hydrocarbon molecule will drastically alter its chemical and physical properties including boiling point, water solubility, and vapor pressure, among others. Halides bound to hydrocarbons typically impart heat resistance to the molecule and greatly restrict biodegradation.

3.2.4 Alkenes

An alkene, also known as an olefin, contains at least one double bond between adjacent carbons and is considered *unsaturated* (i.e., not all C atoms are 'saturated' with H atoms). The general formula for alkenes is C_nH_{2n}. The name of the alkene ends in *-ene* or *-ylene*.

$H_2C=CH_2$ is ethene (or ethylene)

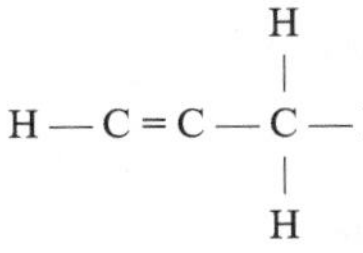

$$\begin{array}{c} \text{H} \\ | \\ \text{H}-\text{C}=\text{C}-\text{C}-\text{H} \text{ is 1-propene} \\ | \\ \text{H} \end{array}$$

$$\begin{array}{c} \text{H} \quad \text{H} \\ | \quad | \\ \text{H—C=C—C—C—H is 1-butene} \\ | \quad | \quad | \quad | \\ \text{H} \quad \text{H} \quad \text{H} \quad \text{H} \end{array}$$

$CH_2CH=CHCH_2$ is 2-butene.

Alkenyl halides contain at least one halogen atom and one carbon-carbon double bond. Examples include trichloroethylene (TCE) and tetrachloroethylene (or perchloroethylene, PCE), which are common industrial solvents:

$$\begin{array}{cc} \text{Cl} \quad \text{Cl} & \text{Cl} \quad \text{Cl} \\ \diagdown \quad \diagup & \diagdown \quad \diagup \\ \text{C=C} & \text{C=C} \\ \diagup \quad \diagdown & \diagup \quad \diagdown \\ \text{H} \quad \text{Cl} & \text{Cl} \quad \text{Cl} \\ \text{trichloroethylene} & \text{tetrachloroethylene} \end{array}$$

3.2.5 Alkynes

An alkyne contains at least one triple bond in the molecule. The general formula is C_nH_{2-n}. The triple bond is a high-energy bond; thus, alkynes are often used for high-temperature flames as in the case of ethyne (acetylene) in welding. Examples of alkynes include:

$$\begin{array}{cc} \text{H—C≡C—H} & \text{H—C≡C—CH} \\ \text{ethyne} & \text{propyne} \end{array}$$

3.2.6 Aromatic Hydrocarbons

Aromatic hydrocarbons comprise a unique category of hydrocarbons in that all such molecules possess the C_6H_6 (benzene) ring, represented as:

Single and double bonds alternate between adjacent carbons. Furthermore, these bonds are not permanent; rather, they are perpetually alternating at extremely high speeds. This continuous transition between single and double bonds is termed *resonance*, which imparts a high degree of stability to aromatic compounds. As we shall see in subsequent chapters, aromatic hydrocarbons are often difficult to decompose by chemical or biological processes.

Examples of aromatic compounds are naphthalene, anthracene, cumene, and benzo[a]pyrene:

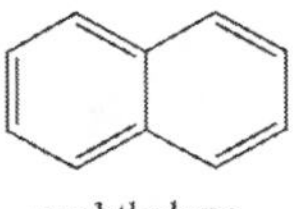

naphthalene 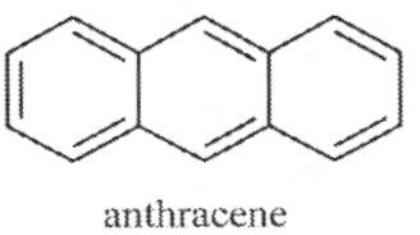 anthracene 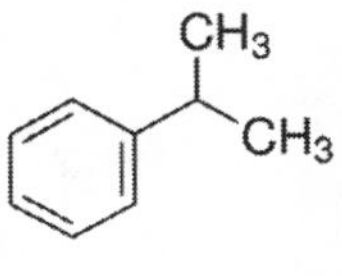 cumene 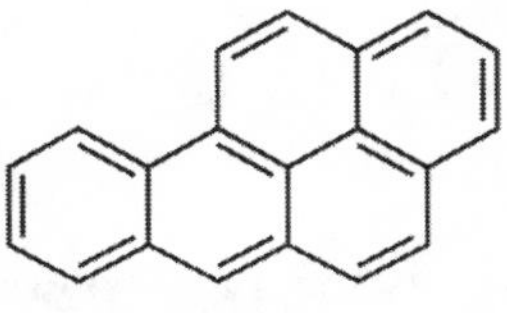 benzo[a]pyrene

3.2.7 Alkylaromatics

This group consists of aromatic compounds that contain an alkyl molecule in place of a hydrogen atom attached to a ring carbon. Examples of alkylaromatics are shown below:

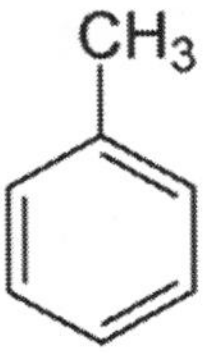

methylbenzene 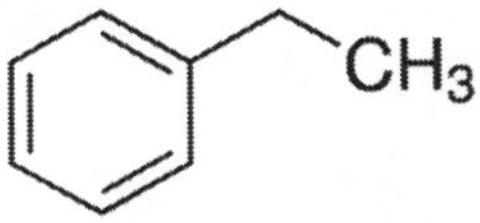 ethylbenzene

Ethylbenzene is a common component in gasoline blends.

3.2.8. Polycyclic Aromatic Hydrocarbons

Polycyclic aromatic hydrocarbons (PAHs), or polynuclear aromatic hydrocarbons (PNAs), are characterized by two or more fused benzene rings. Naphthalene, anthracene, and benzo[a]pyrene, shown above, are examples of PAHs. These compounds are ubiquitous in various products and wastes as well as in combustion gases. Diesel fuels and exhausts from gasoline and diesel engines contain PAHs. PAHs also occur in tobacco smoke. Some PAHs are known to be potent human carcinogens:

- benz[*a*]anthracene
- benzo[*b*]fluoranthene
- benzo[*j*]fluoranthene
- benzo[*k*]fluoranthene
- benzo[*a*]pyrene
- indeno[*1,2,3-cd*]pyrene
- dibenz[*a,h*]anthracene
- chrysene

PAHs are not effectively degraded by microbial action and are strongly adsorbed to soil colloids, particularly clays and organic matter; therefore, they are difficult to remediate. However, due to adsorption processes combined with their very low water solubilities and vapor pressures, they are not highly mobile in soil or groundwater. Several studies have shown that certain PAHs can be decomposed under anaerobic conditions (Robinson et al. 2006; Ambrosoli et al. 2005).

3.2.9 Substituted Aromatics

A number of substituents in the form of organic groups or inorganics—for example, halogens—may be bound to a benzene ring and impart unique properties (e.g., altered water solubility, toxicity). Some common substituted aromatics include:

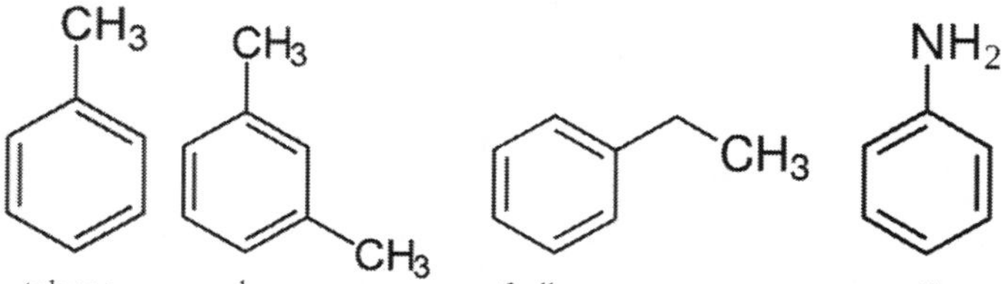

toluene   xylene   ethylbenzene   aniline

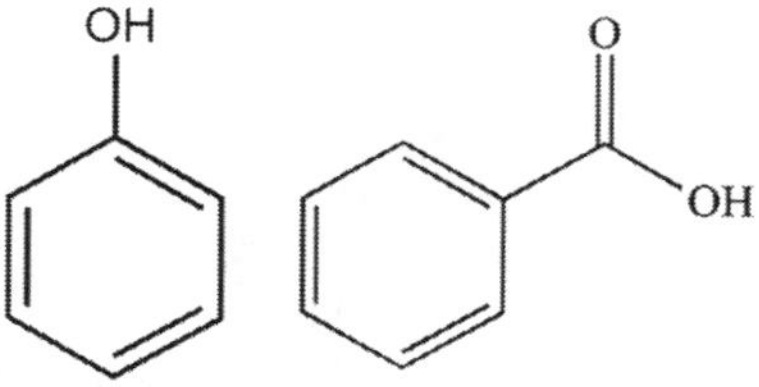

phenol   benzoic acid

The substituents on the benzene ring are numbered from 1 to 6, as in the case of 1,2,3-trichlorobenzene and 2,4,6-trichlorophenol:

1,2,3-trichlorobenzene   2,4,6-trichlorophenol

As with the numbering of substituents on aliphatic chains, those on aromatic molecules are numbered to provide the lowest values. If there is only one additional functional group attached to an already substituted ring, the isomer is instead named by using the prefixes *ortho-*, *meta-*, and *para-* (*o-*, *m-*, and *p-*, respectively).

o-, *m-*, and *p-* positions to a ring.

p-nitrophenol   *m*-chlorobenzoic acid   *o*-bromotoluene

3.3 PROPERTIES OF HYDROCARBON FUELS

Industrialized societies are strongly dependent on petroleum products as fuels and in the production of countless consumer, industrial, agricultural, defense, and other products. The most common uses include fuels for vehicles and industry, heating oils, lubricants, as raw materials in manufacturing petrochemicals and pharmaceuticals, and solvents. Gasoline comprises about 47 percent of total U.S. petroleum consumption (EIA 2018). Thus, petroleum products are extremely common and are the source of many soil and water contamination events.

3.3.1 Composition

Crude petroleum is a complex mixture of thousands of compounds, most of which are hydrocarbons. The detailed composition of crude petroleum depends on the origin and location of the petroleum fields; however, there are many similarities among diverse sources (table 3.1).

Light crudes tend to contain more gasoline, naphtha, and kerosene fractions, and the heavy crudes more gas oil and residue. Typical fractions recovered from the distillation of a crude petroleum sample are shown in table 3.2. The most abundant hydrocarbon is usually a composite of alkanes designated by the formula C_nH_{2n+2}. In petroleum the alkane composition varies from methane, CH_4, to molecules having about 100 carbons. Most petroleum

Table 3.1 Characteristics of crude petroleum components as a function of source

Characteristic or Component	Prudhoe Bay	Louisiana, USA	Kuwait
API gravity (20°C) (°API)	27.8	34.5	31.4
Sulfur (wt %)	0.94	0.25	2.44
Nitrogen (wt %)	0.23	0.69	0.14
Naptha fraction (wt %)	23.2	18.6	22.7
Paraffins	12.5	8.8	16.2
Naphthenes	7.4	7.7	4.1
Aromatics	3.2	2.1	2.4
Benzenes	0.3	0.2	0.1
Toluene	0.6	0.4	0.4
C_8 aromatics	0.5	0.7	0.8
C_9 aromatics	0.06	0.5	0.6
C_{10} aromatics	–	0.2	0.3
C_{11} aromatics	–	0.1	0.1
High-boiling fraction (wt %)	76.8	81.4	77.3
Saturates	14.4	56.3	34.0
n-paraffins	5.8	5.2	4.7
Isoparaffins	–	14.0	13.2
1-ring cycloparaffins	9.9	12.4	6.2
2-ring cycloparaffins	7.7	9.4	4.5
3-ring cycloparaffins	5.5	6.8	3.3
4-ring cycloparaffins	5.4	4.8	1.8
5-ring cycloparaffins	–	3.2	0.4

Source: Clark and Brown 1977; reproduced with kind permission of Elsevier Press, Inc.

Table 3.2 Hydrocarbon products from petroleum

	Number of Carbon Atoms	Boiling Point (°C)	Uses
Petroleum gases	1 to 4	<5	Heating fuel, petrochemicals
Petroleum gases	4 to 10	35–80	Solvents, petrochemicals
Gasoline	5 to 13	20–225	Fuel
Kerosene	10 to 16	180–260	Fuel, lighting
Lubricating oils	20 to 50	350–600	Lubrication
Paraffin	23 to 29	50–60 (m.p.)	Wax
Asphalt	>100	Viscous liquids	Paving, roofing materials
Coke	>100	Solid	Solid fuel

alkanes are of two types: a long, continuous chain of carbons or one chain with short branches, for example, 2-methylhexane:

2-methyhexane

Petroleum also contains cycloalkanes, mainly those with five or six carbons per ring:

methylcyclopentane   cyclohexane

Petroleum contains aromatic hydrocarbons such as benzene and derivatives where one or two hydrogens have been replaced by methyl or ethyl groups (e.g., toluene and ethylbenzene). An additional, albeit unwanted, component of petroleum is sulfur (S). This element occurs in organic compounds arising from the original depositional environment. Sulfur is notorious as an environmental pollutant and is associated with acid precipitation. The S atom is covalently bound to hydrocarbon molecules; therefore, it is not readily removed from petroleum. Petroleum is commonly labeled as being *sweet* or *sour* crude, which is a reflection of its S content: low-S (about 0.5% S) is sweet, and high-S (2.5% or more) is sour. Finally, a range of metals, some of which are potentially hazardous to human health and the biosphere, occur in crude petroleum (again a product of the environment in which the petroleum has formed). Most of these metals, however, typically are present in concentrations less than 1 mg/L (Speight 1980).

3.3.2 Refining

Crude petroleum must be refined to separate its diverse constituents into useful fractions. These fractions are often further processed to produce a specific commercial product. Hydrocarbons in crude petroleum possess unique boiling points according to the number and arrangement of carbon atoms in their molecules. Fractional distillation, accomplished in a distillation tower (figure 3.1), uses the difference in boiling points to separate hydrocarbons in crude oil. Crude petroleum is pumped into a furnace where it is heated to approximately 750°F. The fractionating column is cooler at the top than at the base, so the vapors cool as they rise. An array of perforated caps is situated within the tower. The perforations are fitted with cooled condensers allowing vapors to condense to a liquid. The vapors migrate upward and pass through the openings, while condensed liquids fall and are collected.

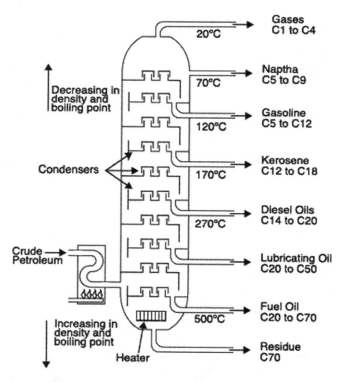

Figure 3.1 Schematic diagram of a distillation tower used to refine petroleum.

The highest boiling vapors condense first and are collected as a liquid. Lower and lower boiling compounds are collected successively up the tower. The highest boiling fractions are heavy, high molecular weight hydrocarbons that have properties suitable for lubricants and heating oils. Asphalts or paraffins also occur depending on the source of the petroleum.

Vapors condensing in the middle range, or 'middle distillates,' are used for heavier fuels, diesel and jet fuels, and other commercial products such as kerosene. The next-lighter hydrocarbons occur in the C_4 to C_9 range and are used in gasoline. The lightest hydrocarbons, approximately C_1 to C_3, are gases at room temperature and are used in heating gas mixtures (figure 3.2).

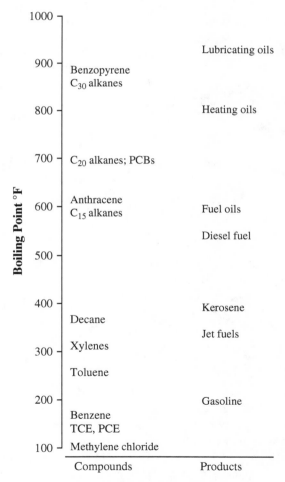

Figure 3.2 Boiling point distribution of petroleum products.

Lowering the pressure over a liquid will lower its boiling point. Heavy residues from the fractionating column are subsequently distilled under a vacuum. As a result the heavier fractions can be further separated without high temperatures that might break them down. These residues are then passed on to a cracking unit, or used to obtain lubricating oils or blended into industrial fuels.

3.3.3 Gasoline

Fuels are complex mixtures of as many as several hundred compounds with molecular weights ranging between C_4 and C_{12}. Most gasoline blends contain from 50 to 150 compounds that are specifically formulated for burn rate (octane number), volatility (for consistent starts and performance in hot and cold weather), and emission control (by the addition of oxygenated compounds) (ATSDR 2015b; Speight 2005). Table 3.3 provides a typical distribution of aliphatic and aromatic compounds for automotive gasoline.

Gasoline fuels are formulated to burn at a certain rate, termed octane number. Power output is maximized by optimizing the rate at which a fuel burns within the cylinders of an engine. The octane scale designates n-heptane an octane number of 0, and iso-octane an octane number of 100. Lead-containing compounds such as tetramethyllead and tetraethyllead had traditionally been used to increase a fuel's octane number. These were convenient additives, as they did not affect any other fuel properties, for example vapor pressure. As a fuel burns, tetraethyl lead converts to lead(II) oxide:

$$2Pb(C_2H_5)_{4(l)} + 27O_{2(g)} \rightarrow 2PbO_{(s)} + 16CO_{2(g)} + 20H_2O_{(l)} \quad (3.1)$$

The lead(II) oxide is produced in the cylinder and is released with the exhaust gases (Meyer 1989). Other hazardous compounds, including ethylene dichloride (EDC) and ethylene dibromide (EDB) were added as Pb scavengers to prevent buildup of lead oxide deposits. In the cylinder, EDC combines with Pb to produce lead chloride, $PbCl_2$, a volatile compound that is carried out of the engine with the exhaust (Sigel et al. 2017). As a result of public health concerns related to leaded compounds, EPA banned the use of Pb additives in fuels in 1973.

Aromatic compounds also increase the octane rating of gasoline blends. Aromatics including benzene and alkylbenzenes are now used in place of Pb to increase octane number. Benzene, toluene, ethylbenzene, and xylene are collectively termed the BTEX compounds, and some premium blends contain up to 50 percent aromatics. BTEX is more reactive than the alkanes typically occurring in fuels and are known to result in the formation of

Table 3.3 Major components of gasoline

Component	Percent by Weight	Component
n-alkanes		Octane enhancers
C_5	3.0	methyl t-butyl ether (MTBE)
C_6	11.6	t-butyl alcohol (TBA)
C_7	1.2	ethanol
C_9	0.7	methanol
C_{10}–C_{13}	0.8	
Total n-alkanes	17.3	Antioxidants
Branched alkanes	2.2	n,n'-dialkylphenylenediamines
C_4	15.1	2,6 dialkyl and 2,4,6-
C_5	8.0	trialkylphenols
C_6	1.9	butylated methyl, ethyl and
C_7	1.8	dimethyl phenols
C_8	2.1	triethylene tetramine di
C_9	1.0	(monononylphenolate)
C_{10}–C_{13}		Metal deactivators
		n,n'-disalicylidene-1,2-ethanediamine
		n,n'-disalicylidenepropanediamine
		n,n'-disalicylidene cyclohexanediamine
		disalicylidene-n-methyldipropylenetriamine
Total branched	32.0	Ignition controllers
cycloalkanes		tri-o-cresylphosphate (TOCP)
C_6	3.0	Icing inhibitors
C_7	1.4	isopropyl alcohol
C_8	0.6	Detergents/dispersants
		alkylamine phosphates
Total	5.0	Poly-isobutene amines
cycloalkanes		
Olefins		long-chain alkyl phenols
C_6	1.8	long-chain alcohols
Total olefins	1.8	long-chain carboxylic acids
		long-chain amines
		Corrosion inhibitors
Aromatics	7.6	carboxylic acids
benzene	3.2	phosphoric acids
toluene	4.8	sulfonic acids
xylenes	6.6	
ethylbenzene	1.4	
C_3-benzenes	4.2	
C_4-benzenes	7.6	
Others	2.7	
Total aromatics	30.5	

Source: Agency for Toxic Substances and Disease Registry 2016a.

photochemical smog (Rappenglück et al. 2000). In countries where catalytic converters are not required in automobiles, increased BTEX concentrations in air may occur. Aromatic compounds are hazardous to human health; therefore, BTEX and cumenes are listed under RCRA Subtitle C regulations as hazardous substances (40 CFR 2015). The octane value of a fuel can be enhanced via blending in other organic substances having high octane numbers such as MTBE, which was introduced into fuels in 1979. Methanol is used to produce the oxygenated additive MTBE, methyl tert-butyl ether:

$$H_3C-O-\underset{\underset{CH_3}{|}}{\overset{\overset{CH_3}{|}}{C}}-CH_3$$

MTBE

Ethanol is used to produce the corresponding ethyl tert-butyl ether that is also blended with gasoline. Both ethers have octane numbers over 100. In addition to improving fuel combustion, the use of MTBE reduces generation of carbon monoxide. MTBE is an oxygenated fuel and produces less CO during combustion than do hydrocarbons. Over the past two decades controversy has surrounded MTBE use, as this water-soluble compound has been linked with contamination of drinking water supplies via leaking USTs, pipelines, and marine engines. By the late 1990s, in many communities where MTBE was used, drinking water supplies were found to contain detectable levels of MTBE. As part of the Energy Policy Act of 2005, Congress voted to remove the oxygen content requirement for reformulated gasoline. Since then, companies have switched from MTBE to ethanol as a gasoline additive. According to EPA, MTBE has not been used in significant quantities in gasoline since 2005. Regardless, groundwater in some parts of the United States might still contain MTBE. Furthermore, MTBE continues to be used as a gasoline additive in other parts of the world.

There is a conflict between the EPA's toxicity characteristic rule and the petroleum exemption under RCRA and CERCLA. Gasoline is obviously hazardous as a result of its flammability and toxicity. Under some conditions petroleum contamination can be hazardous in the workplace since the more volatile products constitute a fire and explosive hazard. Some petroleum compounds are toxic, especially those routinely occurring in gasolines, for example, benzene. Despite these obvious hazards, gasoline and gasoline-contaminated debris have been exempted from hazardous regulations by EPA.

3.3.4 Diesel Fuel

Diesel fuels rank second to gasoline as a fuel for internal combustion engines with a demand of about 25 percent that of gasoline. Diesel engines require that a fuel self-ignite during compression in the cylinder. A measure of ignition quality is the *cetane number*, which corresponds to the percent of cetane, or *n*-hexadecane ($C_{16}H_{34}$) in a mixture of cetane and heptamethylnonane. The *cetane number*, partly a function of the paraffin, olefin, naphthalene, and aromatic composition, is also a measure of the tendency of a diesel fuel to 'knock' in an engine. Paraffins have low self-ignition temperatures; this makes them desirable in a diesel fuel. Diesel fuels contain a heavier range of hydrocarbons compared to gasoline, and are in the C_{10} to C_{18} range. Diesel fuel blends are a compromise between ease of starting (high volatility) and good fuel economy (low volatility).

3.3.5 Kerosene

Kerosene originated as a straight-run petroleum fraction that boiled between about 400°F and 500°F and has been used as a fuel oil since the petroleum refining industry began. Kerosene is defined as a petroleum distillate and must be free of aromatic and unsaturated hydrocarbons, which are potential smoke producers. The desirable components are saturated hydrocarbons, primarily C_{12} or higher.

3.3.6 Fuel Oil

Domestic fuel oils are those commonly used in homes and include stove oil and furnace fuel oil. These are so-called distillate fuels because they are vaporized during the refining process and thus possess a distinct boiling range. The fuel oil range hydrocarbons possess distinct characteristics that make them desirable for heating oil; for example, they possess a higher calorific content than the lighter hydrocarbons such as kerosene or naphtha (Speight 2005).

3.3.7 Lubricating Oils

Lubricating oils are distinguished from other fractions of crude oil by their high (>750°F) boiling point and high viscosity. Hydrocarbons suitable for production of lubricating oils contain as many as forty carbons per molecule. In these oils there is a predominance of normal and branched paraffins. There are also polycycloparaffins, whose rings are commonly condensed. Mono-, di-, and trinuclear aromatics, for example, naphthalene and phenanthrene, are the main component of the aromatic portion:

naphthalene   phenanthrene

Lubricating oils typically possess a high additive content. These compounds are included in an oil blend to improve both physical and chemical properties. Some of the more important additives include antiknock agents, oxygenates, fuel stabilizers, detergents, and dispersants.

Enormous quantities of used lubricating oils are generated from maintenance of motor vehicle engines and industrial machinery. Used oil is a public health and environmental concern because many are known to be contaminated with a suite of metals, other inorganics, and hazardous organic compounds. For example, some used automobile oils are known to contain Pb, As, and Ba, as well as PAHs, many of which are known carcinogens (ATSDR 2015b). These oils are not, unfortunately, managed as RCRA hazardous wastes in the United States as a result of the EPA petroleum exemption. They are managed under other regulations, however, which were formulated to restrict their release to the biosphere. For example, the EPA Used Oil Management Standards, enacted in 1992, prohibit the use of waste oil for dust control or weed control (U.S. EPA 1994). Unfortunately, however, 200 million gal of used oil are estimated to be improperly disposed each year in the United States alone (U.S. EPA 2018).

Aliphatic compounds comprise about 73 to 80 percent of the total weight of used motor oil. This fraction is composed of alkanes and cycloalkanes of one to six rings. Monoaromatics and diaromatics make up another 11 to 15 percent and 2 to 5 percent of the weight, respectively (Vazquez-Duhalt 1989).

3.4 PETROCHEMICALS

Natural gas and crude distillates from petroleum refining are used as feedstocks to manufacture a variety of petrochemicals that are subsequently used in the manufacture of consumer goods.

The basic petrochemicals manufactured by cracking, reforming, and other processes include alkenes (e.g., ethylene, propylene, butylenes, butadiene) and aromatics (e.g., benzene, toluene, xylenes). Some petrochemical plants may also have additional (e.g., alcohol) compound manufacturing units on-site. These petrochemicals or products derived from them, along with other raw materials, are converted to a wide range of products. Common examples include (Ranken Energy 2017; Burdick and Leffler 2010):

- plastics such as low-density polyethylene, high-density polyethylene, polypropylene, polystyrene, and polyvinyl chloride
- synthetic fibers such as polyester and acrylic
- engineering polymers such as acrylonitrile butadiene styrene (ABS)
- rubbers including styrene, butadiene rubber, and polybutadiene rubber
- solvents
- industrial chemicals, including those used for the manufacture of detergents, coatings, dyes, agrochemicals (pesticides), pharmaceuticals, and explosives

Some single-carbon compounds manufactured at petrochemical plants include methanol, formaldehyde, and halogenated hydrocarbons. Alkenes (olefins) are typically manufactured from the steam cracking of hydrocarbons such as naphtha. Major alkenes produced include ethylene, propylene, butadiene, and acetylene. Benzene is generally recovered from cracker streams at petrochemical plants. The major aromatic hydrocarbons manufactured include benzene, toluene, xylene, and naphthalene. The alkanes, alkenes, and aromatics produmanufactured are used in the manufacture of a wide range of products and are shown in table 3.4.

Table 3.4 Uses of alkanes, alkenes, aromatics, and related hydrocarbons in petrochemical applications

General Category	Compound	Applications
C-1 alkanes	Formaldehyde	Manufacture of plastic resins including phenolic, urea, and melamine resins. Bakelite, Formica, methanol
	Halogenated	Manufacture of solvents, refrigerants and degreasing hydrocarbons agents.
Alkenes	Ethylene	Low-density polyethylene, high-density polyethylene, polystyrene, polyvinyl chloride, ethylene glycol, ethanolamines, nonionic detergents
	Butadiene	Manufacture of nitrile rubber and styrene butadiene rubber

(Continued)

Table 3.4 (Continued)

General Category	Compound	Applications
	Butanol	Additive in hydraulic and brake fluids and perfumes; manufacture of solvents (methyl ethyl ketone)
	Various	Acetone (solvent)
		Acrylonitrile (manufacture of acrylic fibers and nitrile rubber)
		Ethanol amines (solvents)
		Polyisoprene (for synthetic rubber manufacture)
		Polyvinyl acetate (used in plastics)
		Polypropylene
		Isopropanol (solvent and in pharmaceuticals manufacturing)
		Propylene glycol (used in pharmaceuticals manufacturing)
		Polyurethane
Aromatics	Benzene	Solvent; manufacture of phenol, styrene, aniline, nitrobenzene, detergents, pesticides (e.g., hexachlorobenzene), cyclohexane (intermediate in synthetic fiber manufacture), caprolactam (used in the manufacture of nylon)
	Toluene	Solvent in paints, rubber, and plastic cements
		Feedstock in the manufacture of organic chemicals, explosives, detergents, and polyurethane foams.
	Xylenes	Manufacture of explosives (TNT), alkyd resins, plasticizers
	Naphthalene	Manufacture of dyes, pharmaceuticals, insecticides, mothballs, phthalic anhydride (used in the manufacture of alkyd resins, plasticizers, and polyester)
	Phenol	Thermoset plastics
		Solvent; manufacture of pesticides, pharmaceuticals, dyestuffs
	Styrene	Manufacture of synthetic rubber and polystyrene resins.
	Phthalic	Manufacture of alkyd resins and plasticizers (e.g., anhydride phthalates)
	Maleic anhydride	Manufacture of polyesters and alkyd resins, malathion
	Nitrobenzene	Manufacture of aniline, benzidine, dyestuffs; solvent in polishes
	Aniline	Manufacture of azo dyes, and rubber chemicals such as antioxidants

Adapted from: World Bank Group 1998.

3.5 CHEMICAL AND PHYSICAL PROPERTIES OF FUELS AND PETROCHEMICALS

From the standpoint of delineation of a contaminant plume and prognosis for success in remediation, essential physical properties of petroleum compounds include solubility in water, specific gravity, viscosity, and vapor pressure (table 3.5).

3.5.1 Solubility

Petroleum hydrocarbons are, with few exceptions, insoluble in water. The maximum solubility of benzene is 1,750 ug/L (1,750 ppb) of water. That amount is, however, sufficient to be harmful to human health. The EPA limit for benzene in groundwater is 5 ug/L (table 3.6). In most cases, solubility of a compound is inversely proportional to its molecular weight; lighter hydrocarbons are more soluble in water than are higher molecular weight compounds. Lighter hydrocarbons (C_4 to C_8 including the aromatics) are relatively soluble. Gasoline is the only petroleum product in common use that contains constituents that are sufficiently soluble in water to cause health problems. The aromatic compounds (benzene and alkyl benzenes) are the primary concern.

Solubility is also a practical concern from a remediation perspective. Contaminants must be in soluble form in order for microorganisms to attack molecules for catabolism (degradation) and respiration. Nonpolar (i.e., uncharged) compounds tend to be hydrophobic and partition into the organic component of a soil. The result is that nonpolar compounds are less mobile in soil and groundwater compared to polar compounds.

3.5.2 Specific Gravity

Specific gravity is defined as the density of a substance compared with the density of water. Petroleum products are less dense than water and will float. For example, gasoline has a specific gravity of approximately 0.73, and no. 2 fuel oil about 0.90. Petroleum products are designated as LNAPLs (light nonaqueous phase liquids). A halogenated hydrocarbon, for example, CCl_4, has a specific gravity of 1.59 and is labeled a DNAPL (dense nonaqueous phase liquid). LNAPLs are relatively easy to locate and recover in the subsurface environment because they tend to float on the water table. In contrast, DNAPLs sink to bedrock and are neither easy to locate nor recover.

Table 3.5 Physical and chemical properties of selected hydrocarbons

Product	Vapor Pressure mm Hg 20°C	Vapor Density Air = 1	Flashpoint °C	Flammability Percent by Volume LE	Flammability Percent by Volume UEL	Solubility in H_2O 20°C, mg/l
Gasoline	450	3–4	−30–43	1.4	7.6	50–100
Benzene	100	2.8	−11	1.3	7.9	1,790
Toluene	36	3.1	4	1.2	7.1	515
Ethylbenzene	10	3.7	18	0.8	6.7	75
Xylene	21	3.7	27	1.1	7.0	150
n-Hexane	124	3.0	−40	1.2	7.1	12
Jet fuel JP-4	103–155	5.5	−10–35	1.3	8.0	<1
Diesel	<1	4.5	40–65	1.3	6.0	<1
Kerosene	1	4.5	40–75	1.4	6.0	<1
Light fuel oil	50	3–7	40–100	nf[1]		<1
Heavy fuel oil	neg	<0.1	65–130	1.0	5.0	<1
Lubricating oil	<0.1	1	50–225	nf		<1 ppb
Used oil	–	–	>100	nf		<1 ppb
Carbon tetrachloride	91	5.3	N/A	nf		neg.
Trichlorethylene	57.8	4.5	N/A	12.5	90	neg.
Chloroform	167	4.1	N/A	nf		1.8 g/100 ml
Pentachlorophenol	40@ 221°C	9.2	–	nf		neg.
Methylene chloride	400@75°C	2.9	N/A	nf		1.32%

[1]Relatively nonflammable.

Table 3.6 Maximum Contaminant Levels (MCLs) of selected organic compounds in drinking water

Compound	Empirical Formula	Molecular Weight	Solubility* mg/L	MCL mg/L
Benzene	C_6H_6	78.1	1,800	0.005
Toluene	C_7H_8	92.2	500	2
Xylenes	C_8H_{10}	106.2	198	10
Ethylbenzene	C_8H_{10}	106.2	150	0.7
Pentachlorophenol	C_6OHCl_5	266.3	20	0.2
Carbon tetrachloride	CCl_4	153.8	800	0.005
Trichloroethylene	C_2HCl_3	131.4	1.1	0.005
Ethylene dibromide	$C_2H_4Br_2$	187.9	4,000	0.00005
Tetrachlorodibenzo[p]dioxin	$C_{12}H_2O_2Cl_4$	322	0.0002	0.00000005
Vinyl chloride	C_2H_3Cl	62.5	2,792	0.002

*In water, 20°C.
Source: 40 CFR 2015.

3.5.3 Viscosity

Viscosity of a hydrocarbon is a measure of its resistance to gravity flow. In the context of site remediation, viscosity indicates the ease with which the compound flows through soil. Only gasoline is of sufficiently low viscosity to migrate rapidly in most soils. Diesel and jet fuel will migrate, but more slowly than gasoline. A fuel oil release can thus be treated much less aggressively than, say, a gasoline release because the fuel oil is not going to migrate from the site as readily.

3.5.4 Vapor Pressure

Vapor pressure, or volatility, is the tendency of a molecule to rise from the surface of a liquid (figure 3.3). This property is approximately the inverse of boiling point; likewise, it is the inverse of molecular weight. Some refined petroleum products, especially gasolines, vaporize readily and have flashpoints at room temperature or below. At the other end of the spectrum are the heavy viscous products such as lubricating oils and fuel oils that vaporize minimally (Leffler 2008; Speight 2005).

3.5.5 Explosive Limits

Gasoline is explosive when its vapors are present in the range of 1.4 to 7.6 percent by volume in air. The LEL (lower explosive limit) is defined as the lowest percent by volume of a mixture of explosive gases in air that will propagate a flame at 25°C and atmospheric pressure. At concentrations

Figure 3.3 There are right ways and wrong ways to assess a compound's vapor pressure! *Source*: U.S. Environmental Protection Agency.

less than 1.4 percent gasoline vapor will not explode as the air-vapor mixture is considered 'lean.' At concentrations greater than 7.6 percent, the vapor will not explode as the mixture is considered 'rich.' The 7.6 percent threshold is the UEL (upper explosive limit), defined as the maximum concentration of a gas above which the substance will not explode when exposed to a source of ignition. The explosive hazard range occurs between the LEL and the UEL. It must be noted that at gasoline concentrations above the UEL an explosion may still be possible, and asphyxiation will occur. In addition, a sudden dilution of gasoline vapors in the local atmosphere can bring the mixture back within the explosive range (figure 3.4).

3.5.6 Flashpoint

When a hydrocarbon liquid burns, it is actually *the vapor above the liquid* that combusts. Whereas explosive limits address concentration, flashpoint relates to temperature—that is, the lowest temperature at which a flame will propagate through the vapor of a combustible material. In other words, flashpoint is the minimum temperature at which the liquid produces a sufficient concentration of vapor that it forms an ignitable mixture with air. The source of ignition need not be an open flame, but could be static electricity or other electrical discharge (figure 3.4).

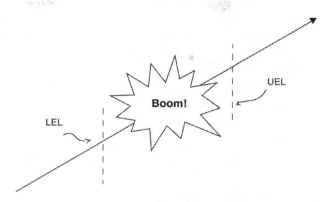

Figure 3.4 Schematic representation of LELs and UELs.

QUESTIONS

1. Choose the correct answer. A hydrophobic compound: (a) has a defatting effect on biological tissue; b) is soluble in fats; (c) is soluble in oils; (d) is insoluble in water; (e) is nonpolar; (f) all of the above.
2. Heavy (high molecular weight) petroleum hydrocarbons are usually insoluble in water. True or false?
3. Choose the correct answer. A substance which is immiscible: (a) will not dissolve in water; (b) is heavier than water; (c) is insoluble in fats.
4. Define a fuel's octane number.
5. Discuss some of the major octane boosters used in fuels. Discuss also any known or suspected health and environmental threats.
6. During a UST excavation, would a hydrocarbon with a vapor density >1.25 be expected to rise, sink, or not migrate at all?
7. PCBs are readily absorbed into the body but are slowly metabolized and excreted. Draw its structure and explain why this occurs.
8. Hydrocarbons are good solvents for fatty substances. They are also good degreasers. True or false?
9. All other factors being equal, why are aromatic hydrocarbons more difficult to chemically degrade as compared with an alkane of similar number of carbon atoms?
10. All other factors being equal, why is a large PAH molecule (ten fused rings) more difficult to decompose as compared with a mixed aromatic-alkane of similar molecular weight?
11. How do flashpoint and explosive limits differ for a hydrocarbon substance?

12. A field technician must enter a closed UST in order to remove petroleum sludges. The UST had formerly stored diesel fuel. What hazards should the worker be aware of prior to tank entry? Consider both immediate and long-term effects.

REFERENCES

Agency for Toxic Substances and Disease Registry. 2016a. *Toxicological Profile Information Sheet. Gasoline.* https://www.atsdr.cdc.gov/ToxProfiles/tp.asp?id=468&tid=83.

———. 2016b. *Toxicological Profile Information Sheet. Used Mineral-based Crankcase Oil.* https://www.atsdr.cdc.gov/ToxProfiles/tp.asp?id=667&tid=123.

Ambrosoli, R., F. A. Marsan, L. Petruzzelli, and J. L. Minati. 2005. Anaerobic PAH degradation in soil by a mixed bacterial consortium under dentrifying conditions. *Chemosphere* 60 (9): 1231–36.

Burdick, D. L., and W. L. Leffler. 2010. *Petrochemicals in Nontechnical Language*, 4th ed. Tulsa, OK: PennWell Corp.

Clark, R. C., and D. W. Brown. 1977. Petroleum: Properties and analyses in biotic and abiotic systems. In *Effects of Petroleum on Arctic and Subarctic Marine Environments and Organisms*, edited by D. W. Brown, R. C. Clark Jr., N. L. Karrick, and W. D. MacLeod. New York: Academic Press.

Cole, G. M. 1994. *Assessment and Remediation of Petroleum Contaminated Sites.* Boca Raton, FL: CRC Press.

40 CFR. 2015. *Part 141.50. Maximum Contaminant Level Goals for Organic Contaminants.* Washington, DC: U.S. Government Printing Office.

———. *Part 261. Identification and Listing of Hazardous Waste.* Washington, DC: U.S. Government Printing Office.

Leffler, W. 2008. *Petroleum Refining in Nontechnical Language*, 4th ed. Tulsa, OK: PennWell Corp.

Madrigal, A. 2018. The world has never seen an oil spill like this. *The Atlantic Monthly.* https://www.theatlantic.com/technology/archive/2018/01/the-oil-spill-that-wasnt/550820/.

Meyer, E. 1989. *Chemistry of Hazardous Materials*, 2nd ed. Englewood Cliffs, NJ: Prentice Hall.

Ranken Energy. 2017. *A Partial List of Products Made from Petroleum.* https://www.ranken-energy.com/index.php/products-made-from-petroleum/.

Rappenglück, B., P. Oyola, I. Olaeta, and P. Fabian. 2000. The evolution of photochemical smog in the metropolitan area of Santiago de Chile. *Journal of Applied Meteorology* 39 (3): 275–90.

Robinson, J., R. Kalin, R. Thomas, S. Wallace, and P. Daly. 2006. In situ bioremediation of cyanide, PAHs and organic compounds using an engineered Sequenced Reactive Barrier (SEREBAR). *Land Contamination and Reclamation* 14 (2): 478–82.

Sigel, H., A. Sigel, and R. O. Sigel. 2017. *Lead: Its Effects on Environment and Health*. Berlin: De Gruyter Pub.

Speight, J. G. 2005. *Environmental Analysis and Technology for the Refining Industry*. New York: Wiley Interscience.

U.S. EIA (U.S. Energy Information Administration). 2018. *What are Petroleum Products, and What is Petroleum Used For?* https://www.eia.gov/tools/faqs/faq.php?id=41&t=6 (Accessed January 22, 2018).

U.S. EPA (U.S. Environmental Protection Agency). 2018. *Used Oil Basic Information*. https://archive.epa.gov/wastes/conserve/materials/usedoil/web/html/oil.html (Accessed January 22, 2018).

———. 1994. *Environmental Regulations and Technology. Managing Used Motor Oil*. EPA/625/R-94/010. Washington, DC: Office of Research and Development.

Vazquez-Duhalt, R. 1989. Environmental impact of used motor oil. *Science of the Total Environment* 79: 1–23.

World Bank Group. 1998. *Petrochemicals Manufacturing. Pollution Prevention and Abatement Handbook*. ifcln1.ifc.org/ifcext/enviro.nsf/AttachmentsByTitle/gui_petrochem_WB/$FILE/petrochm_PPAH.pdf#search=%22petrochemicals%22.

Chapter 4

Subsurface Properties and Remediation

We know more about the movement of celestial bodies than about the soil underfoot.

—Leonardo da Vinci

4.1 INTRODUCTION

At a site contaminated with organic or inorganic chemicals, the importance of understanding the physical and chemical properties of the soil and subsurface materials cannot be overstated. Soil properties play critical roles in: (1) determining the potential for contaminant migration, including vertical and horizontal distribution; (2) altering the toxicity of contaminants; and (3) planning and implementation of a remediation program. Soils are highly variable with depth, over large geographic regions, and even over short distances. Soils vary significantly in their capacity for mobilizing solutes, for adsorbing organic and inorganic constituents, and for chemical and biological degradation of contaminants. In order to optimize remediation success, therefore, it is essential to possess a thorough knowledge of the subsurface conditions present in the affected area.

Soils are composed of inorganic solids (e.g., clay, silt, sand, and gravel), organic matter, and void space. The inorganic solids form as a result of long-term weathering of the parent geologic material by physical (e.g., freeze-thaw, sedimentation) and chemical (e.g., dissolution, precipitation) forces. Soils are classified by texture, which is a function of their sand, silt, and clay content. Many classification schemes exist for particle size analysis (figure 4.1). The organic matter content may range from less than 1 percent, common

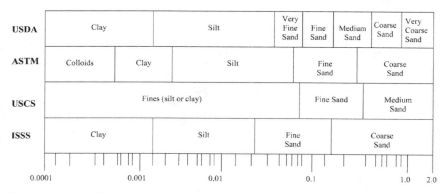

Figure 4.1 Classification schemes for soils based on particle size distribution. USDA = United States Department of Agriculture; ASTM = American Society for Testing and Materials; USCS = Unified Soil Classification System; ISSS = International Society of Soil Science.

for soils in arid regions, to more than 50 percent in peat deposits. Air and water occupy the pore spaces between particles. Pore space may comprise from 25 percent to 60 percent of noncompacted soils.

4.2 RELEVANT CHEMISTRY AND MINERALOGY

4.2.1 pH

Soil pH strongly influences the availability of nutrients and the mobility of metals and other elements. It therefore follows that pH is directly associated with plant and microbial growth and activity. The optimum pH range for most crop plants, and also for microbial populations, is approximately 5.5 to 8. Soils become acidic where there is sufficient precipitation to leach exchangeable base-forming cations (Ca^{2+}, Mg^{2+}, K^+, Na^+) from the profile. Acid rain contributes to lowering soil pH, as does application of certain fertilizers. The two cations that are primarily responsible for soil acidity are H^+ and Al^{3+}. In cases of extreme acidity, aluminum solubilizes into the soil solution as Al^{3+}, which contributes to acidity via hydrolysis reactions.

$$Al^{3+} + H_2O \rightarrow Al(OH)^{2+} + H^+ \qquad (4.1)$$

The H^+ ions are released into the soil solution, thus lowering pH.

Under mildly acidic conditions, Al may occur as aluminum hydroxyl ions, for example, $Al(OH)^{2+}$. In the soil solution they produce H^+ ions via hydrolysis.

$$Al(OH)^{2+} + H_2O \rightarrow Al(OH)_2^+ + H^+ \quad (4.2)$$

$$Al(OH)^{2+} + H_2O \rightarrow Al(OH)_3 + H^+ \quad (4.3)$$

Soils are alkaline when colloids experience a high degree of saturation with basic cations. Most metals precipitate, and are therefore less available, at neutral to high pH. Soil pH can be lowered by the addition of mineral acids, for example sulfuric or nitric; ferrous or aluminum sulfate; or elemental sulfur. Conversely, pH can be increased by addition of limestone ($CaCO_3$), CaO and $Ca(OH)_2$.

Soils possess a resistance to pH change known as *buffering capacity*. In a soil, an equilibrium exists between residual, exchangeable, *and* soluble forms of acidity:

$$\underset{\text{residual}}{[Colloid\text{-}Al]} \leftrightarrow \leftrightarrow \underset{\text{exchangeable}}{[Colloid]\text{-}Al^{3+}} \leftrightarrow \leftrightarrow \underset{\text{soluble}}{Al^{3+}} \quad (4.4)$$

If a small amount of base is applied to the soil to neutralize the H^+ ions in solution, the acid cations are replenished from exchangeable and possibly residual forms. The overall result is a negligible change in soil pH. Soil pH change will be appreciable only when sufficient base is added to neutralize the H^+ and Al^{3+} occurring in the exchangeable *and* residual storehouses.

It follows that soils with a high CEC (i.e., a high number of sites for cation sorption) tend to have the greatest buffering capacities. A number of simple laboratory tests are available to measure soil buffering capacity.

4.2.2 Oxidation-Reduction Status

The redox potential of the soil (oxidation-reduction potential, Eh) is directly related to the concentration of O_2 in both gas and liquid phases within the void spaces. The O_2 concentration is a function of the rate of gas exchange with the atmosphere, and the rate of respiration by soil microorganisms and plant roots. Respiration may deplete O_2, lowering redox potential and creating anaerobic (i.e., reducing) conditions. Such an environment will restrict aerobic reactions and initiate anaerobic processes such as denitrification, sulfate reduction, and fermentation. Polyvalent metal cations may be reduced and become more soluble (and thus more mobile) than their oxidized forms. Well-aerated soils have an Eh of about 0.8 to 0.4 V; moderately reduced soils measure about 0.4 to 0.1 V; reduced soils are about 0.1 to –0.1 V; and highly reduced soils about –0.1 to –0.3 V.

4.3 THE COLLOIDAL FRACTION

Soils are capable of adsorption of nutrients, contaminant elements, and other charged components due primarily to the presence of colloidal particles of varied composition. A *colloid* is defined as a particle measuring less than 1 μm across. By virtue of their extremely small size, soil colloids possess an enormous external surface area per unit weight. The external surface area of 1 g of colloidal clay is about 1,000 times greater than that of 1 g of coarse sand. Some colloids, especially certain silicate clays, possess extensive internal surfaces as well. These occur between the crystal units of each particle; their surface area often exceeds external surface area. The total surface area of soil colloids ranges from 10 m^2/g for clays with only external surfaces, to more than 800 m^2/g for clays with internal surfaces (Weil and Brady 2016).

Colloidal surfaces, both external and internal, are extremely reactive chemically as a result of the presence of permanent electrical charges. For most soil colloids in temperate regions, electronegative charges predominate. The presence and intensity of these influence the attraction and repulsion of particles, thereby influencing a wide array of physical and chemical properties. Surface charges also play a critical role in adsorption or repulsion of contaminant elements, especially metallic cations.

The four major types of colloids present in soils are layer silicate clays, iron and aluminum oxide clays, allophane and similar amorphous clays, and humus.

4.3.1 Silicate Clays

Layer silicate clays occur as a result of the weathering of primary minerals such as pyroxenes and amphiboles, felspars, muscovite, olivine, and volcanic ash (silicate glass). Silicate clays are the preeminent inorganic colloids in soils of temperate regions and occur in soils of the tropics as well. Repeating, crystalline layers characterize these minerals. Each particle contains a series of layers composed of horizontally oriented sheets of silicon, aluminum, magnesium, and/or iron atoms bound together by oxygen and hydroxy groups. The general composition of these clays is illustrated by the formula for the clay kaolinite, $Si_2Al_2O_5(OH)_4$. The exact composition and internal arrangement of atoms in a crystal determine its surface charge and other properties, for example, the capacity to retain and exchange ions. Physical properties will also be affected; for example, certain colloidal clays experience extensive shrinking and swelling with changes in moisture status while others are much less affected. The most common and important silicate clays are the phyllosilicates. These are composed of two distinct sheets, one dominated by silicon, the other by aluminum and/or magnesium.

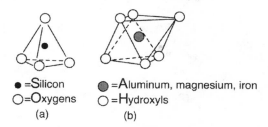

Figure 4.2 The basic structural units of 1:1- and 2:1-type clays: (a) silica tetrahedron; (b) aluminum octahedron.

The basic building block for the silica-dominated sheet is a four-sided unit (tetrahedron) composed of one silicon atom surrounded by four oxygen atoms (figure 4.2). An interlocking array of these tetrahedra creates the tetrahedral sheet. Aluminum and/or magnesium ions are the key cations in the second type of sheet. An aluminum or magnesium ion is surrounded by six oxygen atoms or hydroxy groups, resulting in an eight-sided unit, or octahedron (figure 4.2). Numerous octahedra linked together form the octahedral sheet. The tetrahedral and octahedral sheets comprise the fundamental structural units of silicate clays. The so-called 1:1-type minerals are composed of one Si tetrahedral sheet attached to one Al octahedral sheet. The 1:1 structure is fixed; that is, no expansion occurs between layers when the clay is wetted. Cations and water molecules cannot penetrate between the structural layers of a 1:1 clay particle; the surface area of these clays is thus limited to the external surface. In contrast with other categories of silicate clays, the 1:1 types demonstrate very little plasticity (capability of being molded), cohesion, shrinkage, or swelling. In soil, kaolinite is the dominant member of the 1:1 clay minerals, which also include halloysite, nacrite, and dickite.

The crystal layers of 2:1 clays are characterized by an Al octahedral sheet sandwiched between two Si tetrahedral sheets (figure 4.3). The smectite group experiences significant interlayer expansion when the minerals are wetted. Water can enter the interlayer space and force layers apart. Montmorillonite is the most prominent member of the smectite group. Beidellite, nontronite, and saponite also occur in this category. The smectites possess properties of high plasticity and cohesion, and marked swelling and shrinkage between wetting and drying. Wide cracks form upon drying of smectite-dominated soils (e.g., Vertisols) (figure 4.4). Dry aggregates are very hard, making soils difficult to work in the field.

Vermiculites are also 2:1-type minerals. Water molecules, as well as Mg and other ions, are strongly adsorbed within the interlayer space of vermiculites; however, they act as bridges holding the units together. The extent of swelling is, therefore, less for vermiculites as compared to smectites.

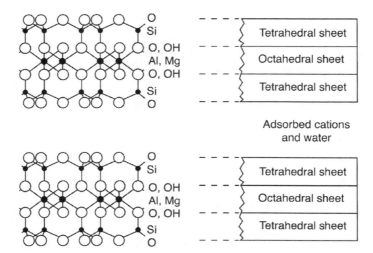

Figure 4.3 Schematic of a 2:1 silicate clay showing two Si tetrahedral sheets surrounding one Al octahedral sheet.

Figure 4.4 Vertisol showing deep cracks. *Source*: USDA-NRCS.

4.3.2 Hydrous Oxides

Hydrous oxides of iron and aluminum are dominant colloids in the highly weathered soils of the tropics and semitropics and also occur in significant quantities in some temperate region soils. Examples of common iron and aluminum oxides are gibbsite ($AlOH_3$) and goethite (FeOOH). Some of these colloids possess distinct crystalline structures whereas others are amorphous. Hydrous oxides possess a significant pH-dependent charge; in other words, the exchange capacity is strongly influenced by soil pH. Strongly acidic soil tends to impart a positive charge to hydrous oxides because H^+ and Al^{3+} ions saturate the colloidal surfaces. Under more alkaline conditions, these cations are stripped away, revealing a significant CEC.

4.3.3 Allophane and Other Amorphous Colloids

In many soils significant quantities of amorphous colloidal matter occur. One of the more significant amorphous aluminum silicate colloids is allophane. This mineral is prevalent in soils developed from volcanic ash and has the general composition $Al_2O_3 2SiO_2 H_2O$. Allophane has a high capacity to absorb cations as well as anions.

4.3.4 Humic Colloids

Soil organic matter is generally composed of 25 to 35 percent polysaccharides and protein-like compounds that are readily decomposed by microorganisms and therefore experience a short half-life in soils. About 65 to 75 percent is composed of humic materials, which are a diverse assortment of secondary organic molecules formed from the decomposition of plant and animal tissue deposited in soil. Humic colloids are complex mixtures of high molecular weight hydrocarbons which are resistant to decomposition. There is no single chemical formula or structure which accurately describes the colloidal structure and organization of humus. Humus is not a single compound; rather, it is a collection of organic molecules having drastically differing chemistries and molecular weights (figure 4.5). Put simply, humus colloids are highly charged organic and amorphous rather than crystalline micelles.

Soils high in organic matter are capable of adsorbing significant quantities of organic contaminants. The contaminant is attracted to the humic compound as both possess a nonpolar nature. Such adsorption slows contaminant movement and, in many cases, results in the incorporation of the contaminant over the long term. Soil organic matter usually has a substantial CEC and may have a significant anion exchange capacity as well. Anion exchange capacity, however, is usually much lower than CEC. Increased soil organic levels favor microbial activity due to increased nutrient supply (especially N, P, and S), CEC, tilth, water-holding capacity, and available carbon.

The negative electrical charges occurring in humus arise from partially dissociated hydroxyl (–OH), carboxyl (–COOH), and phenolic groups (table 4.1). These groups are associated with sheets of benzene rings or

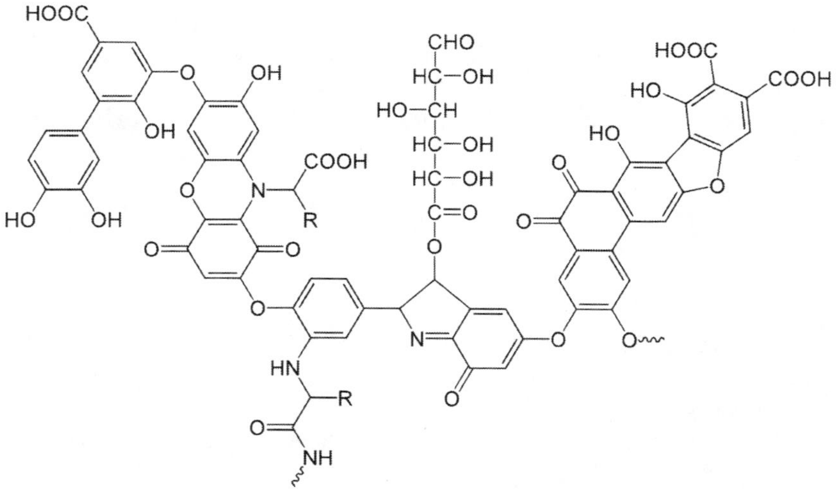

Figure 4.5 Hypothetical structure of a humus molecule.

carbon chains of varying size and complexity (figure 4.4). As is the case for Fe and Al oxides, the negative charge associated with humus is highly dependent on soil pH. Under very acid conditions the negative charge is relatively low because H^+ ions saturate the negative charges. With an increase in pH, however, H^+ ions dissociate from the carboxyl groups and subsequently from the hydroxyls and phenolics. This dissociation imparts a greatly increased negative charge to the colloid. Under neutral to alkaline conditions, the CEC of humus per unit weight greatly exceeds that of the silicate clays.

4.4 THE BIOLOGICAL COMPONENT

Soil materials possess a significant biotic component. Many groups of microorganisms occur; however, the most significant include bacteria, actinomycetes, fungi, algae, and protozoa. The first three groups comprise the microbial types most responsible for transformations of organic matter, including hydrocarbon contaminants, in soil. These organisms are essential to nutrient cycling of ecosystems including C, N, P, and S. Active microorganisms decompose plant and animal tissue, converting them to cellular biomass, secondary materials (humus), CO_2, and other gases.

Table 4.1 Functional and structural groups of humus

Group	Structure
Alcohol	$R\text{-}CH_2OH$
Aldehyde	$R\text{-}\underset{\|}{C}=O$ with H
Amino	$R\text{-}NH_2$
Amine	$R\text{-}CH_2\text{-}NH_2$
Amide	$R\text{-}C(=O)\text{-}NH_2$
Carboxyl	$R\text{-}C(=O)\text{-}OH$
Carboxylate	$R\text{-}C(=O)\text{-}O^-$
Enol	$R\text{-}CH=CH\text{-}OH$
Ether	$R\text{-}CH_2\text{-}O\text{-}CH_2\text{-}R'$
Ester	$R\text{-}C(=O)\text{-}O\text{-}R'$
Ketone	$R\text{-}C(=O)\text{-}R'$
Keto acid	$R\text{-}C(=O)\text{-}COOH$
Quinone	cyclohexadiene-1,4-dione
Hydroquinone	4-hydroxycyclohexadienone
Peptide	$^+H_3N\text{-}CH_2\text{-}C(=O)\text{-}NH\text{-}CH_2\text{-}COOH$

4.4.1 Bacteria

Bacteria are essential to nutrient cycles of ecosystems including C, N, P, and S. Cells decompose plant and animal tissue, converting them via mineralization, immobilization, and humification reactions. Other microbial groups carry out many transformations similar to those of bacteria; however, bacteria stand out due to their ability to multiply rapidly combined with their vigorous decomposition of a wide range of substrates.

Bacteria are the preeminent group of microorganisms in a bioremediation program (see chapter 11, "Microbial Remediation") because of their physiological diversity and, furthermore, because they are the most abundant group among all microbial populations in a soil. In a healthy—that is, non-contaminated and fertile soil—soil, bacterial biomass is substantial: 1 g may contain several hundred million cells. Estimates of bacterial numbers vary according to the means of determination, however. Thousands of bacterial species have been identified in soils throughout the world. (Some microbiologists put the put the upper limit at ten million species.) The numbers of species and individual cells are a function of soil characteristics and environmental conditions such as moisture content, pH, and temperature.

Bacteria possess widely varying morphology as well as physiology (figure 4.6). Among the major bacterial cell types are the bacilli (rod-shaped bacteria), which are the most numerous; cocci (spherical-shaped cells); and spirilla (spirals). The size of individual cells ranges from about 0.2 to 10.0 μm. Bacteria possess an outer layer known as the *capsule, slime* layer or *glycocalyx*, which is composed mostly of polysaccharides or polypeptides. This capsule measures from 100 to 300 A° thick and may protect the cell from engulfment by protozoa (Paustian 2018). Beneath the capsule is a rigid cell wall, a cell membrane that encapsulates the cytoplasm, a nucleus, and numerous other organelles. Some bacilli persist in stressed environments by the formation of endospores, which endure because of their resistance to both desiccation and high temperatures. The endospore can persist in a dormant state long after the death of vegetative cells. When conditions are again adequate for vegetative growth, the spore will germinate. Spore-forming genera are present among both aerobic and anaerobic bacteria.

Bacteria are typically not free in the soil water, as most cells adhere to clay particles, soil organic matter, and other colloidal surfaces. The environmental factors that influence soil bacterial numbers and activity include soil organic matter content, temperature, pH, nutrient supply, moisture level, and aeration. Additional variables such as cultivation, season, and depth have also been documented.

Community size in mineral soils is in part a function of organic matter content. Thus, humus-rich locations have high bacterial numbers, a result

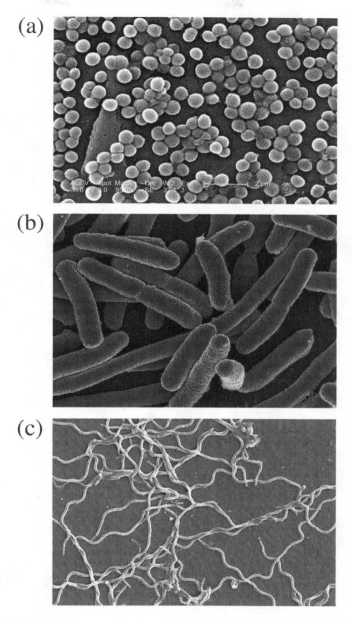

Figure 4.6 The varying morphologies of bacterial cells: (a) cocci; (b) bacilli; and (c) spirilla.

of greater root density and a greater supply of organic matter available from the decomposition of roots and plant debris. The addition of carbonaceous material strongly influences bacterial numbers and activities; for example, the plowing under of crop residues promotes a rapid response. This stimulation is most pronounced during the first months of decomposition and disappears after a season (Weil and Brady 2016).

Bacterial growth is strongly influenced by temperature. Certain species develop best at temperatures below 20°C and are termed *psychrophiles*. True psychrophilic bacteria are not common in soil, however. Thermophiles grow readily at temperatures of 45°C to 65°C. Most microorganisms, however, are mesophiles with optima in the range of 25°C to 35°C and a capacity to grow from about 15°C to 45°C. The majority of soil bacteria are mesophiles. Temperature regulates the rate of biochemical processes carried out by bacteria and increases the rate of reaction up to the point of optimum temperature for biochemical reactions.

The optimum pH regime for most bacterial species is near neutrality. Highly acid or alkaline conditions inhibit many common bacteria. There are exceptions, however; for example, *Thiobacillus ferrooxidans*, an autotrophic iron-oxidizing bacterium, catalyzes the oxidation of ferrous iron with the consequent production of sulfuric acid and can survive at pH 3.0 or lower. In most cases, however, the greater the hydrogen ion concentration, the smaller the size of the bacterial community. This may be a function of direct toxicity of the H^+ ion, the increased solubilization of metals in the growth medium, or a combination of the two. Liming of acid environments increases bacterial abundance.

Inorganic nutrients are required for optimal bacterial growth; therefore, flora are often affected by application of inorganic fertilizers. Cultivation practices also exert direct and indirect biological effects. Plowing and tillage usually result in marked bacteriological fluctuations. Changes will vary with the type of operation (i.e., moldboard plowing vs. harrowing), soil depth, and type of residues that are turned under. The effects occur from improving the soil structure and porosity, encouraging air movement, altering the moisture status, and exposing organic-bound nutrients to bacterial action (Weil and Brady 2016).

4.4.2 Fungi

Fungi are unicellular or multicellular nonphotosynthetic higher protists. They possess cell walls, are nonmotile, and use organic material for both energy and as carbon sources. Fungi produce a filamentous mycelial network composed of individual hyphal strands (figure 4.7). In many cases, the mycelium is divided into individual cells by cross walls or septa. Individual hyphae may be vegetative or fertile, and the fertile filaments produce either sexual or asexual spores. The condia (asexual spores) are widespread and the sexual spores are relatively uncommon. Fungi can be differentiated into genera and species on the basis of morphology; therefore, size, shape, structure, and cultural characteristics are important in taxonomy.

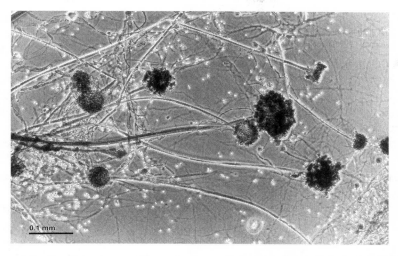

Figure 4.7 Fungi produce a filamentous mycelial network composed of individual hyphal strands.

The filamentous fungi are almost exclusively strict aerobes. Fresh soil samples have been found to contain over 600 mg biomass per kg soil, representing 130 fungal species (Anderson 1984).

Fungi are heterotrophic; that is, their growth is dependent on the availability of oxidizable carbonaceous substrates. The primary environmental influences on soil fungal activity include types and amounts of organic matter, pH, amount of moisture and aeration, temperature, and composition of native vegetation.

Fungal species grow over a wide pH range, although they tend to predominate in acidic pH regimes. In laboratory culture fungi can survive comfortably at pH values as low as 2.0, and some strains are active at pH 9.0 or above. Since bacteria and actinomycetes are intolerant of acid conditions, the microbial community in low pH media is dominated by fungi.

Application of inorganic fertilizers to soil modifies the abundance of filamentous fungi. Such changes are the result of acidification as well as nutrient addition. For example, treatment with fertilizers containing ammonium salts increases fungal numbers because microbial oxidation of the nitrogen leads to formation of nitric acid, which favors fungi and discourages bacteria and actinomycetes. The general reaction:

$$NH_4^+ \rightarrow HONH_2 \rightarrow HONNOH \rightarrow NO_2^- + H^+ \rightarrow NO_3^- \quad (4.5)$$

ammonium　hydroxylamine　hyponitrite　　nitrite acid　　nitrate

In terms of temperature requirements, most fungal species are mesophilic; a limited number of thermophilic strains can be identified in typical soil. Thermophiles are abundant only during heating of compost piles. Thermophiles grow well between 50°C and 55°C but are inactivated at 65°C.

The metabolic processes of fungi are less diverse than those of bacteria. One fungus that has potential for treatment of hazardous organics is *Phanerochaete chrysoporium*, a white rot fungus. This fungus produces an extracellular peroxidase enzyme that decomposes lignin, a complex polymer composed of phenylpropane units. The reaction of interest is relatively nonspecific and has been found to be effective in initiating the degradation of both PAHs and several chlorinated, recalcitrant compounds including pentachlorophenol (PCP), PCBs, and chlorinated dibenzodioxins (Gore 2007; Robles-Hernandez et al. 2007; Andersson et al. 2000; McGrath and Singleton 2000).

4.4.3 Actinomycetes

The actinomycetes are often considered a middle group between the bacteria and the fungi. Like the latter group, most genera of actinomycetes produce slender, branched filaments that develop into a mycelium (figure 4.8). The filament is typically long, and individual hyphae appear morphologically similar to fungal filaments. Actinomycetes are widely distributed in soil, compost piles, river sediments, and other environments. They are prevalent in surface soil and also in the lower horizons to great depths. They are second to bacteria in terms of abundance. Actinomycetes have been shown to decompose aromatics, chlorinated aromatics, steroids, and phenols (Nnamchi et al. 2006; Das et al. 2005; Eweis et al. 1998; U.S. EPA 1983).

The primary environmental influences for actinomycetes include soil organic matter content, pH, moisture level, and temperature. The free-living forms of the actinomycetes are solely heterotrophic. Actinomycetes are strongly affected by the presence of oxidizable carbon and their numbers may be substantial in organic-rich soils. Addition of organic wastes such as crop residues or animal manure to soil increases their abundance. Their populations sometimes reach 10^8 per g of soil with incorporation of crop residue, especially in high-temperature environments (Alexander 1977). Upon organic matter additions the bacterial and fungal flora usually proliferate, and actinomycetes do not respond until later stages of decay. It is therefore likely that a microbial succession is taking place; that is, bacteria and fungi initiate the decomposition reactions due to their rapid growth and physiological diversity. Subsequently, the actinomycetes appear when the readily available compounds have been metabolized and competition for substrates has decreased.

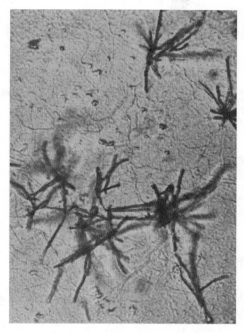

Figure 4.8 Actinomycetes tend to produce slender, branched filaments that develop into a mycelium.

Actinomycetes thrive in neutral pH environments and do not grow well under acidic conditions. In high pH environments, a large proportion of the total microbial community may consist of actinomycetes. Under conditions of waterlogging, for example at 85 to 100 percent water-holding capacity the growth of actinomycetes is limited, as most are strict aerobes. Actinomycetes are not as influenced by dessication as are the bacteria, and they are often favored by low moisture levels both in vegetative development and in formation of reproductive cells. Consequently, as a soil dries, the numbers of actinomycetes remain high while bacteria decrease due to intolerance to arid conditions (Zvyagintsev et al. 2007).

4.5 RELEVANT SOIL PHYSICAL PROPERTIES

Soil and subsurface materials recovered for analysis during a phase II subsurface investigation may have developed naturally from weathering of bedrock or as consolidated or unconsolidated sediments, or from fill materials placed during grading or backfilling an excavation. Whether formed naturally or emplaced, these materials possesses unique physical properties.

Soil physical properties affect the movement, retention, and availability of water and nutrients to microbes and plants; the ability of roots to penetrate the subsurface; and the flow of contaminants and treatment reagents. Physical properties play an essential role in determining soil suitability for engineering uses ranging from conventional waste disposal practices to remediation systems.

Soil physical properties are expressed by characteristics such as texture, structure, particle density, bulk density, and porosity, among others.

4.5.1 Soil Texture

The solids fraction of a typical soil consists primarily of discrete mineral particles. Soil texture, also known as mechanical analysis, is determined by measuring the relative amounts of three groups of soil primary particles (also termed *soil separates*). These are sand, silt, and clay. Texture provides a convenient means to classify soil by measuring the percentages of the three soil particle size ranges.

Soil particles of size less than 2 mm in diameter are included in the classification. These soil separates have the following size ranges:

- Sand = < 2 to 0.05 mm
- Silt = 0.05 to 0.002 mm
- Clay = < 0.002 mm

Sand and silt are considered relatively inactive in soil, as they do not serve to retain soil water or nutrients. These separates are commonly composed of quartz, feldspars and mica, or some other inactive mineral. Clay particles are mainly secondary minerals such as kaolinite, illite, vermiculite, montmorillonite, chlorite, and hydrated oxides of iron and aluminum. Because of its small size and sheet-like structure, clay possesses a vast quantity of surface area per unit mass. The surface charge on clay particles attracts ions and water. Because of this, clay is considered the active portion of soil minerals.

A number of methods are commonly used to determine soil texture. These are presented in table 4.2.

Several systems are available for classification of soil particle distribution. The United States Department of Agriculture (USDA) system, and the International Society of Soil Science (ISSS; renamed as the International Union of Soil Sciences, IUSS), are widely used (figure 4.1). For mineral soils, the proportion of sand, silt, and clay adds up to 100 percent. These percentages

Table 4.2 Selected methods for determination of soil texture

Method	Comments
Texture by feel	Qualitative. Can be effective with experience.
Dry sieving	Mostly for sand fractionation (or between 0.05- and 2-mm diameter particles).
Bouyoucos hydrometer	Based upon upon settling velocities of soil separates in suspension. Relatively rapid.
Pipet	Based upon settling velocities of particles. More accurate than hydrometer method.

Source: Sheldrick and Wang (1993); Gee and Bauder (1986); Day (1965).

are grouped into twelve soil texture classes, which have been organized into a textural triangle (figure 4.9). To illustrate the use of the textural triangle, consider a hypothetical analysis where the soil contains 30 percent sand, 60 percent, silt and 10 percent clay by weight. The relevant lines on the textural triangle intersect in the polygon labeled *silt loam*.

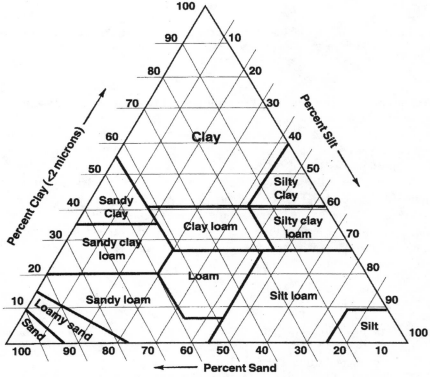

Figure 4.9 The soil textural triangle as applies to the USDA classification system.

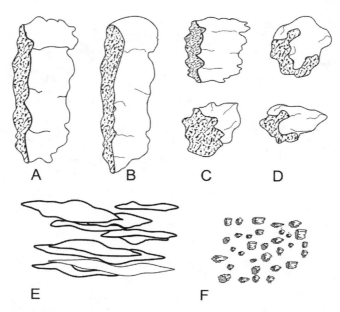

Figure 4.10 The primary classes of soil structure: (a) prismatic; (b) columnar; (c) angular blocky; (d) subangular blocky; (e) platey; (f) single grain.

4.5.2 Soil Structure

Except in very rare cases in nature, soil particles do not exist separately but are bonded into larger units or *aggregates*. The term *soil structure* refers to the arrangement and organization of soil particles into these units. The principal types of soil aggregates are platy, prismatic, columnar, blocky, and granular (figure 4.10).

Soil structure influences water movement, aeration, availability of plant nutrients (as well as of toxic elements), root penetration, microbial activity, and other factors. Strong aggregation decreases detachability and transportability of soil particles and therefore reduces erosion by the action of wind or water. Soil structure is affected by vehicle traffic, tillage, cultivation, application of amendments, and irrigation.

4.5.3 Particle Density

Particle density is the weight of a given soil particle per unit volume. In other words, it is the weight of the particle divided by its volume. In most mineral soils the average particle density is between 2.6 and 2.7 g/cm^3. In contrast, organic matter typically has a particle density of about 0.8 g/cm^3.

4.5.4 Bulk Density

Bulk density of a soil is defined as the dry soil weight divided by the unit volume. A unit volume of soil includes both solids and pore space. Bulk density is important as it reflects the porosity of a soil. A loose, porous soil has a lower bulk density than a tight, compacted clayey soil.

Bulk density provides a measure of how easily a soil can be tilled, how easily water will infiltrate, and its suitability for growing plants. Typical bulk densities for fine sands, silt loams, and silty clay loams are 1.5, 1.35, and 1.25 g/cm^3, respectively.

Stable soil aggregates are important in a soil because they help maintain good soil structure. Good soil structure translates into low bulk densities (1.3–1.5 g/cm^3). The bulk density of soil increases with compaction (e.g., from heavy vehicular traffic). If an aggregate is compacted or crushed, its bulk density increases and pore space decreases. High bulk densities (> 1.6 g/cm^3) can result from compaction. As a rule of thumb, bulk densities greater than 1.7 to 1.8 g/cm^3 impede root penetration.

4.5.5 Porosity

In site assessment and remediation activities it is important for the operator to determine properties of porosity, permeability, and hydraulic conductivity of subsurface materials. These parameters are all closely related. Porosity is the ratio of the pore space of the soil to the total soil volume. A strata having a porosity of 38 percent means that 38 percent of the total volume occurs as open space. Pore spaces may be classified according to size as *micropores* and *macropores*. Soil scientists classify macropores as cavities larger than 75 μm and micropores less than 75 μm. The porosity of sandy soils consists mainly of macropores while clayey soils contain mostly micropores. Air, water, and hydrocarbon vapors and other contaminants occupy pore spaces. The ratio of micropores to macropores influences the movement of soil water and gases. The ratio is also highly relevant to a number of remediation activities; for example, certain aqueous contaminants can be literally 'flushed' out of a soil under the proper conditions. If the soil has adequate porosity, especially as macropores, the flushing solution may be recharged quickly and remediation is rapid, all other factors being equal. In the case of in situ bioremediation the ratios of soil air and water, which are a function of porosity, greatly influence microbial activity.

Typically, more rounded particles such as gravel, sand, and silt, especially those having uniform particle sizes (sorted), have higher porosities than soils containing platy clay minerals. Soils containing a mixture of grain sizes (poorly sorted) also experience low porosities because smaller particles fill void spaces between the larger ones.

4.6 BASIC HYDROGEOLOGY

Soil and groundwater are key avenues for the transport of aqueous and nonaqueous contaminants. Water from direct application, precipitation, or runoff *infiltrate* through the soil surface; some will be stored in the upper horizons while some will *percolate* downward by the force of gravity until impermeable material is encountered. This subsurface material is sufficiently restrictive such that groundwater cannot flow through it. Strata that prohibit the passage of water to a well or spring are termed *aquitards*, *aquicludes*, or *confining layers*. Aquitards generally consist of clay and silt or unfractured,

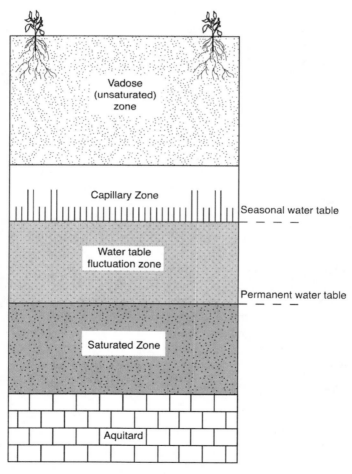

Figure 4.11 Soil profile showing differing moisture regimes.

dense rock. Above this point of confinement, percolating water fills all pore spaces until the lowermost soil is completely saturated. This *saturated zone* rises as water continues to fill the pore spaces. The top of the saturated zone is the *water table*. Above this is a zone in which water is held through adhesion, that is, capillary action between the soil grains. This is the *capillary fringe*. Finally, above the capillary fringe is a zone of unsaturated soil, referred to as the *vadose zone* (figure 4.11).

An *aquifer* is a geologic stratum that has the ability to store and transmit large volumes of water and can supply significant quantities to a well or spring. Aquifers are typically composed of sand and gravel or fractured rock. At the water table, groundwater is subject to atmospheric pressure and the aquifer is termed *unconfined*. Most aquifers impacted by contamination from hazardous substances are shallow, unconfined bodies; that is, they are accessible to the land surface. Municipal water supply wells do not typically withdraw water from unconfined aquifers; however, domestic wells will sometimes tap into these. Groundwater in shallow, unconfined aquifers may discharge into rivers, wetlands, or other surface water bodies (figure 4.12). An aquifer may also outcrop in one area, then dip beneath an impermeable layer, becoming confined at some distance downgradient. Artesian groundwater conditions develop when an aquifer is overlain by an aquitard. The aquitard acts as a rigid cover over the groundwater reservoir, and water will experience pressure beyond solely atmospheric pressure; therefore, when a well is completed in an artesian aquifer, water will rise in the well to some height above the top of the aquifer. These *confined aquifers* are a significant groundwater resource and are typically isolated from contamination.

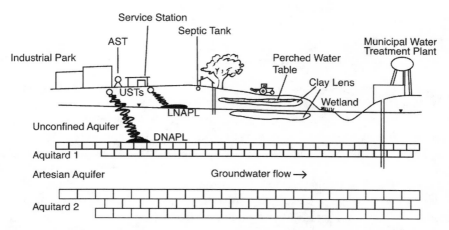

Figure 4.12 Unconfined and confined aquifers and examples of possible contamination sources.

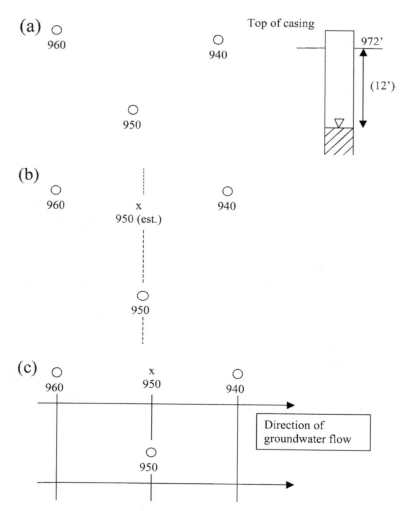

Figure 4.13 Triangulation method for the estimation of groundwater flow direction.

Groundwater is by no means static. The direction of groundwater flow is determined by measuring the elevations of the water table across the area of interest. This is accomplished by installing an array of monitoring wells. The top of the well casing is surveyed; the depth to water in each well is measured and this value is subtracted from the top of the casing elevation to determine water table elevation. The water table elevations are then plotted on a site map to generate a water table contour map (figure 4.13). Water will flow perpendicular to the contour lines from higher to lower elevations. Low-density contaminants (LNAPL) will flow in the same direction as groundwater. High-density organic solvents (DNAPL) will not be significantly affected by groundwater flow gradients; rather, they will flow from higher to lower elevations along the top of the aquiclude. The location of recharge and discharge points partly controls the flow of groundwater. Recharge occurs as rain falls onto the surface, infiltrates, and percolates downward to groundwater. Recharge occurs more readily where surface soil or rock is permeable and surface topography is relatively flat. Determining the depth and type of aquifers, groundwater transport velocities, and proximity of receptors (rivers, lakes, etc.) is essential in the phase II environmental assessment (see chapter 5, "Environmental Site Assessments").

4.6.1 Permeability

Permeability is the ability of the soil or subsurface strata to transport fluids. The more open and interconnected the pores in the rock, the higher the permeability. Permeability is expressed in units of length per time—centimeters per second or feet per day. In general, the larger the grain size, the larger the pore size and therefore the higher the porosity and permeability. Porosity and permeability usually correlate; that is, greater porosity results in greater permeability. There are exceptions to this association, however. Clays which have very small grain size may have porosities of 40 to 50 percent but are generally impermeable.

4.6.2 Hydraulic Conductivity

Hydraulic conductivity is a function of the properties of both the porous medium and the fluid passing through it. Typically, hydraulic conductivity has higher values for gravel and sand and lower values for clay (figure 4.14). Thus, even

Permeability, cm/sec

10^2	10	1	10^{-1}	10^{-2}	10^{-3}	10^{-4}	10^{-5}	10^{-6}	10^{-7}	10^{-8}	10^{-9}
Clean gravel			Clean sands; mixtures of clean sands and gravels			Very fine sands; silts; mixtures of sand, silt, and clay; glacial fill; stratified clays; etc.			Unweathered clays		
10^6	10^5	10^4	10^3	10^2	10	1	10^{-1}	10^{-2}	10^{-3}	10^{-4}	10^{-5}

Permeability, gal/day/ft^2

Figure 4.14 Abilities of different soil size fractions to transmit water. *Source*: U.S. Environmental Protection Agency.

though clay-rich soils have high porosities, they usually have lower hydraulic conductivities because pore diameters in clay-rich soil are much smaller.

Henri Darcy, a French engineer of the nineteenth century, formulated one of the earliest descriptions of groundwater flow. He observed a relationship between the volume of water flowing through sand and properties of the sand:

$$Q/t = KA \cdot dH/dL \tag{4.5}$$

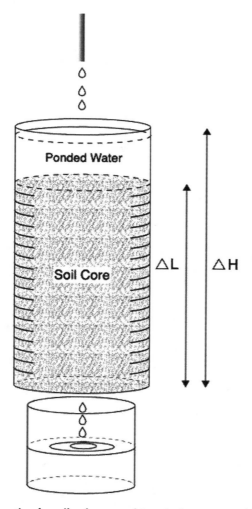

Figure 4.15 Schematic of a soil column used to calculate saturated hydraulic conductivity.

where Q is the volume of flow per unit time, t, through a column of a given cross-sectional area of flow, A. The flow is under a pressure gradient dH/dL, and the change in water level over a given length is L. K is the saturated hydraulic conductivity, a proportionality constant (figure 4.15).

Darcy's law calculates the volumetric flow rate through a unit cross section of the aquifer, not the actual velocity of water movement. The equation is rewritten in order to solve for hydraulic conductivity:

$$K = (Q \cdot \Delta L)/A\Delta H \qquad (4.6)$$

Example 4.1. Soil material is being considered for use as a lagoon liner. An intact core (6 cm diameter) is collected from a nearby site and brought to the laboratory. The soil column measures 20 cm tall. A head of 2 cm water is ponded on the top of the soil column. After the system has reached equilibrium, a total of 172 ml is collected per hour. Calculate the Ks.

$$K = (Q \cdot \Delta L)/A\Delta H$$

$$K = \left[(172\,\text{cm}^3/60/60\,\text{sec}) \times (10\,\text{cm})\right]/\left[28.3\,\text{cm}^3 \times (12\,\text{cm})\right]$$

$$K = 1.4 \times 10^{-3}\ \text{cm/sec}$$

The path of groundwater flow is highly dependent upon the type of material through which the groundwater flows; for example, the pore spaces between sand grains are generally continuously connected. Flow of groundwater through sand or similar porous material is referred to as *porous media flow*. In contrast, in a rock such as basalt, the quantity of pore space is limited and the pores are not interconnected; therefore, flow cannot occur. If the basalt is fractured, water will flow through cracks. In general, porous media do not restrict flow, while fractured media place significant constraints on flow paths.

Hydraulic properties affect the feasibility of adding or removing water, air, nutrients, or other materials to the soil. Saturated hydraulic conductivities of about 1.0×10^{-4} to 1.0×10^{-6} cm/sec are favorable for adding or removing water and/or chemical compounds. Soils with conductivities above this range undergoing a remediation program require careful management to prevent excessive drainage or contaminant migration off-site. Conductivities below this range tend to restrict movement of reagents, air, and water, thus limiting the speed and effectiveness of remediation.

The hydraulic conductivity of saturated soil is dependent on grain size and particle sorting and is relatively stable over time. Hydraulic conductivity in unsaturated soil is also influenced by grain size and sorting, but additionally by water content. At low water content, soil water moves mostly in response to adhesive and cohesive forces, which comprise the soil *matric potential*. Soluble contaminants in unsaturated soil move in the thin films of water surrounding soil particles. The thicker the film of water (i.e., the wetter the soil), the greater the path for contaminant movement and the greater quantity of contaminant that is transported during a given period.

4.6.3 Water-Holding Capacity

The ability of a soil to hold moisture is determined primarily by its proportion of clay, organic matter, and other colloids. These colloids hold more water relative to their volume than do coarse-grained materials such as sands.

The ratio of water and air in soil voids influences a number of chemical and biological processes in the soil. For example, activity of aerobic microorganisms is maximized when the soil moisture level measures about 70 to 80 percent of field capacity. Relatively dry soils tend to adsorb contaminants more strongly than moist soils because water competes with contaminants for adsorption sites. In an unsaturated soil, water and water-soluble compounds move in all directions in response to matric potential (a reflection of adhesion of the water molecule to solids and cohesion of water molecules to each other). In wet soils, water and water-soluble components are most strongly influenced by the force of gravity. NAPLs move through moist (not wet) soils rather quickly. Dry soils tend to adsorb NAPL; water in a completely wet soil will impede NAPL migration.

QUESTIONS

1. The two cations which are primarily responsible for soil acidity are H^+ and Al^{3+}. Explain how these cations are chemically interrelated as regards soil pH.
2. Explain why soil pH measurement with a glass electrode pH meter does not provide a satisfactory estimate of the amount of liming material required to increase soil pH.
3. Choose the correct answer. Clay and humus are significant to soils because of: (a) high CEC; (b) serving as a significant nutrient storehouse; (c) high surface area per unit volume; (d) high water-holding capacity; (e) all of the above.
4. Most soil N, P, and S is held in the soil in organic, that is, humus-incorporated forms. True or false?

5. Humus is not a single molecule; rather, it contains a range of chemical substances and is quite variable. Explain.
6. What type(s) of colloids predominate in soils of the temperate regions? Of the humid tropics? How might the electrical charges differ between the colloids of these two regions?
7. Soil material is being assessed as a possible liner for a waste disposal facility. A soil core was collected and brought to the laboratory. A 10-cm tall section of soil has 2 cm of water continuously ponded on it. The area of the core surface is 78 cm^2. A total of 62 ml H$_2$O is collected per hour. Calculate the Ks. The upper limit for Ks for lagoon liners is 1×10^{-7} cm/sec. Is this soil suitable for a liner?

REFERENCES

Brady, N. C., and R. R. Weil. 2002. *The Nature and Properties of Soils*, 13th ed. Upper Saddle River, NJ: Prentice Hall.

Day, P. R. 1965. Particle fractionation and particle-size analysis. In *Methods of Soil Analysis, Part I. Agronomy*, edited by C. A. Black, et al., 545–67. Madison, WI: American Society of Agronomy/Soil Science Society of America.

Gee, G. W., and J. W. Bauder. 1986. Particle-size analysis. In *Methods of Soil Analysis, Part 1. Physical and Mineralogical Methods*, edited by A. Klute, 383–411, 2nd ed. Agronomy Monograph number 9. Madison, WI: American Society of Agronomy/Soil Science Society of America.

Ghorbani-Nasrabadi, R., R. Greiner, H. A. Alikhani, J. Hamedi, and B. Yakhchali. 2013. Distribution of actinomycetes in different soil ecosystems and effect of media composition on extracellular phosphatase activity. *Journal of Soil Science and Plant Nutrition* 13(1): 223–36.

Gore, A. B. 2007. *Environmental Research at the Leading Edge*. Nova Science Pub.

Mortvedt, J. J., P. M. Giordano, and W. L. Lindsay. 1972. *Micronutrients in Agriculture*. Madison, WI: American Society of Agronomy.

Paustian, T. 2018. Bacterial cells are often covered in glycocalyx. *Through the Microscope*. http://www.microbiologytext.com/5th_ed/book/displayarticle/aid/67.

Phogat, V. K., V. S. Tomar, and R. Dahiya. 2015. Soil physical properties. Chapter 6. In *Soil Science: An Introduction,* edited by R. K. Rattan, J. C. Katyal, B. S. Dwivedi, A. K. Sarkar, T. Bhattachatyya, J. C. Tarafdar, and S. S. Kukal. New Delhi: Indian Society of Soil Science.

Pope, D. F., and J. E. Matthews. 1993. *Bioremediation Using the Land Treatment Concept.* EPA/600/R-93/164. Washington, DC: U.S. Environmental Protection Agency, Office of Research and Development.

Robles-Hernandez, L., A. Cecilia Gonzalez-Franco, D. L. Crawford, and W. W. Chun. 2007. Review of environmental organopollutants degradation by white-rot basidiomycete mushrooms. *Tecnociencia* 2(1): 32–40.

Sheldrick, B. H., and C. Wang. 1993. Particle size distribution. In *Soil Sampling and Methods of Analysis,* edited by M. R. Carter, 499–511. Canadian Society of Soil Science. Ann Arbor, MI: Lewis Publishers.

USDA-NRCS (U.S. Department of Agriculture-Natural Resources Conservation Service). 2018. *Examination and Description of Soil Profiles.* https://www.nrcs.usda.gov/wps/portal/nrcs/detail/soils/ref/?cid=nrcs142p2_054253.

U.S. Environmental Protection Agency. 1989. *Requirements for Hazardous Waste Landfill Design, Construction and Closure.* Seminar publication. EPA/625/4-89/022. Cincinnati: Center for Environmental Research Information.

Weil, R. R., and N. C. Brady. 2016. *The Nature and Properties of Soils,* 15th ed. Pearson.

Zvyagintsev, D. G., G. M. Zenova, E. A. Doroshenko, A. A. Gryadunova, T. A. Gracheva, and I. I. Sudnitsyn. 2007. Actinomycete growth in conditions of low moisture. *Biology Bulletin* 34(3): 242–47.

Chapter 5

Environmental Site Assessments

Dr. Watson: This is indeed a mystery. What do you imagine that it means?
Sherlock Holmes: I have no data yet. It is a capital mistake to draw conclusions before one has data. Invariably one begins to twist facts to suit theories, rather than theories to suit facts.

—Sir Arthur Conan Doyle, "A Scandal in Bohemia,"
The Adventures of Sherlock Holmes

5.1 INTRODUCTION

The purpose of the environmental site assessment (ESA) is to determine the possible presence of hazardous conditions existing on a parcel of land. The ESA is often conducted to protect a property buyer from the liability and costs associated with cleanup of a site later found to be contaminated with toxic materials. Almost all commercial transactions must now be accompanied by at least a so-called phase I environmental assessment. If a site is determined to be in need of remediation, the information compiled during the ESA facilitates the planning process for subsequent sampling, analysis, and remediation activities.

The ESA provides a comprehensive assessment of on-site conditions and operations: site conditions are compared with regulatory requirements to determine compliance; hazardous substances and conditions are identified. The potential for human exposure and any risk and liability involved for the owners, operators, and other relevant parties should be included.

Finally, recommended future courses of action (i.e., remedial action) if any, for the site are provided. Ultimately, the ESA process will produce a detailed report of findings. This will serve as a guide for future decisions as to the disposition of the property.

The ESA process is achieved via a review of regulatory and technical records, on-site inspection, mapping, sampling and analysis, interviews with site owners and managers, and finally, report preparation. The ESA conventionally has been divided into three phases. Phase I is a general problem identification; this is where historical records and documents are gathered, integrated, and analyzed. The phase I is a period of speculation; that is, the assessor keeps in mind that any scenario is possible for the site in question. The end result includes declaring a probability for contamination. Sampling is relatively uncommon during phase I with the possible exceptions of asbestos in building materials, radon sampling in buildings, and lead sampling in drinking water.

A detailed site characterization is central to the phase II ESA. The basic purpose is to confirm or deny any suspicions that may have arisen during the phase I assessment. Additional historical research and information gathering takes place during phase II. Sampling and analysis may be extensive (figure 5.1). Soils may be sampled at the surface or to great depths; likewise, geologic strata may be recovered from core sampling. Surface water and groundwater may be collected and analyzed for concentrations of suspected contaminants. During phase II there is often a thorough survey for possible presence of asbestos in building materials. Other specialized activities may involve analysis of a wildlife habitat by a biologist, audits of facility records of product use and waste management, and even tracking of suspicious activities by a private investigator (Hess 1997). The phase II ESA, by virtue of the time and resources involved, will prove to be a much more costly undertaking than the phase I.

A phase III ESA occurs after suspicions have been confirmed. By this point contamination has been identified and the assessor must now determine its extent. This may be accomplished by more extensive sampling of soil, strata, and water resources on and adjacent to the property. Often, the phase III becomes incorporated within the actual remediation activity. For example, containment barriers may be constructed and groundwater extracted while additional monitoring wells are installed and soil cores collected.

There is often no clear delineation between the above three phases. For example, some modified phase I ESAs may involve extensive soil sampling and analysis. The decision as to degree of assessment is arranged between client and assessor.

The driving force behind ESAs is the Superfund Amendments and Reauthorization Act (SARA), which provides an incentive for environmental diligence in commercial property transactions. The so-called innocent purchaser or innocent landowner defense has placed much emphasis on identifying "recognized environmental conditions" in such transactions. "Recognized

Figure 5.1 The phase II environmental site assessment involves extensive data collection for soil, water, building materials, and potentially hazardous materials at a site.

environmental conditions" are defined by ASTM (American Society for Testing and Materials 2005) as

> the likely presence of any hazardous substances or petroleum products on a property under conditions that indicate an existing release, a past release, or a material threat of a release of any hazardous substances or petroleum products into structures on the property or into the ground, groundwater, or surface water of the property. The term includes hazardous substances or petroleum products even under the conditions in compliance with the laws.

Petroleum products are included in this definition because courts have extended CERCLA liability to petroleum-contaminated sites.

Standards for the phase I and phase II ESAs have been established by the ASTM to address the 'All-Appropriate Inquiry' (AAI) aspect to the CERCLA. A phase I ESA should be prepared in accordance with the requirements of ASTM E-1527, *Standard Practice for Environmental Site Assessments, Phase I Environmental Site Assessment Process* (2013). Likewise, the phase II should follow ASTM E1903–11, *Standard Practice for Environmental Site Assessments, Phase II Environmental Site Assessment Process* (2011). The assessor must be aware of state, local, or federal regulations outside of CERCLA that impose other site assessment requirements and liability protections.

Clients of ESAs are primarily buyers of property (about 75% of all cases). Other clients include the property seller, the lending institution, and corporate shareholders. All these parties are concerned as to whether the property may be a liability due to the presence of hazardous substances.

The degree or extent of a site assessment is often a function of land use. For example, if the assessor is reasonably confident that the site has been undeveloped or used solely as agricultural land for the previous five decades, then a somewhat less rigorous approach regarding the extent of a records search and site reconnaissance may be justified. Conversely, however, if the site is a brownfield with a long and not-so-stellar history of heavy industry and documented releases in the area, the assessment may be slow, cautious, and meticulous.

5.2 THE PHASE I ESA

Some of the key requirements of phase I assessments are to (O'Brien 2011; Hess 1997):

- identify potential environmental liability on a property scheduled for a transaction;
- identify hazardous materials that might be disturbed and released during subsequent construction activities. Hazardous substances pose a reduced risk if left undisturbed. Early identification allows alternative plans for excavation;
- provide information about environmental conditions during planning stages, so that alternative designs can be formulated to avoid complications and reduce costs;
- ensure worker and public safety via limiting exposure to soil, water, and air contamination;
- demonstrate that due diligence was exercised in case of unexpected problems and/or litigation. This aspect is critical in property transfers.

The specific components of the ESA will vary as a function of the site in question, client needs, and possible costs. However, the assessors should, in all cases, search and review the following:

5.2.1 Physical Setting

Topographic characteristics provide important information about surface drainage as well as possibilities for contaminant transport. A USGS topographic quadrangle map indicates surface contours, the presence of waterways, roads, and rail lines, and several other man-made features (e.g., quarries and large buildings) (figure 5.2). A topographic map will help the assessor predict directional movements of contaminants, thus identifying possible receptors (reservoirs, wetlands, homes, etc.) affected by contaminant transport. The topographic map will also provide information on proximity to population centers. It must be remembered that the land parcel in question is not an isolated area; activities on neighboring properties may be affecting your site. Likewise, your parcel may be impacting neighboring land, populations, and other receptors.

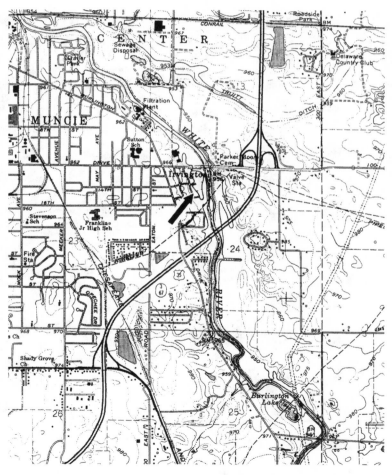

Figure 5.2 Portion of a USGS topographic map showing area of study (arrow).

5.2.2 Geologic Characteristics

USGS and USDA provide maps on soil, surficial deposits, and bedrock types for a selected area (figures 5.2 and 5.3). These data offer additional clues as to the possibility for contaminant migration; for example, do fractures or solution cavities occur that will cause rapid transport, or is the substrata relatively level and impermeable? Are local soils sandy loam or dense clay? Is the water table near the surface or at significant depth? In many soil surveys so-called 'made land,' which is soil composed of fill material of uncertain origin, may be indicated on the soil survey (figure 5.3). USGS, state natural resources, and other maps will provide hydrogeologic data, for example, depth to groundwater and presence of perched water tables (figure 5.4).

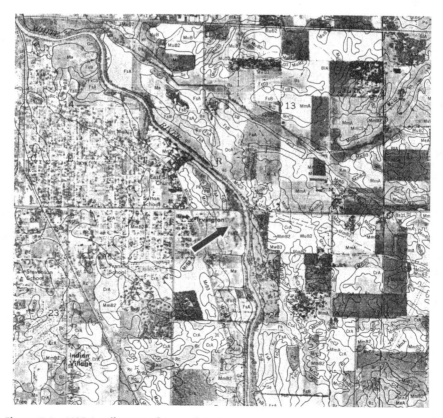

Figure 5.3   USDA soils map of area of study.

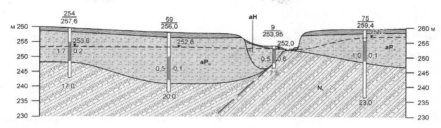

Figure 5.4 Sample geologic profile of an area of study.

5.2.3 Site History

Learning the identity of the previous site owners can provide the assessor with knowledge of the range of commercial and industrial activities carried out. The assessor must therefore review the chain of title. All ESAs require a thorough title search; the process involves visits to the county assessor's office to determine the chronological list of individuals and companies that owned and operated the property. Past or present operations at the site (i.e., petrochemical manufacturing versus a woodlot) will influence the extent to which the assessor must review. If the site occurs in an area of heavy industry (e.g., a mid-western automotive or metalworking area), it may be necessary to search the title of ownership as far back as one hundred years. Title searches can be complicated at times; for example, if the property frequently changes hands, the property is part of several deeds, or the property is part of more than one county (table 5.1).

Table 5.1 Examples of sites and facilities posing significant environmental risk

Site or Facility	Possible Hazards
Petroleum and petrochemical facilities	Oils, fuels, resins, solvents, corrosives, chlorine
Manufacturing plants	Solvents, corrosives, reactives, resins, baghouse dusts
Waste accumulation sites	Solvents, used oil, metals, acids
Metal plating	Cr, Cd, Cu, Ni, Pb, Zn, acids, alkalis, cyanide
Metal smelters, foundries	Metals, corrosives, paint, baghouse dust, resins, binders
Automobile battery recycling	Lead, other metals, acids
Automobile salvage yards	Waste oil, fuels, BTEX, MTBE, ethylene, glycol, tanks (propane, oxygen)
Uncontrolled or private dumps	Organic and inorganic solids and liquids
Agricultural operations	Pesticides, fertilizers, unregulated dumps
Gasoline service stations	Gasoline, diesel fuel, BTEX, used oil
Dry cleaners	Solvents
Older buildings	Asbestos, lead-based paint
Coal and metal mines and spoils	Metals, acids

5.2.4 Aerial Photographs

The photo review is another important tool for determining prior land usage (figure 5.5). Aerial photographs are available on the internet from state natural resources and environmental agencies. The assessor should seek one aerial photo for the site for every five to ten years, going back at least fifty years. The site is to be scrutinized for changes in topography, which may imply a disposal area. Also, the existence of man-made objects such as drums, ASTs, USTs, and buildings should be noted (figure 5.6). A man-made object will usually appear as a distinct entity, as most possess sharp corners or straight edges. A good rule of thumb for identification of a natural versus a man-made

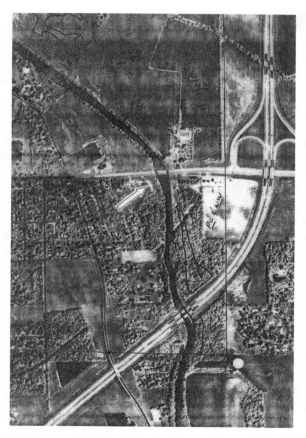

Figure 5.5 Aerial photograph of area of study, an abandoned excavation site converted to a waste disposal site.

Figure 5.6 Aerial photograph of a derelict petroleum refinery showing storage tanks and buildings.

structure is that natural formations almost always possess rounded, irregular, or diffuse shapes (figure 5.6).

5.2.5 Fire Insurance Maps

Sanborn Fire Insurance Maps were originally formulated to assist insurance agents in assessing fire risk and the potential for the spread of fire. The maps depict more than 12,000 American towns and cities—they show the size, shape, and construction materials of dwellings, commercial buildings, factories, and other structures. The maps indicate the width of streets, show property boundaries, and how individual buildings were used. They also provide locations of water mains, fire alarm boxes, and fire hydrants. Other useful information that may be gleaned from the maps includes type of construction (brick, asbestos, etc.), storage of raw materials, products manufactured or stored, and presence of wells (figure 5.7).

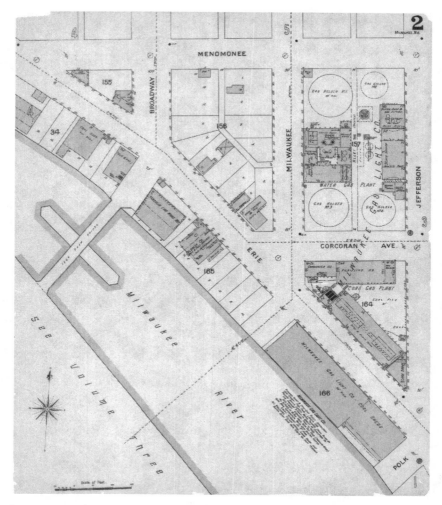

Figure 5.7 Sanborn Fire Insurance Map.

The Library of Congress has placed online nearly 25,000 Sanborn Fire Insurance Maps. The online collection now features maps published prior to 1900. By 2020, Fire Insurance Maps from all states will be online, showing maps from the late 1880s through the early 1960s.

5.2.6 Additional Resources

Other resources that a site assessor should consider in the historical records search include architectural drawings of structures, local street directories, zoning records, building permits, and commercial directories.

Relevant regulatory agency listings must be scrutinized to ascertain the documented presence of hazards or violations by federal, state, or county agencies. These listings are on file in each state's office of environmental protection. Such lists are usually available online and are updated regularly. Records of highest priority for the assessor include NPL (Superfund), CERCLIS, and RCRA lists; state hazardous waste site lists, registered UST and LUST lists, and county health department violations lists (table 5.2).

Table 5.2 Relevant data for the phase I historical and records review

Federal Level
- National Priorities List (NPL) facilities list
- Comprehensive Environmental Response, Compensation, and Liability Information System (CERCLIS) list
- RCRA treatment, storage, and disposal (TSD) facilities list
- Other RCRA lists
  - Large quantity hazardous waste generators
  - Small quantity hazardous waste generators
  - Hazardous waste transporters
- SARA Title III Reports
- Emergency Response Notification System list

State Level
- State priority site list
- Unplanned release and spill databases
- Air emissions records
- Notices of violation or similar citations
- Hazardous waste site list
- Landfill/solid waste disposal site list
- Registered UST lists
- LUST lists

County Level
- Historical maps and plans
- Property surveys
- Building permits
- Landfill sites list
- Well records

Other Local Agencies
- Department of Health records
- Fire department hazardous materials emergency responses
- Sanitation department records
- Water quality agency records

Published Sources
- Aerial photos
- Fire Insurance directories (Sanborn Directories)
- City directory listings
- USGS topographic maps
- State or USGS groundwater survey maps
- State or USGS subsurface geology maps

While visiting state agencies and websites, the assessor should determine the history, if any, of permits or citations for the site. For example, have there been citations for improper disposal? Has a previous owner requested a permit for waste disposal? Has an NPDES permit been provided? This is an opportunity to confirm compliance with the state and federal occupational safety and health statutes and to describe any record of worker- or workplace-related environmental incidents.

5.2.7 The Site Reconnaissance

Efforts to determine possible site hazards by reling solely opreviously documented information are not adequate for a site assessment. A visit to the site is required in order to obtain firsthand observations of past or ongoing activities and occurrences. The purpose of a site walkover, then, is to locate specific problems at the site, if any, such as contaminated areas that pose a liability to those with a financial interest in the property.

The visual search is conducted for evidence of:

5.2.7.1 Current Use of the Site

Is the area undeveloped? Is it commercial, light industry, or heavy industry? The degree of development is related to degree of potential hazard. For example, light manufacturing may generate small quantities of relatively nonhazardous solid waste. In contrast, an established industrial facility may be responsible for the generation of solid and hazardous waste, emission control residues (e.g., baghouse dust), fugitive emissions, and process wastewater. Spills and other releases to soil and paved areas may have occurred.

5.2.7.2 Prior Use

Look for abandoned buildings and sheds that may be used to store hazardous materials. There are many documented cases where containers in old buildings have leaked as a result of corrosion and vandalism.

5.2.7.3 Hazardous Materials Storage and Use Areas

Hazardous waste may be stored in drums, bulk corrugated containers, tanks, or simply in lagoons, a common practice prior to enactment of RCRA. Look for waste accumulation areas, storage sites, and disposal areas. This is also an opportunity to assess the overall condition of containers and storage areas; for

Figure 5.8 Open and unlabeled drums of hazardous waste found during an environmental site assessment.

example, note if drums are open or unlabeled, if they are rusting, and if there is evidence of previous spills (figure 5.8). Is the waste storage area suitably covered to protect containers from precipitation? Is secondary containment available in the event of a release?

5.2.7.4 Underground Storage Tanks (USTs) and Aboveground Storage Tanks (ASTs)

Both USTs and ASTs have been notorious sources of soil and water contamination (figure 5.9). Look for vent pipes, manways, and other covers that may indicate the presence of active or abandoned tanks. A depression within a paved area may indicate the presence of an abandoned UST. Evidence of hydrocarbon residues at or near the ground surface from previous spills and overfills may be visible.

Figure 5.9 USTs being removed from closed facility. *Source*: U.S. Army Environmental Command.

5.2.7.5 Topography, Surface Water, and Wetlands

It is important to delineate drainage patterns in order to predict possible flow paths of surface water and any released contaminants on the land surface. Stormwater drains, surface water, and wetlands are likely receptors. The assessor should look for depressions, ditches, ponds, streams, and other low elevation seeps. Unusual mounds or depressions may indicate the presence of dump areas. A sheen on the water table is a sign that free hydrocarbon product may be present. Product could be released from a UST, loose connections, or spills. Although a sheen on water is generally not an indication of major contamination, it may indicate the need for further assessment as to the lateral extent of contamination, especially if drinking water sources occur locally.

5.2.7.6 Vegetation and Surface Soil

Soil and ground cover will provide subtle insights into activities on the site. Does the soil appear to be native to the area, or has fill material been brought in and graded over the site? The assessor should also look for stressed

Figure 5.10 Soil stained by leaking hydrocarbons at a closed petroleum refinery.

vegetation. Look for bare areas in fields otherwise having healthy vegetation. This may indicate spillage or disposal of toxic compounds. Discolored or stunted plants may simply be a result of inadequate soil fertility; however, they may also indicate the presence of toxic heavy metals and/or hydrocarbons. Hydrocarbons can also cause moisture stress as NAPLs tend to exclude water from soil pores. Stained soil and unusual odors are additional signs of hydrocarbons or other wastes in the soil (figure 5.10).

5.2.7.7 Radon

Radon originates from the natural decomposition of radioactive elements, particularly uranium, in soil and rock. EPA and the Surgeon General's Office have estimated that as many as 20,000 lung cancer deaths are caused each year by radon. This gas is a concern in closed buildings, especially in basements or first floors. Scan the property using radon canisters and report results of testing. If radon levels exceed the EPA action level a recommended response should be included in the site assessment report.

5.2.7.8 Lead-Based Paint

It is estimated that about 25 million older homes in the United States contain LBP (figure 5.11) (US C.D.C. 2014). Some LBP materials contain up to 50 percent Pb. About 75 percent of housing units in the United States built prior to 1980 were coated with LBP. In 1978 the manufacture and use of paints

Figure 5.11 Lead-based paint showing characteristic 'alligator' pattern.

containing more than 0.06 percent Pb by weight on the interiors and exteriors of residential surfaces, toys, and furniture were banned.

The most common methods for testing LBP in buildings include laboratory analysis of paint chips and use of portable X-ray fluorescence analyzers. For laboratory analysis, paint samples from several surfaces should be collected throughout the structure(s). Samples are removed by a heat gun, placed into plastic bags, labeled, and shipped to a qualified laboratory for analysis. In the laboratory, samples are digested in concentrated acid and the dissolved extract is analyzed for total Pb content in an atomic absorption spectrophotometer. If the results exceed 0.5 percent Pb by weight, the assessor's report should include a recommended response. More rapid, i.e., real-time methods exist to determine if Pb is present in paint, such as portable anodic stripping voltammetry and X-ray fluorescence detectors (figure 5.12). Both devices provide reasonably accurate, sensitive, and rapid values for Pb content on an intact surface. Both are, however, rather expensive and require adequate training for proper use.

5.2.7.9 Lead in Water

Lead pipes occur in buildings constructed prior to the 1920s. Copper pipes and Pb solder (50% Sn, 50% Pb) came into common use in the 1950s. Lead solder has also been used in the construction of cisterns. The most common source of water contamination in structures is via Pb leaching from Cu pipes

Figure 5.12 An XRF detector is a useful and accurate means of field testing for the presence of lead. Thermo Scientific™ Niton™ XLp 300 Series XRP Analyzers. *Source:* Reproduced with kind permission of Thermo Fisher Scientific.

with Pb-soldered joints. In 1986, the use of Pb in public drinking water distribution systems was banned and the Pb content in brass was reduced to 8 percent.

Human exposure to Pb-contaminated water systems is influenced by corrosiveness (i.e., pH, salt content) of the water, age of the Pb component (newer ones are more prone to leaching), quantity and surface area of the Pb-containing materials, and standing time and temperature of the water.

The assessor should collect one water sample near a main to determine the Pb content entering the property and one sample of drinking water.

Outlets for collection must be inactive for at least six to eight hours before testing. (Overnight is best.) A 'first draw' of 250 ml is collected from each outlet. The 'first draw' is the first water released from the tap after the period of inactivity. If lead is suspected in the system, a thirty-second 'flush' sample should be collected. Samples are to be shipped to a laboratory which is certified to test lead in drinking water (U.S. EPA 2006).

5.2.7.10 Asbestos-Containing Materials

Between 1900 and 1980 more than 30 million tons of asbestos was used in the United States. By some estimates 75 percent of all residential properties more than 40 years old, or those heated with steam or hot water, contain asbestos. Although banned for three decades, some asbestos-containing building materials may still be available (figure 5.12). There is no argument about the utility of asbestos materials for building systems: it is fibrous (and therefore can be woven into fabric), crystalline, thermally insulating, electrically insulating, thermally stable, and chemically stable. The primary concern with asbestos is possible harm to the respiratory tract when it is released to air.

The assessor should look for asbestos in a number of systems including (figure 5.13):

- coverings for hot water pipes and boilers
- heat reflectors such as those on wood stoves
- vinyl floor tiles and linoleum
- acoustical ceiling tiles
- siding on residential and commercial buildings
- roofing shingles and felts
- spray-on insulation or paint on walls and ceilings
- fuse boxes
- linings of air ducts

The asbestos-containing material (ACM) of greatest concern is that occuring in a physical condition termed *friable*. Asbestos is friable if it crumbles easily when subjected to hand pressure. The common types of ACMs that can contain friable asbestos include:

- a fluffy, sprayed-on material used for fireproofing, or a sprayed- or troweled-on material that resembles a cement-like plaster and is used for fireproofing and soundproofing;
- nonfriable asbestos wall board with sprayed- or troweled-on insulating

Figure 5.13 Unopened case of asbestos floor tiles discovered during site assessment.

Figure 5.14 Anthophyllite asbestos as viewed under a scanning electron microscope. *Source*: U.S. Geological Survey Microbeam Laboratory, Denver, CO.

material behind;
- asbestos-based pipe or boiler insulation that may appear felt-like or cement-like.

The assessor should sample suspect materials, send samples to a qualified laboratory, and include all results in the final report.

The presence of asbestos in building material can be detected by three methods of microscopy. The most popular is polarized light microscopy (PLM). A small quantity of the sample is crushed and placed on a microscope slide with refractive index oil. Polarized light is used to observe its optical properties and to distinguish asbestos from non-asbestos fibers. A polarizing filter on a light microscope forces light to vibrate in one particular plane. The polarizing filter, along with other components of the microscope such as the analyzer, a 530-nm waveplate, and a dispersion-staining lens, allow for observation of the properties of light and crystals. Asbestos analysis by PLM is the least expensive and most rapid method to identify and quantify asbestos fibers.

Scanning electron microscopy (SEM) and transmission electron microscopy (TEM) are also used to identify asbestos in samples of building materials. Both microscopes create the image by using a beam of electrons in a vacuum chamber. The SEM produces an image of the topography of the sample. In TEM the electron beam passes through the sample and an image is displaced onto a screen. Using an energy dispersive X-ray (EDX) device, information about the fiber's elemental properties are obtained. With knowledge of the ratios of the elements in the fiber along with other analytical information, the microscopist can distinguish asbestos fibers from non-asbestos fibers, and one type of asbestos fiber from another (EMLab 2008) (figure 5.14).

5.2.7.11 Miscellaneous

Additional considerations for the phase I walk-through include:

- business activities of nearby sites
- accessibility to the site
- suspicious features
- signs of misuse
- air emissions
- toxic chemicals
- PCB leakage from electrical equipment

It is important to note that the surrounding property should also be surveyed for hazardous materials, as a neighbor may have inadvertently or deliberately allowed contaminants to enter your site or facility. A phase I site reconnaissance checklist appears in table 5.2.

5.2.8 Interviews

The phase I ESA should include interviews of personnel on-site regarding current and prior use of the property. The assessor should inquire as to types of activities, management of products and wastes, and regulatory compliance of operations. Typical individuals to interview include the property owner, on-site personnel, managers, occupants, local governments, and, if possible, neighbors.

The facility assessor should inspect a number of key records including registrations for USTs; Safety Data Sheets; Community Right-to-Know Plans; safety plans, Spill Prevention, Countermeasure, and Control Plans; notices of violation; and hazardous waste generator notices or reports.

5.2.9 Site Map

One product of the site reconnaissance is a detailed sketch of the property location. This should be prepared on computer-aided software (e.g., ArcGIS™, Adobe illustrator™) whenever possible. A USGS topographic map can serve as the base map for the site. Relevant details including drainage, structures, and potential hazards are sketched onto the map. As emphasized earlier, it is necessary to examine site usage outside the immediate property boundaries; however, the assessor must set a reasonable limit as to how far this will be. Obtain a street map and a property record ('street lot') map for the site and decide on a total search radius. For a phase I assessment the minimum search radius will depend on local conditions. The ASTM standard uses the term "search distance" because few properties have circular boundaries. Factors to consider in determining the minimum search distance include (ASTM 2013):

- density of use (urban industrial, urban commercial, suburban commercial, suburban multifamily, suburban residential, rural)
- potential contaminant migration routes
- property type (e.g., woodland versus heavy industry)
- existing or past uses of surrounding properties
- distance to drinking water sources

In an urban industrial area, a phase I search might extend 1/2 mile or more. The main criterion for determining search distance is to ensure that potential sources of environmental risk have been covered.

5.2.10 The Phase I Report

Once all data is gathered the assessor prepares a report that provides detailed findings of the inspection. The report should include an executive summary; all significant findings of the inspection; topographic, geologic, and surface maps; and other appropriate appendices. All contacted sources should be listed. If phase II testing is recommended, this is to be stated as well. The executive summary should include limitations encountered in addressing any relevant issues. As an example, a facility may have been unwilling to provide manufacturing information due to use of proprietary chemicals. After the assessor speaks with an attorney the report should quantify potential environmental liabilities. If precise calculation is not possible, a range of potential costs should be provided (Sullivan 2018).

Even if no hazardous conditions are detected, the assessor should *never* guarantee that no hazardous substances exist on the property. The assessor is visiting and researching a site only briefly, and is therefore only observing a 'snapshot' of all components. Unless every cubic centimeter of subsurface material is thoroughly analyzed and found to be uncontaminated, a site should not be declared as free from potential hazard—it is always possible that unwelcome surprises might be discovered at some later date.

5.3 BEYOND THE PHASE I

The phase II ESA is initiated only if the phase I has raised significant concerns about possible hazards on the property. The phase II investigation often includes collection of extensive field data on soil, subsurface materials, groundwater, vegetation, and other materials on-site. As a result the phase II can be a costly and time-consuming activity. It is the intent of this chapter to provide only a brief glimpse into some of the more significant components of phase II. The reader is referred to several references at the end of this chapter for additional details.

Many variables influence the behavior and movement of a contaminant plume. The downward and lateral migration of the plume through the subsurface depends on the quantity released, the physical properties of the product, and the structure and physical properties of the soil and rock through which the product is moving. In performing the phase II ESA, it is essential to possess a thorough understanding of characteristics of the release event, of the contaminant, and of the affected site. Since petroleum-related releases are by far the most common source of environmental contamination, this discussion will outline the phase II ESA process using petroleum hydrocarbons as the example contaminant.

5.3.1 Noninvasive Technologies for Site Investigation

So-called *noninvasive* (i.e., no drilling or sampling required) geophysical techniques provide subsurface site characterization and monitoring of contaminant plumes and geologic features in a rapid, cost-effective, safe manner. Geophysical methods are important tools both for data gathering for environmental site assessments and in the remediation design and monitoring phases. When used during a site assessment, geophysical methods are typically best applied in phase II to help focus resources for the remainder of the assessment (U.S. EPA 2016). Numerous technologies are currently available including GPR, electromagnetics, resistivity, seismic methods, and others. Research is ongoing to improve and evaluate resolution over complex geologic formations. Several methods are introduced below.

5.3.1.1. Ground-Penetrating Radar

In GPR, a high-frequency electromagnetic pulse is transmitted from a radar antenna into the subsurface. The radar pulses are reflected back from various interfaces within the ground, which are detected by a radar receiver. Reflecting interfaces may be soil horizons, soil/rock interfaces, the groundwater surface, or man-made objects such as USTs or buried drums (U.S. EPA 2018a).

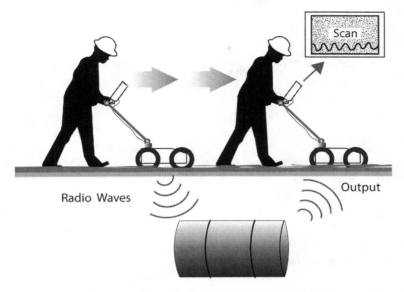

Figure 5.15 Ground-penetrating radar uses a high-frequency electromagnetic pulse transmitted from a radar antenna into the subsurface.

The radar signal is transmitted to the subsurface by an antenna that is on or near the ground surface. Reflected signals are detected by the transmitting antenna or by a separate receiving antenna (figure 5.15). Incoming signals are processed and displayed on a graphic recorder. As the antenna is moved along the surface, the graphic recorder displays results in a cross-section record or radar image of subsurface conditions.

GPR relies upon transmission of short wavelengths of energy—the antenna emits a single frequency between 10 and 3000 MHz. Higher frequencies within this range provide better subsurface resolution at the expense of depth of penetration. Lower frequencies allow for greater penetration depths but sacrifice subsurface target resolution.

Clay materials with a high CEC increase the attenuation and decrease penetration of the electromagnetic pulse. Additionally, the presence of solutes or other substances which increase the electrical conductance of groundwater also cause signal attenuation.

Target detection has been reported under unusually favorable circumstances at depths of 300 ft (100 m) or more. A careful feasibility evaluation is necessary if investigation depths must exceed 30 ft (10 m).

GPR is useful in locating USTs, utilities, and backfilled areas; determining geologic and hydrogeologic conditions; and, occasionally, delineating floating product (NAPL) (U.S. EPA 2016).

5.3.1.2 Electromagnetics

Electromagnetic (EM) methods, also referred to as electromagnetic induction methods, refer to measurement of subsurface conductivities by low frequency electromagnetic induction. A transmitter coil radiates an electromagnetic field, which induces eddy currents within the subsurface. The eddy currents, in turn, induce a secondary electromagnetic field which is subsequently intercepted by a receiver coil. The voltage measured in the receiver coil is associated with the subsurface conductivity. These conductivity readings are then related to subsurface conditions.

EM methods can be used to locate buried objects (metal and nonmetal); obtain geologic and hydrogeologic information; and sometimes delineate residual and floating product.

The conductivity of geologic materials is strongly dependent upon soil water content and concentration of dissolved electrolytes (salts). Clays and silts typically exhibit higher conductivity values because they contain large numbers of ions. Sand and gravel have fewer free ions in a saturated environment and, therefore, will have lower conductivities. Metal objects such as steel USTs display very high conductivity measurements that will ultimately reveal their presence.

The EM receiver and transmitter coils can be configured in several ways, depending on the objectives of the survey (figure 5.16). One common configuration for shallow environmental investigations utilizes transmitter and receiver coils that are attached to the ends of a rigid fiberglass rod at a fixed distance (i.e., fixed-coil separation). The equipment is then moved across the

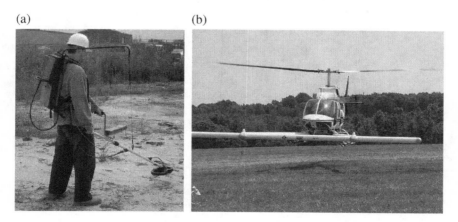

Figure 5.16 Electromagnetic survey (a) on foot and (b) via helicopter. *Source*: US Department of Defense, 2003.

area of investigation. This configuration is particularly suitable for detection of USTs and metal pipes.

EM methods are limited by certain interferences, for example, when used within 5 to 20 ft of power lines, buried metal objects (including rebar), radio transmitters, fences, vehicles, or buildings. In addition, success depends upon contrasts in subsurface conductivity. As an example, the difference in conductivity between a UST and surrounding soil is typically adequate for detection. However, mapping more subtle targets, such as fine versus coarse soil material, is less reliable.

5.3.1.3 Resistivity Methods

Electrical resistivity surveying is based on the principle that the distribution of electrical potential in the ground around a current-carrying electrode depends on the electrical resistivities and distribution of the surrounding soil. In the field, an electrical direct current (DC) is applied between two electrodes placed in the ground. Differences of potential are measured between two additional electrodes that do not carry current.

Data interpretation is usually carried out by qualitative comparison of observed response with that of idealized hypothetical models or on the basis of empirical methods.

Mineral grains are essentially nonconductive, with the exception of metallic ores; therefore, the resistivity of soil and rock is governed primarily by the amount of pore water, its resistivity, and the arrangement of the pores. Differences of lithology are also accompanied by differences of resistivity, so resistivity surveys can be useful in detecting bodies of anomalous materials (e.g., a clay lens) or in estimating the depths of bedrock surfaces (U.S. EPA 2018b).

5.3.2 Release Considerations

An initial step in the phase II investigation is to ascertain what contaminants were released. Knowledge of the type of the product released, its physical and chemical properties, and its key chemical constituents provides insight into subsurface behavior as well as the hazard potential to public health and environmental receptors. Petroleum products include a variety of fuel types, each having unique physical and chemical properties (see chapter 3, "Hydrocarbon Chemistry and Properties"). These properties must be known or estimated to make an accurate judgment as to its mobility—in other words, whether it will spread quickly as free liquid or vapor, its degree of partitioning to various phases, and its potential for degradation.

The assessor must attempt to determine the quantity of product released. This data is used for predicting whether the contaminant has reached the water table, and to estimate the extent of contamination in all affected compartments (e.g., vadose zone, saturated zone). Establishing the time since release is useful for approximating the quantity of product released. The composition and properties of the released product will change over time; volatile compounds evaporate, soluble constituents dissolve in infiltrating rainwater, and some constituents biodegrade. These physical, chemical, and biological changes that occur over time are collectively referred to as weathering. Weathering can result in an old contaminant plume having a markedly different chemical composition compared with that which was originally released.

After release to the soil, petroleum liquids transform between different phases. The most common phases are free product, vapor, dissolved product, and sorbed to soil colloids. The hydrocarbon plume may be present in all four phases at the same time. It is useful to know the relative proportions of contaminant in each phase, as hydrocarbons will vary in terms of mobility and feasibility of remediation depending on phase distribution.

As a petroleum release migrates from the site of origin, it fills soil pores. The plume percolates through the soil and a certain quantity is retained within pore spaces. The product in the soil that remains free-flowing and undissolved is termed *free product*. Because this is a hydrocarbon release, it is also labeled *nonaqueous phase liquid* (NAPL), as most hydrocarbons are nonpolar (i.e., they do not react with water). Free product tends to migrate both vertically and horizontally due to gravity and capillary action, among other forces. The migration occurs by successive permeation of larger areas, depending on quantity of product discharged.

Infiltrating rainwater will dissolve some NAPL, thus promoting leaching through the profile toward the water table. Because of this slow dissolution, free product occurring in soil is a major concern in any site assessment and remediation program. As long as free product remains in soil it is a potential source of contamination. For recent petroleum product releases (less than one year old), hydrocarbons are most likely to occur in the residual liquid (NAPL) phase rather than in vapor and dissolved phases.

When free product reaches the water table it does not dissolve significantly. If the NAPL is less dense (i.e., has a lower specific gravity) than water (LNAPL), it tends to pool on the water table and may be carried in the direction of groundwater flow. If, on the other hand, the NAPL is more dense than water (DNAPL), it will sink through the aquifer and accumulate above an aquiclude such as bedrock or a clay lens (refer to figure 4.9 in chapter 4). The product will flow along the bedrock surface until it penetrates a crevice or is blocked by some obstruction. The rate of free product movement depends on subsurface soil structure as well as the volume of product released. Tighter soils such as clays restrict the flow, whereas sands and gravel allow rapid subsurface transport.

Hydrocarbon odors are a common clue that a release has occurred. The odor originates from vapors generated as the volatile components of petroleum evaporate. Whether emanating from soil pores or floating on the water table, vapors migrate both horizontally and vertically along paths of least resistance. Vapor migration can be blocked by buried structures; however, it will follow pathways through backfill material surrounding structures such as sewer and utility lines. Vapors can then accumulate in basements, underdrains, sewers, and water wells. In buildings the assessor must check for vapors, particularly in basements, that have migrated through cracks in the foundation or through drains entering the building. If vapors are detected when checking storm drains, it is worth testing drains upgradient and downgradient to determine a pattern of migration. If higher concentrations of vapors are found upgradient from the site of interest, the vapors may be originating from some other source. Portable field instrumentation for hydrocarbon vapor monitoring is discussed later in this chapter.

Soluble petroleum constituents dissolve into both rainwater infiltrating through the unsaturated soil zone and in groundwater. Some toxic constituents (e.g., benzene) are soluble in water yet are colorless, odorless, and tasteless in drinking water; therefore, the appropriate instrumentation is required for their detection.

Circumstances regarding a leak or spill will affect the behavior and movement of product. All other considerations being equal, a rapid, high-volume leak can create a very different contaminant plume than will a slow leak. A high-volume release spreads laterally as well as downward; in contrast, a slow release moves downward through macropores. A slow release may also promote partitioning of the product into vapor, dissolved, sorbed, and other phases.

Much release information on the product type and amount released is available through interviews of owners and operators of the facility. Additionally, a review of facility records (e.g., product inventories, purchase and sales records, and tank tightness tests) provides much useful data.

5.3.3 Contaminant Considerations

Numerous physical and chemical properties of the contaminant will dictate the overall mobility and toxicity of the plume. These properties will also influence the feasibility of a selected remediation program. Some of the more critical properties to assess in the phase II ESA are as follows:

- water solubility
- viscosity
- specific gravity
- soil sorption coefficient
- biodegradability index
- vapor pressure
- vapor density

These properties can be measured directly in the laboratory (ASTM 2015) or inferred from published data. In many cases, values may not be available for bulk product; for example, gasoline composition will vary as a function of petroleum source, refinery, and blend. Often, however, one can consult default tables, which provide generalized values for fuels. In some cases it may be advantageous to collect data for those components that pose a significant hazard and/or comprise a large volume of the plume (e.g., benzene or toluene in a gasoline plume). As mentioned in chapter 3, fuels are composed of a complex mixture of hydrocarbons, octane boosters, blending agents, detergents, and so on. For example, the molecular weight fractions of gasoline range from about C4 to C9. Gasoline contains aromatic compounds including BTEX. These more volatile and toxic aromatic constituents are relatively mobile and water-soluble; they dissolve at least partially in infiltrating rainwater, travel through soil, and can mix with groundwater. The most soluble components in gasoline are the oxygenated compounds such as ethanol and ethers, which are used as octane boosters. These latter compounds are added in concentrations as high as 11 percent by volume. MTBE is very mobile, soluble, and easy to detect. It partitions rapidly into the water phase and is not present in the vapor phase. Because of its high solubility, MTBE moves quickly out of the gasoline plume, forming a halo on the outer edges of the BTEX plume (U.S. EPA 2008; NEIWPCC 1990). If MTBE is detected in groundwater, the assessor can infer that gasoline is present nearby because MTBE is used only in gasoline. Middle distillate fuels such as diesel fuel, kerosene, jet fuel, and lighter fuel oils tend to be more dense, less volatile, less mobile, and less water soluble than gasoline. They also contain lower percentages of the more toxic aromatic compounds such as BTEX. Heavier fuel oils and lubricants are even more dense, insoluble, and immobile in the subsurface environment as compared to the middle distillates. The middle distillate and heavier fuels may be less volatile and less mobile, but they can migrate and have the potential to cause substantial environmental contamination.

As gasoline weathers it experiences natural biodegradation, simple chemical oxidation, and transformation due to loss of the lighter, more volatile hydrocarbons (figure 5.17). The heavier hydrocarbons are retained; thus, over time, the gasoline begins to take on the properties (including odors) of

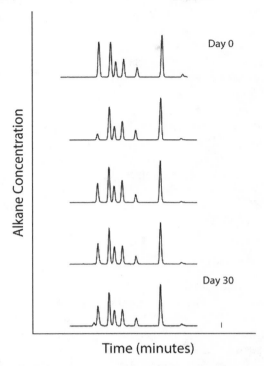

Figure 5.17 Gas chromatogram showing weathering of gasoline. Each peak represents a specific component of gasoline.

fuel oil. These changes will complicate the identification of the type of spill. Aging does not proceed rapidly unless conditions are favorable. The aging process can be limited by such factors as dense soil, barriers to vapor loss (e.g., paving), and/or high petroleum concentrations that inhibit bacterial breakdown.

In general, constituents with low molecular weights migrate from the source more quickly over time. This will occur through volatilization, natural flushing by infiltrating rainwater, and by virtue of lower viscosity. The more stable and less mobile constituents (typically, high molecular weight compounds) remain near the source longer. Contaminant volatility, as measured by vapor pressure, indicates how readily it will transform from liquid to vapor. Contaminants with high vapor pressures partition readily into the vapor phase.

Contaminant solubility is a measure of how readily the contaminant will dissolve in water (rainwater and pore water). Highly soluble contaminants dissolve readily. Vapor analysis and soil sampling can be undertaken to determine the predominant phase(s) in which most of the contamination is present. Soil material is often analyzed for total hydrocarbon concentration, while soil gas is sampled for presence of hydrocarbon vapors.

Viscosity affects the mobility and phase partitioning of a contaminant. A highly viscous constituent is likely to remain in the liquid phase (and perhaps sorbed to soil) rather than volatilize or dissolve. It will also remain in the unsaturated zone longer than would a lower viscosity constituent. The extent to which a contaminant will sorb to soil particles depends on the phase (free liquid, vapor, dissolved) and physical properties of the contaminant. For example, sorption to colloids is inversely proportional to contaminant water solubility (i.e., the higher the solubility, the lower the degree of sorption). For soils with a high organic carbon content (greater than 0.1% by weight), sorption from aqueous solution is almost completely controlled by organic carbon (Delle Site 2001).

5.3.4 Site Considerations

Site information embraces all relevant surficial, geologic, and hydrogeologic characteristics. It is important to be able to predict whether or not the release will migrate significantly and where it may spread. Some of this data may have already been compiled during the phase I ESA as part of the records search.

For a site assessment, published default values are available to enable assessors to estimate critical parameters on a timely basis. Representative geologic data such as that provided on the internet or on file at the local USGS office may suffice to gain an approximate understanding of the subsurface. In most cases, however, field-determined data will greatly improve the precision of the findings (figures 5.18–5.20). Geologic characteristics can vary greatly, even over short distances, making accurate estimates of soil parameters difficult to determine without collecting extensive field data. Additionally, such data may be required to make a choice of corrective action technology.

A thorough understanding of local geologic and soil conditions is a crucial goal in the phase II ESA; the migration path and mobility factors are critical if a tank is located near a surface or groundwater source of drinking water. In a relatively uniform soil formation, product migration is mainly downward. If a tank is located in a coarse-to-medium unlayered sandy soil, a gasoline release is apt to move rapidly toward the water table. In a layered subsurface, product migration has a more significant horizontal component. A subsurface clay lens will limit the downward movement of the product and will redirect its flow laterally (figure 5.21). Migration through heterogeneous formations results in gasoline plumes that vary in shape. Estimated migration should eventually be validated on a site-by-site basis through a

Figure 5.18 Truck-mounted drilling system commonly used for split-spoon sampling of soils and for installation of groundwater monitoring wells.

detailed sampling program that would take place when a remedial action is implemented.

Some key parameters to determine in the phase II site assessment include:

- depth to groundwater
- depth to bedrock
- soil temperature
- moisture content
- particle size distribution/texture
- soil structure
- bulk density
- saturated hydraulic conductivity (Ks)
- unsaturated hydraulic conductivity

Optional parameters include:

Figure 5.19 A split-spoon sample from a soil boring. *Source*: Julian C. Gray.

- organic matter content
- cation exchange capacity
- soil microbial activity
- soil field capacity

Providing the specifics regarding measurement of the above parameters is beyond the scope of this book. The reader is referred to several references on determination of of site characteristics.

A shallow depth to groundwater can result in contaminants escaping the unsaturated zone more rapidly than at a site with a thicker unsaturated zone. The greater the depth to groundwater, the greater the likelihood of detecting significant amounts of contaminants in the vadose zone in all four phases (NAPL, dissolved, vapor, sorbed).

Soil hydraulic conductivity (Ks) affects the contaminant's ability to migrate from the release site as NAPL and dissolved phases. Soil air conductivity, analogous to Ks, affects the mobility of contaminant vapors. Both hydraulic and air conductivity vary from formation to formation in much the same way, with formations of low hydraulic conductivity generally having low air conductivity as well. Low conductivity, either hydraulic or air, indicates a greater probability of encountering contaminants in all four phases close to the release site.

A high rainfall infiltration rate can enhance the transformation of the contaminant plume to different phases. Some hydrocarbons will dissolve in the infiltrating rainwater, thereby reducing the residual NAPL portion of contamination while increasing the quantity dissolved in pore water.

Soil temperature also affects contaminant mobility. Contaminant vapor pressure, and therefore the ease with which contaminants migrate into soil pore spaces, increases with increasing temperature. A greater percentage of the entire contaminant plume is likely to occur in the vapor phase in warmer

(a)

(b)

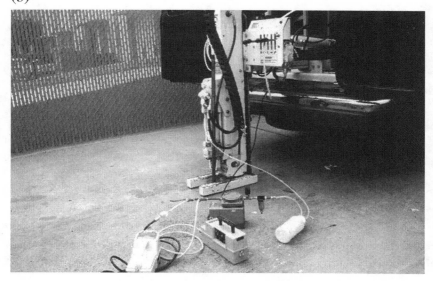

Figure 5.20 An innovative system for sampling the subsurface. A direct push technology Geoprobe™ is pushed into the soil and withdraws vapor samples that are measured on-site for organics via a van-mounted gas chromatograph. The van also houses an inductively coupled plasma-atomic emission spectrometer for analysis of individual elements. *Source*: Julian C. Gray.

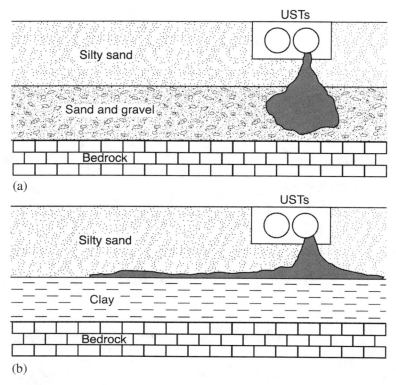

Figure 5.21 Migration of NAPL as a function of soil type: (a) homogeneous sandy loam; (b) sandy loam underlain by a clay lens.

regions or seasons. High soil temperatures also tend to reduce liquid viscosity, increasing the rate of downward movement through soil to the water table and out of the vadose zone. Finally, higher soil temperatures enhance the proliferation of indigenous microbial populations. A more diverse and active microbial consortia is expected to decompose hydrocarbons more rapidly.

5.3.5. A Note on Field Sampling

Proper sample collection, preservation, and storage are essential if soil and water samples are sent to a laboratory for analysis of volatile organic components (VOCs), total petroleum hydrocarbons (TPH), and so on. Vapor losses and biodegradation of the sample must be minimized. Specifically, samples should be:

- collected with minimal disturbance to the soil. A soil sample exposed to air will lose constituents by volatilization, and the introduction of air will encourage biodegradation;
- preserved and sealed in a clean, vapor-tight jar as quickly as possible. Water samples should fill the jar completely, leaving no air bubbles. Samples should be collected in jars containing the appropriate chemical preservative;
- collected to avoid any cross-contamination. Sample collection equipment should be decontaminated between samplings;
- labeled appropriately to correspond to the collection location at the site;
- immediately placed into an ice-filled cooler and transported to a qualified laboratory as soon as possible;
- properly stored at 4°C until analyzed. Holding time should be kept to a minimum. (U.S. EPA 2012)

Sampling of soil can be as simple as random collection of surface material to deep drilling over a surface grid pattern to a number of depths. The boreholes can later be converted to groundwater monitoring wells. Tests to be conducted on the environmental samples will vary with client, regulatory agency requirements, laboratory facilities and expertise, and other factors. Some of the more common test methods are listed in table 5.3.

5.3.6 Testing in the Field

The usefulness of various instrumentation for field measurement is contingent on the use of good field measurement procedures. A variety of field

Table 5.3 Selected test methods for environmental samples

7190	Chromium FLAA
7191	Chromium GFAA
7196A	Chromium, hexavalent (Colorimetric)
7200	Cobalt FLAA
7210	Copper FLAA
7380	Iron FLAA
7420	Lead FLAA
7430	Lithium FLAA
7450	Magnesium FLAA
7460	Manganese FLAA
7470A	Mercury in liquid waste CVAA
7471A	Mercury in solid or semisolid waste CVAA
7480	Molybdenum FLAA
7520	Nickel FLAA
7550	Osmium FLAA
7740	Selenium GFAA

(Continued)

Table 5.3 (Continued)

7760A	Silver FLAA
7770	Sodium FLAA
7780	Strontium FLAA
7840	Thallium FLAA
7950	Zinc FLAA
8011	EDB and DBCP by microextraction and GC
8015B	Nonhalogenated organics using GC/FID
8021B	Halogenated volatiles by GC using PID and ECD in series; capillary column technique
8081A	Organochlorine pesticides by capillary column GC
8082	PCBs by capillary GC
8100	Polynuclear aromatic hydrocarbons
8121	Chlorinated hydrocarbons by GC: capillary column technique
8141A	Organophosphorus compounds by GC: capillary column technique
8240/60	TPH
8240/60	BTEX (GC/MS)
8020 or 8240/60	VOCs
8240	MTBE (GC/MS)
8240	EDB, EDC (GC/MS)
8240	TML, TEL (GC/MS)
8260B	Volatile organic compounds by GC/MS
8270C	Semivolatile organic compounds by GC/MS
8275A	Semivolatile organic compounds (PAH and PCB) in soils/sludges and solid wastes using TE/GC/MS
8280A	Dioxins and furans by HRGC/LRMS
8290	Dioxins and furans by HRGC/LRMS
8310	Polynuclear aromatic hydrocarbons
8515	Colorimetric screening method for TNT in soil
9010A	Total and amenable cyanide (colorimetric manual)
9012A	Total and amenable cyanide (colorimetric automated UV)
9020B	Total organic halides (TOX)
9031	Extractable sulfides
9040B	pH electrometric measurement
9050A	Specific conductance
9056	Determination of inorganic anions by ion chromatography
9060	Total organic carbon (TOC)
9070	Total recoverable oil and grease
9071A	Oil and grease extraction method for sludge and sediment samples
9078	Screening test method for PCB in soil
9100	Saturated hydraulic conductivity, saturated leachate conductivity, and intrinsic permeability

Source: U.S. Environmental Protection Agency 1986.

procedures is currently in use. Details on the types and procedures are described elsewhere.

Active soil vapor sampling and analysis measures the volatile hydrocarbon concentration in a sample that is collected in situ by pumping or withdrawing the sample into a field instrument for analysis (see figure 5.20 above). Soil vapor samples are collected by: (1) using a drill or auger to install a borehole, inserting a portable instrument probe (see below), and taking a reading; (2) driving a hollow steel probe into the soil, collecting a sample using a gas-tight syringe, and injecting into a field instrument for analysis; (3) driving a hollow steel probe into the soil and collecting a sample in a polyethylene bag for analysis with a portable field instrument; or (4) direct in-line sampling with a portable analytical field instrument (i.e., PID or FID) from a driven probe (ASTM 2012; NEIWPCC 1990).

Headspace analysis of soil or water involves collecting a sample, placing in an airtight container, and analyzing the headspace vapor above the soil or water using a portable analytical instrument such as an FID or a PID (see below). This practice has its share of disadvantages—because there is no standard technique, readings may vary significantly with different users. A variation of the technique is the dynamic headspace analysis of soil and water using a polyethylene bag; this technique involves collecting a sample, placing it in a reclosable freezer bag, agitating the sample to release vapors into the bag, then measuring the vapor concentration using analytical field instruments. The procedure includes generating a calibration curve using field standards to determine sample concentrations and conduct a quality control check of analytical results.

In the *Hanby procedure*, aromatic compounds are extracted from a soil or water sample to yield a colorimetric indication of type and concentration of contaminant. Color indicates the type of compound and color intensity indicates the concentration (Hanby 2006). Training and practice are necessary for performing the analysis and for interpreting results.

Many instrument manufacturers calibrate vapor detectors to a standard gas such as benzene and set action levels from 100 ppm to 500 ppm.

Comparing results from field instruments with those from lab analysis is fraught with difficulty. As indicators of contamination, the techniques serve the desired purpose. However, the numbers obtained from these techniques may not be the same because many field instruments test for groups of compounds (e.g., TPH, volatile hydrocarbons), while labs analyze for individual constituents.

The following paragraphs provide a brief description of some commonly used field testing instruments.

5.3.7 Field Testing Instruments

Colorimetric detector tubes are among the simplest field testing tools for environmental contamination. The tubes are designed for the measurement of a single, specific vapor or gas in air. Each tube is a short length of glass tubing that contains a unique chemical packing (figure 5.22). Air is drawn through the tube by a hand pump or mechanical pump. As the air is pulled through the tube the contaminant of interest reacts with the detector reagents to produce a color in the packing material. The length and intensity of the color is proportional to the concentration of the contaminant. The accuracy and detection range of a specific tube are stated in the manufacturer's technical literature. These tubes are relatively inexpensive and easy to transport, use, and interpret.

The photoionization detector (PID) relies on the ionization of hydrocarbons to detect and measure the presence of organic vapors (figure 5.23). An ultraviolet light beam within the instrument is used to ionize organic vapor molecules. The air sample is drawn through the instrument probe past the lamp by an internal pump. If the ultraviolet light can excite the vapor sample and cause it to ionize, a signal is produced on the instrument readout. The strength of the signal is a measure of contaminant concentration. Different UV lamps are used to detect different volatile constituents. All PID lamps have a specific sensitivity to BTEX. Some PIDs have interchangeable UV lamps that are sensitive to different ranges of compounds. The detection range for these instruments ranges from about 0.2 to 2,000 ppm.

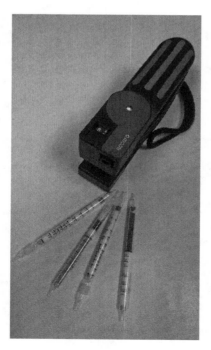

Figure 5.22 Draeger™ tubes for colorimetric analysis of vapors.

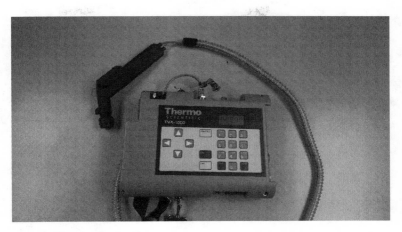

Figure 5.23 Photoionization detector.

Because PIDs do not detect alkanes such as methane, they are useful in detecting aromatic constituents released from USTs in areas where 'natural' methane may exist (e.g., septic fields, sewer lines). Ideal conditions for conducting PID analyses are dry weather and temperatures above 50°F. The responsiveness of PIDs decreases in moist conditions and high relative humidity (i.e., above 90%) (IST 2006; NEIWPCC 1990). The flame ionization detector (FID) also measures the presence of organic gases and vapors including methane. The FID uses a hydrogen flame to ionize molecules of volatile organic constituents present in the sample. The ionized molecules produce a current that is proportional to the total volatile organic vapor concentration in the sample. FIDs are less sensitive than PIDs to relative humidity and temperature; however, excess carbon dioxide and depleted oxygen may extinguish the flame in the instrument. FIDs are also more sensitive to alkanes such as hexane and butane, which make up a higher fraction of gasoline than the aromatics.

A portable gas chromatograph (GC) uses a reactive separation column to isolate and analyze specific constituents in either a liquid or vapor phase in conjunction with a PID or FID detection system. A GC is a substantial improvement over the PID and FID in terms of accuracy, resolution, and technical sophistication. A portable GC consists of a sample injection system, separation column, output detector, and detection system. A GC/FID system contains a combustible gas supply for the flame; a GC/PID system contains a UV lamp.

The vapor sample is injected into the GC and carried through the sample column by an inert carrier gas such as helium. The individual hydrocarbon contaminants travel through the column at unique speeds, thus reaching the

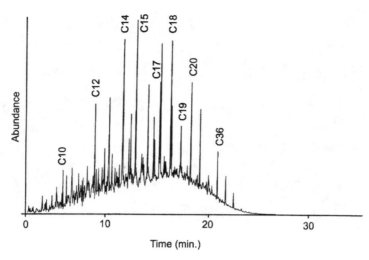

Figure 5.24 Sample gas chromatograph of diesel fuel that had been weathered in soil for 100 days.

detector at different times. Under ideal circumstances each component is separated by the time it enters the detector. Typically, the lightest constituents elute first, followed by progressively heavier constituents. The detection process is translated into a chromatogram, which shows the length of time from injection to maximum peak height. The peak height from baseline to the top of the peak is proportional to the concentration of a constituent (figure 5.24).

The portable GC is an extremely versatile and powerful tool for use in the field. However, performance of the GC is greatly dependent on operator capabilities; the instrument requires a substantial level of skill to operate and to interpret the results.

QUESTIONS

1. Environmental site assessments (ESAs) associated with a property transfer are common in light of CERCLA. List possible clients of an ESA.
2. What specific qualities have made asbestos enormously popular in construction and other applications for centuries?
3. Which of the following is considered a common type of material that can contain friable asbestos and may signal the need for further testing? (a) wall board with sprayed- or troweled-on material on the back that serves as insulation; (b) fluffy, sprayed-on material used for

fireproofing ceilings or walls; (c) oils used to lubricate sliding doors; (d) wall insulation with a foamy texture; (e) a and b only; (f) all of the above.
4. An LNAPL will sink in an aquifer but will not dissolve in the water. True or false?
5. A phase I environmental inspection includes deep subsurface sampling combined with thorough chemical, physical, and biological testing of soils, groundwater, and geologic materials. True or false? Explain your answer.
6. The Hi-Jinx Petrochemical Company is preparing to sell 2.5 acres of its industrial complex east of town. The prospective buyer has hired you to assess the site for possible liabilities. Mark (T/F) those items that are essential to assess *in the phase I ESA*: (a) aerial photographs for the past fifty years; (b) fire insurance maps; (c) NPL, UST, and LUST lists; (d) citations and notices of violation; (e) detailed soil borings and groundwater and geologic analyses; (f) asbestos analysis of floor-, wall-, and pipe-insulation samples.
7. During the site reconnaissance you discover an area of buried drums, most of which appear to be unlabeled, rusting, and possibly leaking. There is a slight odor of solvents. A limited phase II ESA, *at least* analyzing the above area, is strongly recommended. True or false?

There are several structures at the Hi-Jinx site: one production facility, a tank farm, two warehouses, and an unidentified building. All were constructed prior to 1955. Use this information to answer questions 8 to 11.

8. Asbestos-containing material (ACM) should be a concern for the site assessor. True or false?
9. LBP should be a concern, both inside and outside the structures. True or false?
10. It is possible that unregistered USTs may occur on-site. True or false?
11. Will soil acidity/alkalinity significantly affect movement of NAPL? Why or why not?
12. You are investigating a potentially hazardous atmosphere in a UST. Oxygen levels measure approximately 15 percent and there is a known risk from gasoline and benzene vapors at this site. What type of detector would you recommend using for simple field analysis of these vapors? Why?
13. Search the UST and LUST databases for your community. Where are the LUSTs located? What is the status of their cleanup?

14. Locate a commercial or industrial facility in or near your county. Compile the following information for the site and/or surroundings: (a) bedrock geology; (b) surficial geology; (c) soil types and descriptions; (d) presence of one or more aquifers; (e) depth to groundwater. All these data can be found in an internet search.
15. For the site located in question 17, locate: (a) the presence, if any, of USTs; (b) LUSTs; (c) NPDES permits; (d) waste disposal permits; (e) TRI listings; (f) NPL listings. Most or all of this data can be found in an internet search.

REFERENCES

American Society for Testing and Materials. 2015. *ASTM Standards on Environmental Site Characterization*. West Conshohocken, PA: ASTM.

———. 2013. *Standard Practice for Environmental Site Assessments: Phase I Environmental Site Assessment Process*. E1527-13. West Conshohocken, PA: ASTM.

———. 2012. *Standard Practice for Active Soil Gas Sampling in the Vadose Zone for Vapor Intrusion Evaluations*. ASTM D7663-12. West Conshohocken, PA: ASTM International.

———. 2011. *Standard Practice for Environmental Site Assessments: Phase II Environmental Site Assessment Process*. E1903-11. West Conshohocken, PA: ASTM.

Delle Site, A. 2001. Factors affecting sorption of organic compounds in natural sorbent/water systems and sorption coefficients for selected pollutants. A review. *Journal of Physical and Chemical Reference Data* 30(1): 187–439.

EMLab P&K. 2008. *Asbestos Identification and Quantification in Bulk Samples. The Environmental Reporter.* https://www.emlab.com/resources/education/environmental-reporter/asbestos-identification-and-quantification-in-bulk-samples/.

Hanby. 2006. *Hanby Environmental Laboratory Procedures*. www.hanbytest.com.

Hess, K. 1997. *Environmental Site Assessments, Phase I: A Basic Guide*, 2nd ed. Boca Raton, FL: CRC Press.

International Sensor Technology. 2006. *Photoionization Detectors*. www.intlsensor.com/pdf/photoionization.pdf.

Lyman, W., D. Noonan, and P. Reidy. 1990. *Cleanup of Petroleum Contaminated Soils and Underground Storage Tanks*. Park Ridge, NJ: Noyes.

NEIWPCC (New England Interstate Water Pollution Control Commission). 1990. *What Do We Have Here? An Inspector's Guide to Site Assessment at Tank Closure*. Boston: NEIWPCC.

O'Brien, E. F. 2011. Environmental consultants. In *Environmental Aspects of Real Estate and Commercial Transactions,* edited by J. B. Witkin. Chicago: American Bar Association.

Sullivan, E. 2018. Estimating environmental liabilities. *Risk Management*. http://cf.rims.org/Magazine/PrintTemplate.cfm?AID=2804 (Accessed February 9, 2018).

U.S. CDC (Centers for Disease Control and Prevention). 2014. *Lead.* https://www.cdc.gov/nceh/lead/tips.htm.

U.S. Department of Defense. 2003. *Report of the Defense Science Board Task Force on Unexploded Ordnance December 2003.* Washington, DC: Office of the Under Secretary of Defense for Acquisition, Technology, and Logistics.

U.S. EPA (Environmental Protection Agency). 2018a. *Ground-penetrating Radar.* Environmental Geophysics. https://archive.epa.gov/esd/archive-geophysics/web/html/ground-penetrating_radar.html (Accessed February 9, 2018).

———. 2018b. *Resistivity Methods.* Environmental Geophysics. https://archive.epa.gov/esd/archive-geophysics/web/html/resistivity_methods.html (Accessed February 12, 2018).

———. 2016. *Expedited Site Assessment Tools For Underground Storage Tank Sites. A Guide For Regulators.* EPA 510-B-16-004. Washington, DC: Land and Emergency Management.

———. 2012. *Sampling and Analysis Plan. Guidance and Template.* Version 3, Brownfields Assessment Projects. R9QA/008.1. http://www.epa.gov.

———. 2008. MTBE. In *Regulatory Determinations Support Document for Selected Contaminants from the Second Drinking Water Contaminant Candidate List (CCL 2).* EPA Report 815-R-08-012. Washington, DC.

_____. 2006. *3Ts for Reducing Lead in Drinking Water in Schools.* Revised Technical Guidance. Washington, DC.

_____. 1986. *Test Methods for Evaluating Solid Waste, Physical/Chemical Methods.* SW-846. Washington, DC.

Chapter 6

Isolation of the Contaminant Plume

"It makes a difference, doesn't it, whether we fence ourselves in, or whether we are fenced out by the barriers of others?"

—E. M. Forster, *A Room with a View*

6.1 INTRODUCTION

If a plume of hazardous contaminants drifts beyond the vadose zone and contacts groundwater, its rate of migration may accelerate, increasing the threat of public health or environmental damage. Contaminants contact groundwater via several avenues: mobilization by infiltration of precipitation or run-on from the surface; groundwater migration into the contaminated zone; and migration of contaminants in the form of NAPL and dissolved liquids.

Some contaminant hydrocarbons are *immiscible* and less dense than water (e.g., LNAPLs or 'floaters') and will float on the water table. Other contaminants (e.g., hydrocarbon 'mixers' and a number of metals, salts, and anions) readily dissolve in groundwater. Any of these contaminants can be carried along with groundwater, which is by no means a static body. In order to control contaminant migration it may be necessary to contain the plume and prevent additional dissolution of hazardous constituents by water entering the contamination zone. During such an isolation period engineers, scientists and regulatory officials can formulate a detailed remediation program.

Before a specific containment technique is selected and established in the field, it is essential to identify all significant contaminants, thoroughly assess the site hydrogeology, and understand the chemical and physical behavior of the contaminants in soil, groundwater, and other environmental media. A comprehensive environmental site assessment will provide the project manager with an understanding of site soils, geology and hydrology, and the properties and behavior of the contaminants in the subsurface.

A number of systems are available to isolate the affected area such that contaminants are contained, either for permanent isolation or for removal or treatment at a later date. Isolation techniques are commonly used in association with so-called pump-and-treat systems, which are designed to remove contaminated groundwater from the contained area for eventual treatment. Specific techniques that effectively limit the spread of the contaminant include diverting the flow of groundwater; subsurface barriers placed in the direction of flow to control lateral spread; and placement of an impermeable cap on the land surface to reduce infiltration of precipitation and run-on. Each varies in terms of effectiveness, logistics of installation, and overall cost. These techniques all share the common purpose of isolating the contaminant. They do not remediate the site per se; that is, they do not remove or destroy the contaminant.

6.2 DIVERTING GROUNDWATER FLOW

6.2.1 Groundwater Pumping

Vertical or horizontal movement of groundwater can be altered by injection of water into a substratum or by extraction of groundwater from selected locations. This ultimately serves to divert, contain, or remove a contaminant plume. An array of extraction wells is positioned downgradient from a contaminated zone and functions to remove groundwater and contaminants (figure 6.1). Another approach is to inject water directly into the subsurface to literally push the plume away from an area requiring protection, such as a public drinking water source or an ecologically sensitive area (figure 6.2) (U.S. EPA 1989). Pumping techniques comprise an *active* strategy for diverting groundwater. A disadvantage of this technique, then, is that continuous operation and ongoing maintenance are required for the entire period of the remedial activity, which can prove costly for lengthy operations.

6.2.2 Subsurface Drains

In contrast to pumping techniques, subsurface drains comprise a passive system that does not require substantial maintenance. Subsurface drains are

Isolation of the Contaminant Plume 153

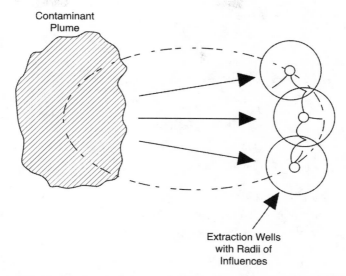

Figure 6.1 Removal and containment of a hazardous plume using extraction wells. *Source*: U.S. Environmental Protection Agency.

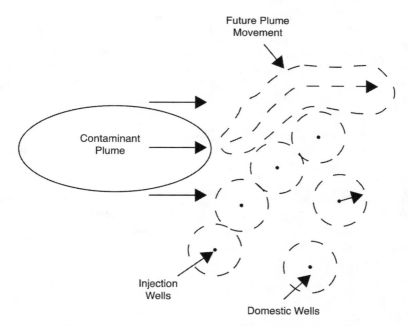

Figure 6.2 Diversion of a contaminant plume by injecting water into the subsurface. *Source*: U.S. Environmental Protection Agency.

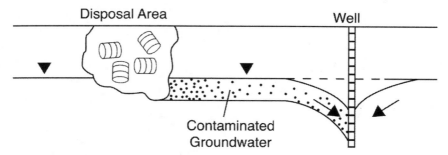

Figure 6.3 The use of subsurface drainage to contain a hazardous plume.

permeable regions established in or near a contaminated zone, having the function of intercepting groundwater flow. Water is collected at a low point and pumped or drained by gravity to a collection basin or storage tank. The collected liquids are subsequently treated and released. Subsurface drains can also isolate a contaminated area by intercepting uncontaminated groundwater before it comes into contact with the site (figure 6.3) (U.S. EPA 1989).

6.3 SUBSURFACE BARRIERS

Low-permeability barriers can serve as a practical and permanent, albeit cumbersome, isolation technique for a contaminated site. Barriers block the flow of uncontaminated groundwater from entering the affected location; additionally, they prevent contaminated liquids from migrating off-site. When enclosed by such a barrier, liquids are recovered under controlled conditions for eventual treatment. Barriers include slurry walls, grout curtains, and sheet piling. These barriers comprise a *passive* means of containment, whereas groundwater pumping actively controls pollutant migration and extracts contaminated groundwater for subsequent treatment. In order to serve as an effective barrier, walls are constructed of low-permeability materials. These can be either soluble or insoluble and are selected based on site considerations, installation logistics, and cost factors. Typical materials include soil-bentonite mixtures, cement-bentonite mixtures, or rigid sheet materials (e.g., steel). Prior to selection of materials, a chemical analysis of wall and contaminant compatibility should be conducted.

The overall effectiveness of barriers can be significantly improved by curbing surface infiltration. This may be accomplished by installing surface caps prior to or during installation of barriers (see below). The placement of a low-permeability barrier usually involves substantial earth moving, is energy- and labor-intensive, and is expensive. However, once installed, it is a long-term, low-maintenance containment system.

6.3.1 Slurry Walls

Slurry walls are a common installation for reducing groundwater flow in unconsolidated material. Slurry walls are rigid underground physical barriers formed by pumping slurry into a deep vertical trench during excavation, allowing the slurry to set, and backfilling with native soil or other suitable fill (figures 6.4 and 6.5). Several materials are suitable for construction of the slurry wall depending on the chemical and physical nature of the contaminants and on the strength required. Slurries are commonly constructed with Portland cement, a cement-bentonite mixture, or a soil-bentonite mix. The trench is excavated by standard construction equipment such as a backhoe; dedicated trenching machines are also available. The slurry hydraulically shores the trench walls to prevent collapse.

Once the lateral extent and properties of the contaminant plume are determined, the positioning of the wall is established. If contaminants are soluble and/or mobile in water—for example, metals, salts, and certain organics (mixers)—a slurry wall can be connected (keyed) to a low-permeability stratum such as bedrock (figure 6.6). If installed properly, this will halt the flow of all liquids; however, it will not be feasible in locations of deep bedrock. In the case of floating contaminants, such as LNAPL, a wall can be constructed 'hanging' (figure 6.7). However, any soluble contaminants in groundwater can, in theory, pass beneath the wall. A keyed slurry wall would therefore be appropriate.

Slurry walls can be placed upgradient or downgradient of the contaminant site, or it can completely encircle the affected area (figures 6.8 and 6.9).

Figure 6.4 A trencher making its first cut into the ground surface. *Source*: U.S. Air Force. https://www.afcent.af.mil/News/Article/221806/trench-highlights-interoperability/.

Figure 6.5 **Hydraulic clamshell bucket digging slurry wall.** *Source*: Cheryl Rusnak.

Circumferential walls are effective for small plumes and offer advantages over linear walls (IDEM 2017)—they reduce the amount of uncontaminated groundwater entering the affected area from upgradient and also prevent downgradient migration of contaminants. Additionally, the use of circumferential walls in association with extraction wells within the perimeter maintains the hydraulic gradient in an inward direction, which prevents escape of contaminants (Smith et al. 1995).

Several practical issues must be addressed in the design of a slurry wall. An initial step is to determine compatibility of wall materials with the contaminant. The presence of reactive or incompatible contaminants in soil may damage a slurry wall and limit its reliability. Soil-bentonite mixtures can be damaged by strong acids or bases, sulfates, strong electrolyte solutions, or concentrated solvents (IDEM 2017; U.S. EPA 1991). Furthermore, in soils contaminated with hydrophobic compounds (NAPL), insufficient contact may occur between slurry material and adjacent soil. It is beneficial to remove as

Isolation of the Contaminant Plume 157

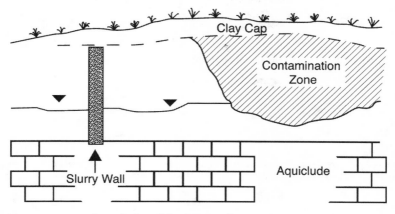

Figure 6.6 Cross section of a keyed-in slurry wall.

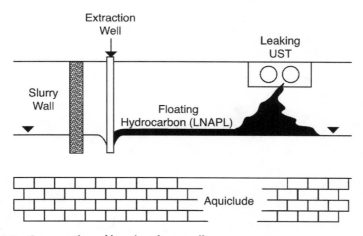

Figure 6.7 Cross section of hanging slurry wall.

much free product as possible prior to installation of the wall. Slurry walls must also be compatible with native soils, geologic strata, and groundwater. Compatibility of the proposed slurry mixture with soil and geologic material should be tested in the laboratory. Site conditions and slurry composition should also be evaluated to determine the optimal configuration of the slurry wall. For example, site topography will affect choice of slurry materials. A soil-bentonite mixture will flow readily; therefore, it should be used at nearly level sites. In contrast, concrete-bentonite slurries, which set rather quickly, are suitable for sites having irregular topography (Smith et al. 1995).

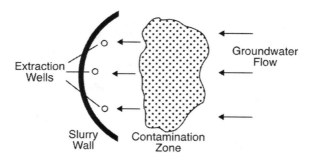

Figure 6.8 **Slurry wall positioned downgradient of a contaminant plume.** *Source*: U.S. Environmental Protection Agency.

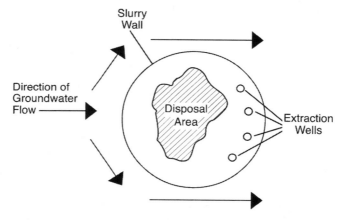

Figure 6.9 **Circumferential slurry wall.** *Source*: U.S. Environmental Protection Agency.

6.3.2 Grout Curtains

Grout curtains are rigid underground barriers formed by injecting grout into porous rock or soil through wells. As with a slurry wall, the grout curtain will contain affected groundwater, divert a chemical plume, and divert groundwater flow around an affected area. Grouts are commonly composed of particulate materials (e.g., Portland cement) or soluble chemicals (sodium silicate). Some grouts are organic and contain polymers or bitumen.

Construction of a grout barrier is accomplished by pressure injecting the grouting material through a pipe into the strata to be sealed. The injection points are usually arranged in lines of primary and secondary grout holes (figure 6.10). During or after drilling, the necessary quantity of grout is injected into one row of holes and allowed to permeate into the surrounding soil. After the grout has had time to set, the secondary holes are injected. The spacing of

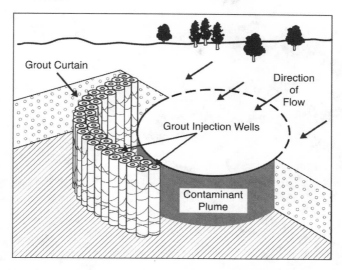

Figure 6.10 Grout curtain configuration. *Source*: U.S. Environmental Protection Agency.

the injection points is calculated based on factors including soil permeability, grout viscosity, and set time of the grout. The points must be spaced to allow the grout injection radii from adjacent injection points to overlap (Quintal and Otero 2018). Each borehole should be drilled into the underlying bedrock if possible, thereby 'locking in' the structure and preventing flow of water-soluble or heavy immiscible contaminants below the curtain.

Grouting tends to carry a higher cost compared to that of slurry walls. Cement has probably been used more than any other grouting material. The addition of clay to the grout can expand its range of usage. Clays have been widely used in grouts because of their relatively low cost.

6.3.3 Sheet Piling

Sheet piling barrier walls involve driving rigid sheets into the ground to form a physical barrier to groundwater movement. These sheets are typically composed of steel or concrete that can be interlocked or sealed to form a continuous impermeable barrier. Steel sheet piling is more commonly used than concrete because of its resistance to shock and physical stress, because it is more effective overall at groundwater cutoff, and because it is less costly.

Sheet piles are seldom used as a groundwater barrier at contaminated sites because the overall effectiveness of the barrier is strongly influenced

Figure 6.11 Steel sheet piling used for containing a plume of no. 2 heating oil at a UST field.

by subsurface conditions. Sheet piling is more commonly employed for temporary dewatering in other construction, for example, removal of a UST (figure 6.11), or for erosion protection. A drawback to sheet piling is that rocky soils and other obstructions (e.g., buried drums) damage and deflect the piles, thus lessening their integrity as a groundwater barrier. Sheet piling may allow leakage through interlocking joints. Such leakage should be taken into consideration when predicting the permeability of the contaminated zone and the adjacent soil.

6.4 CAPPING

At a hazardous site, control of infiltrating water is important in order to prevent solubilization of constituents within the plume as well as to limit plume migration. *Capping systems* are employed to reduce surface water infiltration, to provide a stable surface over contaminated soil, and to improve aesthetics. Several cap materials and designs are in use ranging from simple, single layers of compacted clay to complex multilayer systems. In cases where remedial treatments are not recommended due to cost or feasibility concerns, permanent caps can provide long-term contaminant isolation and prevent mobilization of soluble compounds. Capping may also be used for temporary isolation of contaminants during selection of the final treatment technology. Surface water control practices (ditches, berms, etc.) are often used in association with caps in order to control rainwater drainage and run-on/runoff around the site.

Selection of the capping materials and design depends on remedial objectives, risk factors, and site cleanup goals. Other factors include the local availability and cost of cover materials, the chemical and physical nature of the contaminants being covered, climate, and projected future use of the site. The design of caps usually conforms to the standards in 40 CFR 264, where closure requirements for RCRA landfills are detailed. These standards include minimum liquid migration rates, requirements for maintenance of covers, site drainage, and resistance to settling, and other damage.

6.4.1 Synthetic Membranes

A *synthetic membrane cap* is installed as a series of overlapping sheets over a prepared surface, after which the sheets are sealed together. Membrane materials include polymers, rubber, coated fabrics, and others. Success in use of synthetic membranes depends on compatibility of the polymer with the contaminants, proper installation including seaming and placement to prevent damage, and protection against weathering and root penetration. Seaming of sheets is critical to success as it is a common source of leakage. Edges can be seamed by the use of heat guns (fusion seams), chemical adhesives, and extrusion seams (figure 6.12).

The primary benefits of synthetic membranes are their extremely low permeabilities and variability of composition to match a particular situation. A range of polymer types is available which can be selected for compatibility with a particular contaminant. Common polymers include:

- HDPE (high-density polyethylene)
- VLDPE (very low-density polyethylene)
- PVC (polyvinyl chloride)
- CSPE (chlorosulfonated polyethylene)
- urethane
- polypropylene
- proprietary formulations

Membranes typically vary in thickness from 30 to 45 mils (1 mil = 1/1000 in).

Limitations of synthetic membranes include cost and potential for failure over the long-term due to damage caused by puncturing, tearing, or weathering. These limitations are minimized by proper design and installation.

(a)

(b)

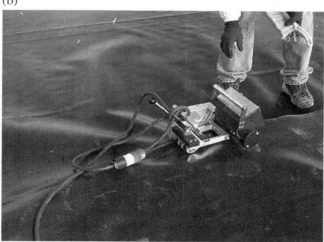

Figure 6.12 Geomembrane cap installation. (a) Rolling out the geomembrane layer. Sandbags hold edges in place against the wind. (b) Wedge seam sealer fuses the seam. The sealer is self-propelling. The seam in advance of the sealer is swept clean to facilitate a quality seal. *Source*: U.S. Environmental Protection Agency 2006.

6.4.2 Low-Permeability Soils

This cap is composed of fine-grained soils (e.g., clays or silty clays) that, when compacted, maintain a Ks of 10^{-6}–10^{-7} cm/sec (0.1 ft/yr) or less. Benefits of such materials are that they may be locally available and therefore relatively inexpensive. Furthermore, because they are natural materials, they will be durable over very long periods.

The cap is installed after the surface is prepared: soil is leveled and compacted, then covered by a clean soil layer followed by topsoil and vegetation. The uppermost soil layer should be limed and fertilized if necessary, then seeded and mulched immediately after placement to prevent erosion. Deep-rooted vegetation such as trees should be avoided, as taproots will eventually compromise the integrity of the liner. The vegetative cover is typically a shallow-rooted grass or grass-legume mixture. The benefit of a mixture of grasses and legumes is that a self-sustaining ecosystem may eventually develop; that is, legumes will provide nitrogen to the grass crop. With time and vegetative turnover from year to year, a minimal degree of maintenance may be required.

6.4.3 Multilayer Cap

The multilayer cap is popular as a barrier against infiltration and for site closure. Such a cap is often required for MSW landfills and secure landfills (i.e., for hazardous waste disposal) as well. The cap minimizes infiltration, diverts water from the site, and acts as a growth medium for vegetation. A typical multilayer cap system as shown in figure 6.13 consists of three layers:

1. Sealing layer. Very low-permeability material is placed directly over the prepared site and is compacted using heavy equipment. This layer serves as the barrier immediately overlying the contaminated material. Its purpose is to prevent downward infiltration, and may be composed of compacted fine-grained soil material or a flexible synthetic geomembrane. If the layer is composed of soil material, a layer of 18 to 24 in is typically required. The sealing layer is protected from the action of weathering and root penetration by the overlying soil and drainage layers.
2. Drainage layer. This is a bed of porous material (e.g., sand, synthetic grid), which promotes drainage of liquids that have penetrated the uppermost layer. This layer is designed and sloped so that flow of infiltrating water will be carried laterally to low points at the edge of the cap, ultimately to be removed either by gravity flow or by a sump. The drainage layer is typically about 18- to 24-in thick.
3. Surface soil. A layer of native soil is placed to a depth of 12 to 24 in. This layer supports vegetation and provides a cover for the drainage layer below. The soil should be of adequate fertility in order to promote long-term vegetative growth with minimal maintenance. It is essential to protect both cap and surface drainage layers from erosion damage. Establishment of appropriate grading for each layer, in addition to mulching and growing dense-rooted vegetation, helps protect the cap against erosion.

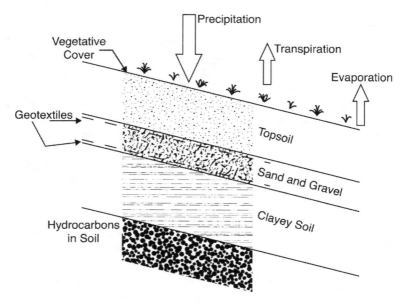

Figure 6.13 Cross section of a typical multilayer cap system.

6.5 OTHER PRACTICAL CONSIDERATIONS

6.5.1 General Design

The design of a containment system must address all vertical and horizontal migration pathways of precipitation, groundwater, and contaminants. Surface water run-on is managed by utilization of diversion channels, drains, berms, and other physical barriers. During the design and construction phases of the containment system, it is important to consider both future use and use restrictions of the land. For example, future operations on a capped area must not damage the integrity of the cap. Inclusion of active gradient controls (i.e., pump-and-treat) required by some installations will add to the long-term operating costs of these systems.

The presence of a barrier such as a cap may actually hinder remediation efforts. For example, if the installation of monitoring wells or exhumation of waste materials should occur after capping, these physical changes will adversely affect the integrity of the cap.

6.5.2 Equipment

Generally, large earth-moving equipment is required for installation of slurry walls and caps. Bulldozers, backhoes, and earth-hauling machines are commonly used, but specialized equipment may also be required (see figures 6.4

and 6.5). Truck-mounted drills and augers are employed for grout curtain installation, and truck-mounted hammers for steel sheet piling.

6.5.3 Monitoring

The ultimate goal of containment is to halt the spread of contaminants in the subsurface following a release. In other words, containment involves securing hazardous materials in place. Monitoring, therefore, becomes important as a means for measuring effectiveness of the containment strategy. Monitoring may be accomplished by groundwater testing to measure water quality and groundwater depth, and by visual inspections of caps for structural integrity. Groundwater monitoring should take place both up- and downgradient of the barrier walls in order to compare soil and water quality in contaminated and presumably non-contaminated zones. Long-term monitoring (i.e., for a decade or longer) may be required to ensure that containment is effective (NRC 2007).

6.6 TREATMENT OF GROUNDWATER (PUMP-AND-TREAT TECHNOLOGY)

Up to this point, containment has implied no treatment of affected groundwater; rather, a plume is simply being isolated or diverted to prevent groundwater contamination or limit additional contamination. An additional objective of groundwater pumping, however, involves removal of dissolved contaminants from the subsurface.

Groundwater remediation became an environmental priority in the United States following the passage of RCRA and CERCLA. At the time, the most commonly used remediation approach was to pump contaminated water to the surface and treat it using carbon filtration and/or air stripping (U.S. EPA 1996)—termed the *pump-and-treat* method.

If groundwater is to be treated, the desired level of finished water quality is determined a priori. Next, the design and implementation of the groundwater pumping system are formulated, based on data evaluated in setting cleanup goals. Criteria for well design, pumping system, and treatment are a function of contaminant and site characteristics. Treatment may include a train of processes such as gravity separation, air stripping, or carbon adsorption for removal of specific contaminants (see below). An associated component of groundwater pump-and-treat is establishing termination criteria. Termination requirements are based on the cleanup goals defined in the initial stage of the remedial process. Termination criteria are also dependent on new discoveries (for example, new sources of contamination) revealed at the site during remedial operations (FRTR 2018).

Groundwater monitoring is a required component of a groundwater extraction program; it is conducted so that the effectiveness of the overall treatment activity can be assessed. Additionally, monitoring data will allow the operator to determine when termination requirements have been met and operations can cease. Groundwater data from wells and piezometers allows the operator to adjust the system in response to changes in subsurface conditions caused by the remediation activity.

The following groundwater treatment technologies are commonly employed following pumping. These are briefly described below; some are described in detail in subsequent chapters.

6.6.1 Bioreactors

Extracted groundwater is transferred to a large vessel where contaminants come into intimate contact with aerobic heterotrophic microorganisms. Common reactor types include suspended growth or attached biological reactors (see chapter 11). In suspended systems, for example the activated sludge process, contaminated groundwater is circulated in an aeration basin. In attached systems, such as rotating biological contactors and trickling filters, microorganisms become established on an inert support matrix. The aerobic microbial consortium decomposes the hydrocarbon contaminants as oxygen and nutrients (if necessary) are introduced.

6.6.2 Constructed Wetlands

Groundwater treatment via *constructed wetlands* (figure 6.14) relies upon natural geochemical and biological processes within an artificial wetland ecosystem to accumulate and remove contaminants from influent waters; for example, microbial activity in the dense rhizosphere (root zone) decomposes hydrocarbons and immobilizes metals. Certain plant types also extract metals from water and soil. Metals may be chemically transformed to less mobile species which precipitate and/or sorb to sediment.

6.6.3 Air Stripping

Air stripping technology involves the mass transfer of volatile contaminants from water to air. Volatile organics are partitioned from groundwater by increasing the surface area of the contaminated water as it is exposed to air. For groundwater remediation, the process is commonly conducted in a packed tower or aeration tank. The typical packed tower air stripper includes a spray nozzle at the top to distribute contaminated water over packing material within the column, a fan to force air countercurrent to water flow, and a sump at the base to collect decontaminated water. Auxiliary equipment

Figure 6.14 Constructed wetlands can be used to remove low concentrations of heavy metals and hydrocarbons from groundwater.

includes an air heater to improve removal efficiencies; automated control systems with safety features such as explosion-proof components; and air emission control and treatment systems such as activated carbon units, catalytic oxidizers, or thermal oxidizers.

6.6.4 Granulated Activated Carbon (GAC) Adsorption

In *carbon adsorption*, groundwater is pumped through a single or multiple canisters or columns containing activated carbon. Both hydrocarbon and metallic contaminants may sorb to carbon internal and external surfaces. When the concentration of contaminants in the effluent from the bed exceeds an established limit, the carbon can be regenerated in place, removed and regenerated at an off-site facility, or removed and disposed. The two most common reactor configurations for carbon adsorption systems are the fixed bed and the pulsed, or moving, bed. Adsorption by activated carbon has a long history of use in treating municipal, industrial, and hazardous wastes.

6.6.5 Ion Exchange

Ion exchange serves as another means of contaminant sorption, where groundwater is pumped through a canister packed with specially formulated resin beads. The beads are typically formulated from synthetic organic polymers that contain ionic functional groups. Inorganic and natural polymeric materials are also available. Contaminant cations or anions become adsorbed to the exchange resin as water passes through. After the resin capacity is exhausted, resins are regenerated for reuse.

6.6.6 Precipitation

Precipitation of metals has long been a method of choice for treating metal-enriched industrial wastewaters. In groundwater treatment, metal precipitation involves conversion of soluble heavy metal ions to insoluble salts that eventually precipitate. The process usually relies upon pH adjustment and/or addition of a chemical precipitant. Typically, metals precipitate from solution as hydroxides, sulfides, or carbonates. The precipitate is removed from the system by physical methods such as sedimentation or filtration. The solubilities of the specific metal contaminants and the required cleanup standards will dictate the process used. In some cases the metallic sludges generated can be shipped to recyclers for metal recovery (FRTR 2018).

6.6.7 Sprinkler Irrigation

The recovered groundwater is collected and distributed over the top of the site. The organic contaminants in the applied water are degraded by microorganisms which occur naturally in the soil (see chapter 11, "Microbial Remediation").

QUESTIONS

1. The purpose of containment is to "buy time" and prevent off-site contaminant migration; containment is not remediation. True or false? Explain.
2. Choose the correct answer. Containment may be needed when there is danger of contamination of: (a) groundwater; (b) drinking water wells; (c) sanitary or storm sewers; (d) sensitive ecological areas; (e) all of the above.
3. List the layers (geologic materials) of a multilayer cap situated over contaminated soil (in order, from the ground *surface* down to the *deepest* layer) and describe each of their functions.

REFERENCES

Clay, L., and J. Pichtel. 2018. Treatment of simulated oil and gas produced water via pilot-scale rhizofiltration and constructed wetlands. *International Journal of Environmental Research.* https://doi.org/10.1007/s41742-018-0165-0

FRTR (Federal Remediation Technologies Roundtable). 2018. *Ground Water Pumping/Pump and Treat Remediation Technologies Screening Matrix and Reference Guide, Version 4.0.* www.frtr.gov/matrix2/section4/4-48.html (Accessed January 26, 2018).

IDEM (Indiana Department of Environmental Management). 2017. *Engineering Control: Slurry Walls.* Office of Land Quality. http://www.in.gov/idem/cleanups/files/engineering_controls_slurry_walls.pdf.

NRC (National Research Council). 2007. *Assessment of the Performance of Engineered Waste Containment Barriers.* Division on Earth and Life Studies, Board on Earth Sciences and Resources, Committee to Assess the Performance of Engineered Barriers. Washington, DC: National Academies Press.

Quintal, D., and M. Otero. 2018. *Vertical Impermeable Barriers (Cutoff Walls).* Geoengineer. http://www.geoengineer.org/education/web-based-class-projects/geoenvironmental-remediation-technologies/impermeable-barriers?showall=1&limitstart=

Smith, L. A., J. L. Means, A. Chen, B. Alleman, C. C. Chapman, J. S. Tixier, S. E. Brauning, A. R. Gavaskar, and M. D. Royer. 1995. *Remedial Options for Metals-Contaminated Soils.* Boca Raton, FL: CRC Press.

U.S. Environmental Protection Agency. 1985. *Handbook: Remedial Action at Waste Disposal Sites (Revised).* EPA/625/6-85/006. Washington, DC: Office of Solid Waste and Emergency Response.

———. 1989. *Corrective Action: Technologies and Applications.* Seminar publication EPA/625/4-89/020.

———. 1991. *Handbook: Stabilization Technologies for RCRA Corrective Actions.* EPA/625/6-91/026.

———. 1996. *Pump-and-Treat Ground-Water Remediation. A Guide for Decision Makers and Practitioners.* EPA/625/R-95/005.

———. 2006. *Cleanup Sites. February 5, 2001 Field Update.* http://www.epa.gov/region5/sites/amerchem/fu20010205.htm.

———. 2018. *Cleanup in Region 10. Kerr-McGee Photo Gallery.* yosemite.epa.gov/r10/cleanup.nsf/9f3c21896330b4898825687b007a0f33/ee48a36956b95202882 56a7800009bce?OpenDocument (Accessed January 26, 2018).

Chapter 7

Extraction Processes

All will come out in the washing.

—Miguel de Cervantes Saavedra

7.1 INTRODUCTION

Extractive processes involve the separation of inorganic and/or organic contaminants from soil for eventual recovery, treatment, and disposal. Also known as *soil flushing, soil washing, chemical leaching, solution mining,* or *hydrometallurgical processes,* a chemical reaction occurs where contaminants are solubilized or similarly detached from soil solids and subsequently recovered. This technology contrasts with physical processes, in which soil fines (silt- and clay-sized particles) are separated in water from coarser material by simple gravity-based settling processes, following which the former fraction is disposed as hazardous waste.

In situ extraction processes are applicable for either the vadose zone or the saturated zone. The flushing solution is applied to the affected site via sprinklers or irrigation, or by subsurface injection. A sufficient period is allowed for the applied reagents to percolate downward and react with contaminants, which are subsequently mobilized by solubilization or formation of emulsions (FRTR 2018). Liquids are collected in appropriately placed wells or subsurface drains, following which they are removed, treated, and/or recycled back to the site (figure 7.1).

Pretreatment steps include preparation of flushing solutions and installation of systems to deliver and recover the flushing solution. Surface grading, caps, or vertical barriers (e.g., slurry walls) may be needed to control flow.

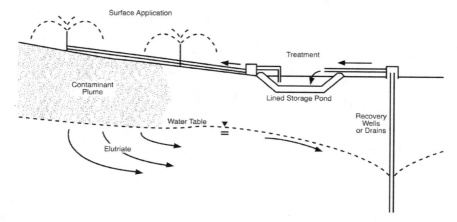

Figure 7.1 Schematic diagram of an in situ flushing process in contaminated strata.

Posttreatment steps may include disposal of treatment residuals, processing and reuse of flushing solution, and recovery of certain contaminants (e.g., metals). Both inorganic and organic contaminants are amenable to soil flushing treatment if they are sufficiently soluble in an inexpensive solvent that is available in large volume. The selection of extracting solution must also conform to regulatory requirements because certain solutions pose public health and/or environmental risks if not properly managed.

An analogous process to soil flushing is its ex situ counterpart, *soil washing*. In this procedure, the contaminated soil is physically removed from the affected area—for example, by backhoes—placed into a reactor vessel, and vigorously mixed with the washing solution. There may be repeated cycles of washing, centrifugation, and recovery of clean soil material (figure 7.2). Eventually the residual extractant is rinsed out of the cleaned soil with H_2O and the soil is returned to its original location. The recovered extractant can be reused for additional washings.

To ensure a thorough level of treatment and compliance with regulatory requirements, extractive processes are often used in conjunction with other treatment technologies such as biodegradation, phytoremediation, reaction with activated carbon, or chemical precipitation (Sung et al. 2011; Kos and Lestan 2003; Smith et al. 1995).

7.2 EXTRACTING SOLUTIONS

Flushing solutions include water, acidic or basic solutions, chelating agents, surfactants, oxidizing or reducing agents, and alcohols. The appropriate solution is chosen based on a variety of considerations including ability to react

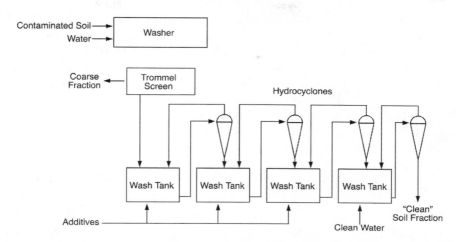

Figure 7.2 Schematic diagram of an ex situ washing process for contaminated soil.

with contaminant, ease of handling, potential environmental and/or health hazards of the formulation, possible adverse effects to the soil, cost, and ease of recycling.

Raw water can be used to extract water-soluble or water-mobile constituents including simple salts and anions, for example, arsenate (AsO_4^{3-}), arsenite (AsO_3^-), cyanide (CN^-), nitrate (NO_3^-), and selenite (SeO_4^{2-}). Acidic solutions are used for recovery of metals and for basic organic constituents including amines, ethers, and anilines. Basic solutions can be used for removal of metals including zinc, tin, and lead, and for some phenols. Complexing and chelating agents are employed for removal of metals, and surfactants for nonaqueous or hydrophobic compounds. In limited cases, surfactants have been successful in the removal of metals (Torres et al. 2012; Doong et al. 1998; Thirumalai Nivas et al. 1996). However, surfactants have shown most success in the solubilization and removal of organic contaminants.

7.3 FLUSHING METALS FROM SOIL

Metals at contaminated sites occur in complex forms, and mobility is controlled by numerous chemical and physical phenomena including soil type, pH, CEC, particle size, and the presence of other inorganic or organic compounds (see chapter 2). Many of these factors are interdependent. Metal removal efficiencies during soil flushing depend not only on soil characteristics but also on metal concentration, chemistry of the metal(s), extractant chemistry, and overall processing conditions.

Metals that are minimally soluble in water often require acids, chelating agents, or other robust solvents for successful flushing. Acids solubilize metals by either simple dissolution reactions of metal-containing solids or by creating metal-hydrogen bonds on a crystal lattice, thereby allowing for metal removal from the solid surface. The most commonly used acids are hydrochloric, sulfuric, nitric, phosphoric, and carbonic acids. High extraction efficiencies, sometimes greater than 90 percent, have been reported with HCl for both artificially and field-contaminated soils (Moon et al. 2012; Moutsatsou et al. 2006; Tokunaga et al. 2005). Using 1M HCl on a contaminated soil, Asel Gzar (2015) measured 98, 94, and 55 percent removal of Cd, Pb, and Ni, respectively. Liu et al. (2010) reported 89 and 96 percent Pb and Cd extraction efficiencies using HCl on soil at a contaminated industrial facility. At a former automobile battery recycling facility Steele and Pichtel (1998) recovered 32 percent soil Pb, and Cd removal was 68 and 98 percent with 0.1 N and N HCl, respectively. The U.S. EPA Acid Extraction Treatment System (AETS), which uses HCl, reduced both the total and TCLP Cd concentrations in soil from the Palmerton (PA) Superfund site to below regulatory limits (U.S. EPA 1994a). Moon et al. (2012) using various washing solutions, reported that HCl, followed by HNO_3, was the most effective option for removal of Zn from contaminated soil.

The use of acid or basic solutions in soil washing presents some disadvantages. The very low or high pH levels created can cause dissolution and leaching of nutrient elements such as Ca, Mg, and K. Extremes in pH will adversely affect the activities of both micro- and macroorganisms.

Chelants generally offer the highest extraction efficiency and are associated with less destruction of soil structure compared to strong acids (Dermont et al. 2008). Chelating agents such as ethylenedinitrilotetraacetic acid (EDTA), diethylene triamine pentaacetic acid (DTPA), nitrilotriacetic acid (NTA) (figure 7.3), and similar agents bond with the metallic cation to facilitate solubilization. Others include:

- N-(acetamido)iminodiacetic acid (ADA)
- cyclohexylenedinitriloetetraacetic acid
- S,S-ethylenediaminedisuccinic acid ([S,S]-EDDS)
- ethyleneglycol-bis (s-aminoethyl ether) $N,N,N'N$-tetraacetic acid
- EGTA (ethylenebis [oxyethylenetrinitrilo] tetraacetic acid)
- ethylene idaminedissuccinate

Chelation mechanisms involve binding with the metal via multiple bonds. Transition metals occurring in nature are typically characterized by the presence of at least one vacant d-orbital; hence, their cationic nature. This orbital

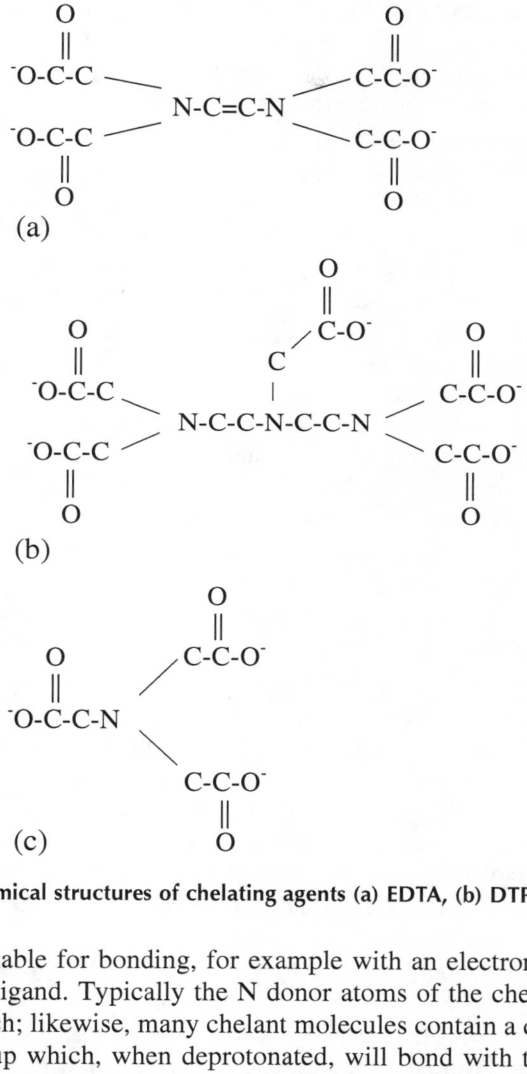

Figure 7.3 Chemical structures of chelating agents (a) EDTA, (b) DTPA, and (c) NTA.

is readily available for bonding, for example with an electron-rich π orbital of an organic ligand. Typically the N donor atoms of the chelant molecules are electron-rich; likewise, many chelant molecules contain a carboxylic acid or similar group which, when deprotonated, will bond with the cation. The ability to form highly stable metal complexes makes chelating agents such as EDTA, DTPA, NTA, and ADA effective extractants for some metal-contaminated soils.

Chelants vary in effectiveness of metal removal, a result of the presence of different numbers of reactive sites and differing degrees of bonding strength. Variability in removal efficiency is also a function of the presence of different solid forms of the metal in the soil, extracting solution pH, and potential interference from other cations that may complex with the chelant. During soil flushing, most of weakly bound metals are first released, and then the rate-limited soil dissolution becomes dominant (Zhang et al. 2013).

Extraction of heavy metals from contaminated soil using EDTA has been reported by numerous researchers. In a study by Asel Gzar and Gatea (2015), metal removal by EDTA was in the order: Cd > Pb > Ni (97%, 88%, and 24%, respectively). In a study of flushing a Superfund soil, EDTA was capable of removing virtually all of the nondetrital forms of Pb when at least a stoichiometric amount (1:1 molar ratio of chelant:metal) was present, and EDTA removed all nondetrital Cd (Steele and Pichtel 1998). Extraction efficiencies between 80 and 100 percent have been reported with EDTA for Cd sorbed to clay, calcium carbonate, and hydrous Fe oxides. Lead and Cd were effectively removed by ADA as well as by PDA (pyridine-2,6-dicarboxylic acid) (Steele and Pichtel 1998). Chen et al. (1995) reported 85 percent Pb extraction efficiency with PDA, and Macauley and Hong (1995) reported 95 percent Pb extraction efficiency from an artificially contaminated soil. Gidarakos and Giannis (2006) measured 95 percent Cd removal from soil using PDA. Hong and Chen (1996) and Chen et al. (1995) noted over 90 percent Cd extraction efficiency with ADA and PDA.

Several researchers have tested naturally occurring compounds and analogs as potential chelating agents. In an artificially contaminated soil, Asel Gzar and Gatea (2015) noted 42 to 100 percent removal of soil Pb using acetic acid. Fischer and Bipp (2004) found, under optimized conditions, the following metal extraction rates achieved with alkaline D-gluconic acid solutions: Ni 43 percent, Cr 60 percent, Cd 63 percent, Zn 70 percent, Pb 80 percent, and Cu 84 percent. Evangelou et al. (2004) investigated the use of humic acids as an alternative to synthetic chelators. Fischer (2002) tested the s-thiol group-containing amino acids L-cysteine and L-penicillamine for their ability to release heavy metals (Cd, Cr, Cu, Hg, Ni, Pb, Zn) from peat, bentonite, and illite.

In some cases surfactant molecules have enhanced metal extraction from soil. In a study by Torres et al. (2012), soil contaminated with arsenic, Cd, Cu, Pb, Ni, and Zn was treated with three surfactants. Collective metal removal efficiencies were 67.1 percent (Tween 80), 64.9 percent (Surfacpol 14104), and 61.2 percent (Emulgin W600). Pichtel and Pichtel (1997), Thirumalai Nivas et al. (1996), and Hessling et al. (1986) found surfactants to be effective to varying degrees in solubilizing soil Cr as the chromate ion. The surfactant molecule, possessing a polar head and a nonpolar tail, has the ability to form micelles around Cr oxyions. Suggested solubilization mechanisms include ion exchange from soil colloids and simple dissolution of insoluble Cr phases (Thirumalai Nivas et al. 1996). In contrast to the above findings, Baziar et al. (2013) determined that Tween 80 and Brij 35 were not effective in removing soil Pb and Cd. Surfactant molecules are discussed in detail in section 7.4.

The applicability of water-soluble polymers as extractants for remediation of heavy metal-contaminated soils has been studied (Sauer et al. 2004). Polyethylenimine (PEI) was functionalized with bromo- or chloroacetic acid to produce an aminocarboxylate chelating group which effectively binds lead. The resulting polymer, PEIC, is believed to have extraction capabilities similar to EDTA. Using soil from a Superfund site in New Mexico that contained approximately 10,000 mg/kg of Pb, the polymers removed more than 97 percent of soil Pb. Subsequent experiments demonstrated that the selective extraction of lead could be controlled by varying polymer functionalization levels.

7.3.1 Chelant Recovery

The eventual separation of metal from the metal-extractant mixture is necessary for a soil extraction system to be complete and cost-effective. Several chemical and physical treatments are in use to recover the metal(s) from the metal-chelant complex, thereby freeing the extracting solution for reuse. Treatment typically involves reaction of the metal-saturated chelant with an element that has a higher affinity for the organic ligand than the contaminant metal(s). For example, if Pb is removed from soil as Pd-EDTA, the Pb is desorbed from the complex by addition of calcium. The Ca^{2+} readily binds ligands at high pH, thus freeing the contaminant metal to form a hydroxide solid. Lead recovery from an EDTA-Pb complex was 70 percent and 62 percent at pH values of 11.0 and 12.0, respectively (Steele and Pichtel 1998). The corresponding recovery of Pb from an ADA-Pb complex and a PDA-Pb complex was over 95 percent. Brown and Elliott (1992) removed 40 percent of Pb from an EDTA-Pb solution at pH 10.8 in the presence of excess Ca^{2+}. Zeng et al. (2005) successfully used Na_2S combined with $Ca(OH)_2$ to precipitate trace metals bound to EDTA, thus allowing it to be reused. Juang and Wang (2000) carried out electrochemical separation of metals and EDTA in a two-chamber electrolytic cell separated with a cation exchange membrane. Pociecha and Lestan (2012) used electrochemical treatment to remove metals from EDTA as insoluble precipitates and electrodeposits on a cathode. The authors assert that EDTA can be recycled in a closed process loop.

Once the metal is displaced from the complex, additional processing will be needed to remove impurities and concentrate the metal in order to convert it to a marketable product. The most commonly used processing methods are precipitation and ion exchange. In some cases the metal salt or complex is reduced to the metallic form. Reduction is accomplished by electrowinning or by using a reducing gas such as hydrogen (Zeng et al. 2005). If the end product is a metal salt, it can be converted to a more marketable oxide or salt form via simple chemical processes.

7.4 FLUSHING ORGANICS FROM SOIL

Soil flushing has been employed for sites contaminated with hydrocarbons including wood preservatives (creosote, pentachlorophenol), PAHs (naphthalene, fluoranthene, pyrene), and some extremely toxic organics (chlorinated dibenzodioxins) (Madadian et al. 2014; Urum et al. 2006; Mulligan and Eftekhari 2003; Leharne and Dong 2002).

As discussed in chapter 3, polycyclic aromatic hydrocarbons (PAHs) are hazardous environmental contaminants which are also quite hydrophobic. Several projects have tested pure solvents or solvent mixtures for extraction of PAHs. Results showed that up to 95 percent of the PAH from contaminated soils can be efficiently removed using solvents such as ethanol, 2-propanol, acetone, and pentanol, and ternary mixtures of solvents such as water/pentanol/ethanol and water/pentanol/propanol (Colombano and Mouvet 2012; Khodadoust et al. 2000; Silva et al. 2005; Gan et al. 2009).

A practical consideration for flushing organics from soil involves the polarity of the contaminants. Many hydrocarbons are completely insoluble in water, whereas others are only slightly soluble. Soluble (hydrophilic) organic contaminants often are removed from soil by flushing with water alone. Organics with octanol/water partition coefficients (K_{ow}) of less than 10 (log K_{ow} < 1) are highly soluble. Examples include low molecular weight alcohols, phenols, and carboxylic acids. The octanol-water partitioning coefficient (K_{ow}) is measured by mixing a test compound, for example a hydrocarbon, with a solution of octanol (a C_8 hydrocarbon) and water and allowing the mixture to equilibrate. Separate aliquots of the octanol and water are removed and the concentration of the contaminant in each phase is measured. The coefficient is calculated as:

$$K_{ow} = [\text{contaminant in octanol}] / [\text{contaminant in water}]$$

Low-solubility (hydrophobic) organics may be removed with the assistance of a surfactant. Examples of hydrophobic compounds include petroleum products (gasoline, jet fuels, oil), aromatic solvents (benzene, toluene, ethylbenzene, xylene), semivolatiles (PAHs), chlorinated pesticides, chlorinated solvents (trichloroethene), and polychlorinated biphenyls (PCBs).

A schematic representation of a reaction of a surfactant molecule with a hydrocarbon contaminant is shown in figure 7.4. Surfactants are useful for solubilization of hydrophobic compounds because of their amphiphilic structure, meaning that one part of the molecule is a polar or ionic group (head) with a strong affinity for water, and the other part is a hydrocarbon group

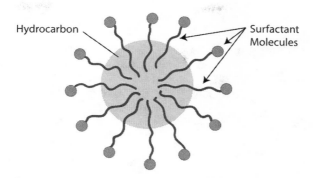

Figure 7.4 Surfactant molecules sequestering contaminant molecules.

(tail) with an aversion to water. An example surfactant molecule may appear as shown below:

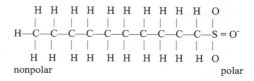

In the presence of water-insoluble hydrocarbons, the tail of the anion associates with the hydrocarbon molecule whereas the head remains in aqueous solution. Thus, the surfactant emulsifies, or suspends, hydrocarbons in water. Colloidal surfactant micelles are formed, which can be removed in the process water.

Baziar et al. (2013) measured 70 to 80 percent and 60 to 65 percent TPH (total petroleum hydrocarbon) removal from soil using nonionic surfactants (Tween 80, Brij 35), respectively. Numerous studies have addressed anionic and nonionic surfactants for enhancing removal of PAHs from soil (Rodriguez-Escales et al. 2012; Ahn et al. 2005; Deshpande et al. 1999). Using Tween 40, Tween 80, Brij 30, and Brij 35, Ahn et al. (2005) determined 33.9 percent extraction efficiency of phenanthrene. Madadian et al. (2013), using Brij 35, measured highest removal efficacies for anthracene, naphthalene, fluorene, and benzo(a)pyrene in coarse and fine soil fractions to be 76, 86, 79, and 86 percent; and 58, 63, 61, and 80 percent, respectively.

Some researchers state that surfactants will sorb to soil particles, which could lead to a capture of PAHs by nonmobile surfactants, thus enhancing adsorption of PAHs to soil (Colombano and Mouvet 2012; Ahn et al. 2008).

Extraction of organics can be combined with other technologies such as bioremediation. For example, the BioGenesis™ soil washing technology involves

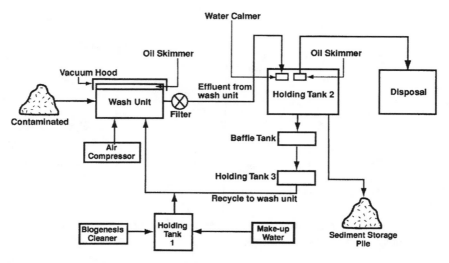

Figure 7.5 Schematic diagram of the BioGenesis™ soil washing technology. *Source*: U.S. Environmental Protection Agency.

mixing excavated contaminated soil in a mobile washing unit. In the first stage, a proprietary surfactant solution is used to transfer hydrocarbons from the solid matrix to the liquid phase. The second stage involves biodegradation of residual hydrocarbons in the soil and solubilized mixture (Biogenesis 2017) (figure 7.5).

7.5 TECHNICAL/PRACTICAL CONSIDERATIONS IN SOIL FLUSHING

The feasibility of soil flushing technology (and the level of treatment) is variable, depending on the appropriateness of solution for the contaminants, the ability to flood the soil with the flushing solution, degree of contact of flushing solution with waste constituents, hydraulic conductivity of the soil, and the installation of collection wells or subsurface drains to recover all applied liquids. Provisions must also be made for ultimate disposal of the elutriate. Some critical variables are outlined below.

7.5.1 Contaminant Forms and Concentrations

It is essential to determine the total concentrations of the contaminant compounds present in the affected soil as well as their chemical forms. Typically, some detective work is required, that is, gleaning data from past disposal records, waste manifests, and so on, and amassing field analytical data. Furthermore, numerous published methods are available for the analysis of inorganics and organics present in soil and water (see chapter 6). The methods selected will be a function of regulatory requirements and client needs.

For many contaminants it is essential to know not only the total concentration, but which chemical species of contaminants occur as well. Partitioning of contaminant metals or organics occurs after release to the soil and involves the chemical redistribution of the contaminant within the soil matrix. The degree and range of redistribution is a function of the contaminant(s), the soil chemical and physical environment, and time. For example, after a soluble metal enters soil it may adsorb to soil clay, organic matter, and hydrous oxides, and a portion may remain dissolved in the soil water. If the soil is in an oxidized state (i.e., high redox potential), the metal may crystallize as a hydroxide or carbonate, thus rendering it highly insoluble (and hence not amenable to extractive processes). In contrast, a reducing environment (e.g., a soil which is frequently saturated) will allow severalmetals to remain in solution. Such metal fractions are, therefore, readily leachable by extractive processes. If the plume is composed of hydrocarbons the effects of weathering (i.e., partitioning into soluble, vapor and sorbed phases) must be considered. A chemical analysis of the affected soil is therefore recommended, especially for aged releases. Contaminant forms also depend somewhat on the source of contamination. For example, at a contaminated lead battery disposal facility, did soil Pb originate as soluble Pb salts such as $PbSO_4$, or as PbO or metallic Pb? For $PbSO_4$, a relatively soluble salt, soil flushing is a relatively straightforward process. For the latter two species, however, soil extractive processes would be prohibitively slow and expensive.

Knowledge of metal partitioning (and ultimately how much metal is readily extractable) can be gained from laboratory tests that separate a soil metal into distinct fractions as a function of solubility in an appropriate solution. For example, researchers (Sposito et al. 1982) developed a technique to determine exchangeable, adsorbed, organic-bound, carbonate-bound, and residual forms of a soil metal. The method involves sequential extraction of a soil sample with 0.05 M KNO_3, deionized H_2O, 0.5 M NaOH, 0.05 M Na_2EDTA, and hot (80°C) 4 M HNO_3. If soil metals are found to occur mostly in the KNO_3 and H_2O fractions, they are considered exchangeable and water-soluble, respectively, and extraction may be rapid and simple. If, however, most occurs in the HNO_3 fraction (as may be the case for Pb from spent battery casings), removal by extractive processes is probably not feasible.

Numerous other fractionation techniques for soil metals are available. Tessier at al. (1979) described a five-step sequential extraction protocol for exchangeable, carbonate, reducible, and organic matter-bound and residual metals. In a study by Keon and others (2001), arsenic-bearing phases in sediment samples were quantitatively recovered by the following extractants in a sequential extraction procedure: adsorbed on goethite, 1 M NaH_2PO_4; arsenic trioxide (As_2O_3), 10 M HF; arsenopyrite (FeAsS), 16 N HNO_3; amorphous As sulfide, 1 N HCl, 50 mM Ti-citrate-EDTA, and 16 N HNO_3; and orpiment (As_2S_3), hot concentrated HNO_3/H_2O_2. In a study by Ahnstrom and Parker (1999), cadmium was partitioned into five operationally defined fractions: 0.1 M $Sr(NO_3)_2$ (soluble-exchangeable); 1 M Na acetate, pH 5.0 (sorbed-carbonate); 5 percent NaOCl, pH 8.5 (oxidizable); 0.4 M oxalate + 0.1 M ascorbate (reducible); and 3 HNO_3:1 HCl (residual).

Metal concentration and overall concentration gradients affect extraction success. At low metal concentrations, lower extraction efficiency may occur as a result of stronger binding, because binding energies associated with low sorption densities are substantial (Benjamin and Leckie 1981). The concentration gradient of the contaminant between soil and solution will also play a role in the extraction process. If the extracting solution is regularly removed and replaced, allowing for a greater concentration gradient, metal extraction is enhanced.

Contaminated soils that are amenable to soil flushing are those preferably containing one principal metal that is readily soluble and extractable in an aqueous medium (> 80% extraction efficiencies) (Griffiths 1995).

7.5.2 Effects of Flushing Agent on Soil Properties

The addition of a flushing solution to a site requires knowledge and meticulous management of reactions that may adversely affect the soil system. For example, application of sodium as dilute NaOH to a soil may drastically restrict soil permeability. The Na^+ ion greatly expands the soil-diffuse double layer (figure 7.6), resulting in the dispersal of soil particles and ultimately a complete loss of saturated hydraulic conductivity, K_s. Therefore, extractive processes would no longer be applicable to this site. Similar dispersive effects may occur with excessive use of surfactants.

The addition of a fairly concentrated acid (e.g., 0.1 N HCl) will result in the dissolution of soil particles (particularly fines) with consequent loss of soil structure; eradication of the indigenous microbial populations that are responsible for nutrient turnover and the ultimate development of a stable soil; and dissolution of native soil metals including nutrient bases (Ca^{2+}, Mg^{2+}, K^+, etc.). The solids that remain do not comprise a 'soil' per se, but rather a mass of inert solids. The dissolved soil solids will also pose a significant wastewater treatment issue. Laboratory tests, although serving as only a crude approximation of events in the field, should assess the impact of a flushing solution on soil properties.

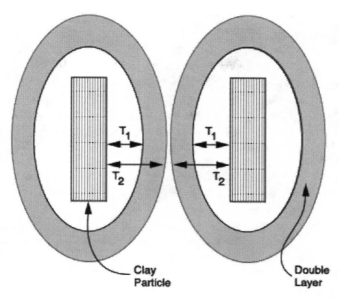

Figure 7.6 The diffuse double layer surrounding a clay particle. T1 = thickness of the layer with an adsorbed divalent cation; T2 = thickness with an adsorbed monovalent cation.

7.5.3 Vertical and Lateral Distribution of Contamination

This data is obtained via use of soil borings, noninvasive detection techniques (e.g., electromagnetic detection, ground-penetrating radar), groundwater analyses, and other field activities during the phase II environmental site assessment. Success of solubilization and removal with depth is a function of site characteristics including K_s as well as engineering limitations such as depth of well installation. If the plume is relatively water-soluble and very deep within the profile, soil flushing may not be feasible because control of flow will be difficult. Furthermore, construction of deep barrier walls will prove costly.

7.5.4 Suitability of Site to Flooding and Installation of Subsurface Drains

For in situ flushing to be effective, favorable hydrogeologic conditions are required, where the flushing solution, with its dissolved and suspended components, can be readily applied, contained, and recovered. It is essential that site managers possess a thorough knowledge of both surface and subsurface properties prior to flushing. When working with dense soils, surface runoff of extracting solution is a possibility. Conversely, in excessively drained soils, migration of the flushing solution may continue through uncontaminated material and into groundwater, creating a much larger contamination problem than that which was originally encountered.

In surficial layers, high clay content, high humic content, or high soil CEC (> 5 to 10 cmol(+)/kg) not only restricts site flooding but interferes with contaminant desorption. Soil K_s is a key physical parameter for determining feasibility of soil flushing. Soils with low K_s (< 1 x 10^{-5} cm/s) will limit the ability of flushing solutions to percolate through soil within a reasonable time frame. Flushing technology is most applicable to permeable soils (Ks > 1 x 10^{-4} cm/s) (approximately 30% or less silts + clays), because desorbed metals will be more quickly removed, allowing for a greater concentration gradient with fresh flushing solution.

For the subsurface, basic requirements include locating impermeable layers, areas of high K_s (e.g., sand or gravel lenses, fractures, discontinuities), and buried wastes, drums, USTs, and other large anthropogenic debris. This data is obtained by use of well logs, geologic and hydrogeologic maps, soil borings, noninvasive detection techniques (e.g., electromagnetic detection, ground-penetrating radar), groundwater analysis, and other routine components of the phase I and phase II environmental site assessments. It is essential to know the location of any utilities (electric, fiber optics, gas, water, steam, telephone, etc.) as such channels may have provided a path of rapid flow of the contaminant, and may likewise cause unwanted dispersal of the flushing solution. Information on the locations of these conduits is available from local utilities.

Prior to field implementation of soil flushing, a thorough groundwater hydrologic assessment should be carried out. This should include information on seasonal fluctuations in water table, direction of groundwater flow, porosity, vertical and horizontal K_s, and infiltration rate.

7.5.5 Trafficability of the Site

Soils high in clay will form a 'smeared' surface if vehicle traffic is heavy, and will limit downward transport of surface-applied flushing solution. Excessively wet sites will not support heavy traffic and may require drainage.

7.5.6 Degree of Site Preparation

Many contaminated sites have served as waste dumps and may contain unconsolidated and large wastes (e.g., machine parts, steel drums, USTs). In order for flushing to be successful, contact between soil fines and extracting solution must be maximized. Furthermore, paths of free flow should be eliminated; therefore, large debris should be removed from the profile if at all possible. Surface drainage and controls of run-on and runoff may be needed.

7.5.7 By-Products

In a system involving vigorous mixing and/or aeration (i.e., soil washing in reactor vessels) of hydrocarbon-contaminated media, volatilization of vapors is inevitable. Biological attack of hydrocarbon molecules will also promote weathering and subsequent vapor release. Off-gas collection and treatment (with permit) may therefore be necessary.

7.5.8 Environmental Impacts

Soil flushing chemicals may themselves become environmental problems if not properly managed. Many must be treated adequately before they can be safely discharged. EDTA, the most commonly employed chelating agent, is toxic (cytotoxic but not carcinogenic) (Lanigan and Yamarik 2002), especially in its free form (Sillanpaa and Oikari 1996; Dirilgen 1998). Furthermore EDTA is poorly photo-, chemo-, and biodegradable in the biosphere (Nörtemann 1999). In recent years, biodegradable chelating agents such as EDDS ([S,S]ethylenediaminedisuccinic acid), IDSA (iminodisuccinic acid), and MGDA (methylglycinediacetic acid) have been studied as alternatives to conventional agents (Nowack and VanBriesen 2005; Tandy et al. 2004). EDDS is considered a promising substitute for EDTA in soil remediation. It has been found that EDDS is fully degraded in wastewater treatment processes (Vandevivere et al. 2001) and in soil as well (Tandy et al. 2006; Wang et al. 2007; Meers et al. 2008).

7.6 THE GOOD AND THE BAD

Soil flushing offers many advantages in site remediation. The technology results in contaminant removal, which reduces future liability for the site operator and owner. In some cases economic benefits from recovery of recyclable metals (radioactives, precious metals) are possible. Metals can be preconcentrated for recovery. In addition, the required equipment is relatively easily constructed and operated, and conventional fluid-processing equipment can be included. This includes irrigation systems, sprinklers, pumps, and extraction wells.

Soil flushing does not involve excavation and soil treatment and disposal, which can be expensive and is associated with certain hazards (e.g., wind dispersal of soil particles, mechanical hazards, etc.).

Soil flushing has its share of costs, however. Flushing will drastically alter the chemical, physical, and biological properties of a soil, associated strata, and aquifers. A solution composed of a fairly concentrated extractant is being forced through a site with the intention of contacting as much of the solid matrix as possible. When using an acidic extracting solution, soil pH will be affected, the saturation of the CEC by bases (Ca, Mg, K) will decrease, soil structure may be destroyed, and microbial activity will decline. The continued saturation of the soil may result in a predominantly reducing environment.

The extracting solution may itself become a pollutant. Flushing solutions may have toxic and other environmental impacts on soil and water receiver systems. It is critical, at the completion of the remedial activity, to remove as much of the extracting solution as possible so that it does not contaminate the treated zone and does not migrate off-site. It may be useful to install subsurface containment barriers, such as sheet piling or slurry walls in order to control the flow of the flushing solution and to create an essentially closed system (see chapter 6).

Remediation times will be long (from one to several years) due to the slowness of diffusion processes in the liquid phase. In order for the entire program to be cost-effective and to reduce environmental impact, the flushing solution should be regenerated on-site and reused. Chelating agents are often expensive and difficult to recover from the flushing solution.

Bacterial fouling of infiltration and recovery systems and treatment units may be a problem, especially if high Fe concentrations are present in groundwater or if biodegradable reagents are being used. Flushing additives may interfere with downstream wastewater treatment processes. For example, surfactants will inactivate common heterotrophic bacterial populations that are needed at wastewater treatment plants for BOD removal.

Complex waste mixtures (e.g., several different metals or metals with organics) make the formulation of flushing solutions difficult. Each waste component may respond to treatment differently.

Several naturally occurring substances may interfere with soil washing and reduce extraction efficiency. For example, soil high in limestone will interfere with acid extraction because $CaCO_3$ will neutralize the acid reagent. A high soil organic matter content will hinder extraction by bases, and $Fe(OH)_3$ and $CaCO_3$ will interfere with extraction by chelating agents. Since the process is highly waste-specific, treatability studies in the laboratory should be conducted prior to applying a procedure in the field.

Specialized acid-resistant equipment must be used for the acid leaching process. Workers must be trained in safe handling procedures for flushing solutions as many are potentially toxic, corrosive, or reactive.

7.7 CASE HISTORY

The King of Prussia Technical Corporation Superfund site, located in south-central New Jersey, began operations as an industrial waste recycling facility in 1971. Six lagoons were constructed for waste treatment and approximately 15 million gal of acids and alkaline aqueous wastes were processed. Facility operations were halted and the site was abandoned in 1973 or 1974. Subsequent to closure, illegal dumping of solid and hazardous wastes apparently occurred (U.S. EPA 2018; ART 2006).

The site, measuring approximately 10 acres, is bordered mainly by pine forest. Site runoff flows via a swale toward nearby Great Egg Harbor River. Soil material is derived from the Outer Coastal Plain and is very sandy.

Soil and sediment at the site were found to be contaminated with several heavy metals including beryllium (Be), Cu, Cr, Ni, and Zn. Remedial alternatives considered for the site included:

- no action
- limited action (site restrictions; installation of additional fencing)
- complete removal and off-site disposal of contaminated media
- limited excavation of sediments and soil followed by capping
- stabilization/solidification, either in situ or ex situ, of contaminated media followed by capping
- excavation of contaminated soils, sediments, and sludges followed by soil washing and return of 'clean' soil to the site

After numerous technical reviews and pilot studies, soil washing was selected as the remedial option. Pilot studies using 990 tons of contaminated media showed that soil washing could effectively remove the contaminants. Furthermore, the treated soil could be returned to its original location to restore site topography (Mann n.d.).

The highest concentration of surface (< 2 ft deep) contamination was located in sediments of the drainage swale. Maximum concentrations of Cr, Cu, and Hg were 8,010, 9,070, and 100 mg/kg, respectively. Subsurface (2–20 ft) contamination was highest in a zone of sludge-like material adjacent to the lagoons. Highest metal concentrations in the sludge material were Cr at 11,300 mg/kg, Cu at 16,300 mg/kg, Pb at 389 mg/kg, and Ni at 11,100 mg/kg. The soils had low concentrations of volatile and semivolatile organic compounds (U.S. EPA 1994a; ART 2006, 1993; Mann n.d.).

Selective excavation of metals-contaminated soils was conducted (figures 7.7 and 7.8). A total of 40,000 tons of material was excavated; however, only

20,000 tons exceeded cleanup levels and required treatment. Selective excavation of soil and sludge involved the following steps:

1. Excavation of clean overburden soil and transport to stockpile area;
2. Excavation and transportation of contaminated soil to the screening and blending area;
3. Analysis of contaminant levels (using an X-ray fluorescence detector) in the trench bottom soils;
4. Backfilling of the clean trench bottom with clean material.

The washing system was designed to physically separate the metal-enriched soil fines from the larger particles, which were then reacted with water and surfactant. The system included the following processes:

1. Screening out the oversize fraction from soil to be treated using vibrating screens. Materials greater than 8 in. across, typically concrete, tree stumps, branches, and so on are periodically removed and stockpiled. The material that passed through the screens is directed to another mechanical screening unit to remove particles greater than 2 in. Materials passing through the screens (< 2 in.) are subjected to wet screening with high-pressure water sprays. The wet screening breaks up clods and forms a slurry.
2. Separating the screened soil/water slurry into coarse- and fine-grained material via a set of hydrocyclones (figure 7.9). Multiple cyclones achieve

Figure 7.7 Plant assembly at the King of Prussia site. *Source*: Reproduced with kind permission of ART Engineering, LLC.

Figure 7.8 Soil feeding. *Source*: Reproduced with kind permission of ART Engineering, LLC.

a separation efficiency of > 99 percent of the sands and fines. The hydrocyclone cut point is set at 40 microns, determined via a treatability study. The flow containing coarse-grained material is directed to a froth flotation stage, while fine-grained material, presumably silt- and clay-sized, is processed into a sludge cake.
3. Froth flotation involves removing the contaminants from the coarse-grained material using air-flotation treatment units, specifically, an air-flotation tank equipped with mechanical aerators. The coarse-grained particles are pumped into the tank, where a surfactant is added. The contaminants float within a froth and are removed from the surface of the tank and transferred to the sludge management process. The cleaned underflow sands are sent to a cyclone and dewatering screens. Approximately 85 percent of the processed material (clean sand) from the site was used as backfill, while the water was recycled back to the wet screening process.
4. Sludge management involves treating the overflow from the hydrocyclones, that is, fine-grained material and water. The overflow was pumped to clarifiers and a polymer added. The clarified solids were sent to a sludge thickener and finally to a filter press where the 15 to 20 percent solids influent was converted to a 50 to 60 percent dry solids filter cake. The filter cake was disposed off-site as nonhazardous waste. The water from the sludge management process was returned to the wet screening phase

Figure 7.9 Hydrocyclone separation of contaminated soil. *Source*: Reproduced with kind permission of ART Engineering, LLC.

for reuse (U.S. EPA 1995; ART, Inc. 1993). The soil washing unit can process 25 tons/hour.

The process oversize and clean sand from the soil washing unit met the cleanup levels established for this case (table 7.1). The average concentrations of Be, Cu, Pb, Ni, and Zn in the clean sand and process oversize were at least an order of magnitude lower than the required cleanup levels. Cadmium, Hg, Se, and Ag were not detected in any process oversize samples; and As, Hg, Se, and Ag were not detected in any clean sand. Chromium, Cu, and Ni were concentrated in the sludge cake with individual contaminants measured at levels greater than 2,000 mg/kg.

More than 19,000 tons of soil contaminated with heavy metals were successfully treated during the soil washing operation (U.S. EPA 2018). Approximately $7.7 million was expended on the soil washing activity including off-site soil disposal costs.

Table 7.1 Treatment performance data for the King of Prussia superfund site

Contaminant	Cleanup Level	Untreated Feed Soil (Range)	Clean Oversize Product (Average)	Clean Sand Product (Average)	Sludge Cake (Average)
			mg/kg		
As	190	n/a	0.62	nd	n/a
Be	485	n/a	5.9	1.9	n/a
Cd	107	n/a	nd	0.64	n/a
Cr	483	500–5,000	172	73	4,700
Cu	3,571	800–8,000	350	110	5,900
Pb	500	n/a	6.5	3.9	n/a
Hg	1	n/a	nd	nd	n/a
Ni	1,935	300–3,500	98	25	2,300
Ag	5	n/a	nd	nd	n/a
Zn	3,800	n/a	48	16	n/a

nd = not detected; n/a = samples were not collected.
Sources: U.S. Environmental Protection Agency 1995; 1994b; ART, Inc. 1993.

QUESTIONS

1. Discuss the benefits of conducting laboratory tests of an extracting solution prior to its application in a field situation.
2. Discuss three potential adverse effects from washing a soil with a moderately concentrated acid; with a surfactant solution.
3. List and discuss two soil components or compounds that may interfere with soil washing using EDTA.
4. Draw a surfactant molecule. How is it used in site remediation (on what types of contaminants)? How, specifically, does it function?
5. A permeable soil is contaminated with a hydrophobic (nonpolar) contaminant, for example, PCBs mixed with waste oil. Which would be an appropriate extracting solution? (a) water; (b) linear alkyl sulfate $CH_3[CH_2]_nCHOSO_3^-$ (c) dilute HNO_3; (d) dilute KOH.
6. The addition of dilute sodium hydroxide (NaOH) to a soil may be used for extraction of certain pollutants. What is one significant drawback to its use?
7. A soil flushing solution may become a pollutant if it is used improperly and/or excessively. Explain and provide an example.
8. Explain how humus compounds may be responsible for metal chelation and complexation.
9. Soil flushing is an ideally suited process for a soil contaminated with several metals and mixed oils; all can be removed with one to two wash cycles. True or false? Justify your answer.

REFERENCES

Ahn, C. K., Y. M. Kim, S. H. Woo, and J. M. Park. 2008. Soil washing using various nonionic surfactants and their recovery by selective adsorption with activated carbon. *Journal of Hazardous Materials* 154(1–3): 153–60.

Ahnstrom, Z., and R. Parker. 1999. Development and assessment of a sequential extraction procedure for the fractionation of soil cadmium. *Soil Science Society of America Journal* 63(6): 1650–58.

Alternative Remedial Technologies, Inc. 1992. *Soil Washing Demonstration Run for the King of Prussia Technical Site, December 14, 1992*. Land O Lakes, FL.

———. 1993. *Site Operations Plan, the King of Prussia Technical Corporation Site, Winslow Township, New Jersey. July 26, 1993*. Land O Lakes, FL.

———. 2006. *Soil Washing at King of Prussia (KOP) Superfund Site*. See: image s.google.com/imgres?imgurl=http://www.art-engineering.com/Projects/KOP-Soil /Image25.jpg&imgrefurl=http://www.art-engineering.com/Projects/KOP.Soil/ Photos.htm&h=185&w=271&sz=16&hl=en&start=10&tbnid=AmMlm3Irh2_Qv M:&tbnh=77&tbnw=113&prev=/images%3Fq%3Dking%2Bof%2Bprussia%2 Bsuperfund%2Bsite%26svnum%3D10%26hl%3Den%26lr%3D%26sa%3DN.

Asel Gzar, H., and I. M. Gatea. 2015. Extraction of heavy metals from contaminated soils using EDTA and HCl. *Journal of Engineering* 21: 45–61.

Baziar, M., M., Reza Mehrasebi, A. Assadi, M. M. Fazli, M. Maroosi, and F. Rahimi. 2013. Efficiency of non-ionic surfactants - EDTA for treating TPH and heavy metals from contaminated soil. *Journal of Environmental Health Science Engineering* 11(1): 41.

Bell, C. F. 1977. *Principles and Applications of Metal Chelation.* Oxford Chemistry Series. Oxford: Oxford University Press.

Benjamin, M. M., and J. O. Leckie. 1981. Multiple-site adsorption of Cd, Cu, Zn, and Pb on amorphous iron oxyhydroxide. *Journal of Colloid and Interface Science* 79(1): 209–21.

Betteker, J., J. Sherrard, and D. Ludwig. 1993. Solidification/stabilization of contaminated dredged material. *Hazardous and Industrial Wastes* 25: 93.

Biogenesis.com. 2017. *Breakthrough Remediation Technologies.* https://biogenesis.com/services/ (Accessed December 4, 2017).

Bricker, T. J., J. Pichtel, H. J. Brown, and M. Simmons. 2001. Phytoextraction of Pb and Cd from a Superfund soil: Effects of amendments and croppings. *Journal of Environmental Science and Health, Part A* 36(9): 1597–610.

Brown, G. A., and H. A. Elliott. 1992. Influence of electrolytes on EDTA extraction of Pb from polluted soil. *Water, Air, and Soil Pollution* 62(1–2): 157–65.

Chen, T. C., E. Macauley, and A. Hong. 1995. Selection and test of effective chelators for removal of heavy metals from contaminated soils. *Canadian Journal of Civil Engineering* 22(6): 1185–97.

Clément, T., E. Mousset, Y. Pechauda, D. Huguenot, E. D. van Hullebusch, G. Esposito, and M. A. Oturana. 2016. Removal of hydrophobic organic pollutants from soil washing/flushing solutions: A critical review. *Journal of Hazardous Materials* 306: 149–74.

Cline, S. R., and B. E. Reed. 1995. Lead removal from soils via bench-scale soil washing techniques. *Journal of Environmental Engineering* 121(10): 700–5.

Cline, S. R., B. E. Reed, and M. Matsumoto. 1993. Efficiencies of soil washing solutions for the remediation of lead contaminated soils. In: *Hazardous and Industrial Wastes. Proceedings of the 25th Mid-Atlantic Industrial Waste Conference*, edited by A. Davis. Lancaster, PA: Technomic.

Colombano, S., and C. Mouvet. 2012. *Remediation Technologies for PAH Contaminated Soils Review Report within the SNOWMAN Project PACMAN (SN-03-11).* Orleans, France: BRGM.

Davis, A. P., and I. Singh. 1995. Washing of zinc(II) from a contaminated soil column. *Journal of Environmental Engineering* 121(2): 174–85.

Dermont, G., M. Bergeron, G. Mercier, and M. Richer-Laflèche. 2008. Soil washing for metal removal: A review of physical/chemical technologies and field applications. *Journal of Hazardous Materials* 152(1): 1–31.

Deshpande, S., B. J. Shiau, D. Wade, D. A. Sabatini, and J. H. Harwell. 1999. Surfactant selection for enhancing ex situ soil washing. *Water Research* 33(2): 351–60.

Dirilgen, N. 1998. Effects of pH and chelator EDTA on Cr toxicity and accumulation in *Lemma minor*. *Chemosphere* 37(4): 771–83.

Doong, R. A., Y. W. Wu, and W. G. Lei. 1998. Surfactant enhanced remediation of cadmium contaminated soils. *Water Science and Technology* 11(8): 65–71.

Ehsan, S., S. O. Prasher, and W. D. Marshall. 2006. A washing procedure to mobilize mixed contaminants from soil. II. Heavy metals. *Journal of Environmental Quality* 35(6): 2084–91.

Elliott, H. A., and G. A. Brown. 1989. Comparative evaluation of NTA and EDTA for extractive decontamination of Pb-polluted soils. *Water, Air, and Soil Pollution* 45(3–4): 361–69.

Evangelou, M., H. Daghan, and A. Schaeffer. 2004. The influence of humic acids on the phytoextraction of cadmium from soil. *Chemosphere* 57(3): 207–13.

Farrah, H., and W. F. Pickering. 1978. Extraction of heavy metal ions sorbed on clays. *Water, Air, and Soil Pollution* 9: 491–98.

Fischer, K. 2002. Removal of heavy metals from soil components and soil by natural chelating agents. Part I: Displacement from clay minerals and peat by L-cysteine and L-penicillamine. *Water, Air, and Soil Pollution* 137(1–4): 267–86.

Fischer, K., and H. P. Bipp. 2004. Removal of heavy metals from soil components and soils by natural chelating agents. Part II: Soil extraction by sugar acids. *Water, Air, and Soil Pollution* 138: 271–88.

FRTR (Federal Remediation Technologies Roundtable). 2018. *Ex Situ Physical/Chemical Treatment (Assuming Excavation)*. Remediation Technologies Screening Matrix and Reference Guide, Version 4.0. https://frtr.gov/matrix2/section4/4-19.html (Accessed February 5, 2018).

Gan, S., E. V. Lau, and H. K. Ng. 2009. Remediation of soils contaminated with polycyclic aromatic hydrocarbons (PAHs). *Journal of Hazardous Materials* 172(2–3): 532–49.

Gidarakos, E., and A. Giannis. 2006. Chelate agents enhanced electrokinetic remediation for removal cadmium and zinc by conditioning catholyte pH. *Water, Air and Soil Pollution* 172(1–4): 295–312.

Grčman, H., D. Vodnik, S. Velikonja-Bolta, and D. Lestan. 2003. Ethylenediaminedissuccinate as a new chelate for environmentally safe enhanced lead phytoextraction. *Journal of Environmental Quality* 32(2): 500–6.

Griffiths, R. A. 1995. Soil-washing technology and practice. *Journal of Hazardous Materials* 40(2): 175–89.

Hessling, J. L., M. P. Esposito, R. P. Traver, and R. H. Snow. 1986. *Metals Speciation, Separation, and Recovery*, Vol. 2. Edited by J. W. Patterson and R. Passino. Chelsea, MI: Lewis.

Hong, A., and T. C. Chen. 1996. Chelating extraction and recovery of cadmium from soil using pyridine-2,6-dicarboxylic acid. *Water, Air, and Soil Pollution* 86(1–4): 335–46.

Hsieh, H., M. Barnes, and E. Z. Aldridge. 1989. *Physiochemical and Biological Detoxification of Hazardous Wastes*. Lancaster, PA: Technomic.

Isoyama, M., and S. I. Wada. 2007. Remediation of Pb-contaminated soils by washing with hydrochloric acid and subsequent immobilization with calcite and allophanic soil. *Journal of Hazardous Materials* 143(3): 636–42.

Juang, R.-S., and S.-W. Wang. 2000. Electrolytic recovery of binary metals and EDTA from strong complexed solutions. *Water Research* 34(12): 3179–85.

Kayser, A., K. Wenger, A. Keller, W. Attinger, H. R. Felix, S. K. Gupta, and R. Schulin. 2000. Enhancement of phytoextraction of Zn, Cd, and Cu from calcareous soil: The use of NTA and sulfur amendments. *Environmental Science and Technology* 34(9): 1778–83.

Keon, N., H. Hemond, C. Swartz, D. Brabander, and C. Harvey. 2001. Validation of an arsenic sequential extraction method for evaluating mobility in sediments. *Environmental Science and Technology* 35(13): 2778–84.

Khodadoust, A., K. Reddy, and K. Maturi. 2005. Effect of different extraction agents on metal and organic contaminant removal from a field soil. *Journal of Hazardous Materials* 117(1): 15–24.

Khodadoust, A. P., R. Bagchi, M. T. Suidan, R. C. Brenner, and N. G. Sellers. 2000. Removal of PAHs from highly contaminated soils found at prior manufactured gas operations. *Journal of Hazardous Materials* B80(1–3): 159–74.

Kos, B., and D. Lestan. 2003. Induced phytoextraction/soil washing of lead using biodegradable chelate and permeable barriers. *Environmental Science and Technology* 37(3): 624–29.

Lanigan, R. S., and T. A. Yamarik. 2002. Final report on the safety assessment of EDTA, calcium disodium EDTA, diammonium EDTA, dipotassium EDTA, disodium EDTA, TEA-EDTA, tetrasodium EDTA, tripotassium EDTA, trisodium EDTA, HEDTA, and trisodium HEDTA. *International Journal of Toxicology*. doi: 10.1080/10915810290096522.

Lau, E. V., S. Gan, H. K. Ng, and P. E. Poh. 2014. Extraction agents for the removal of polycyclic aromatic hydrocarbons (PAHs) from soil in soil washing technologies. *Environmental Pollution*. doi: 10.1016/j.envpol.2013.09.010.

Leharne, S., and J. Dong. 2002. *Investigations of Surfactant Facilitated Removal of Coal Tar Contaminants from Manufactured Gas Works Soils.* IAHS-AISH. Groundwater Quality: Natural and Enhanced Restoration of Groundwater Pollution. Proceedings of the Groundwater Quality 2001 Conference held at Sheffield, UK June 2001. Pub. No. 275.

Leštan, D., C.-L. Luo, and X.-D. Li. 2008. The use of chelating agents in the remediation of metal-contaminated soils: A review. *Environmental Pollution* 153(1): 3–13.

Liu, L., S. P. Hu, Y. X. Chen, and H. Li. 2010. Feasibility of washing as a remediation technology for the heavy metals-polluted soils left by chemical plant. *Ying Yong Sheng Tai Xue Bao* 21(6): 1537–41.

Lo, I. M. C., D. C. W. Tsang, T. C. M. Yip, F. Wang, and W. H. Zhang. 2011. Influence of injection conditions on EDDS-flushing of metal-contaminated soil. *Journal of Hazardous Materials* 192(2): 667–75.

Luo, C. L., Z. G. Shen, X. D. Li, and A. J. M. Baker. 2006. Enhanced phytoextraction of Pb and other metals from contaminated soils through the combined application of EDTA and EDDS. *Chemosphere* 63: 1773–84.

Macauley, E., and A. Hong. 1995. Chelation extraction of lead from soil using pyridine-2,6-dicarboxylic acid. *Journal of Hazardous Materials* 40(3): 257–70.

Madadian, E., S. Gitipour, L. Amiri, M. Alimohammadi, and J. Saatloo. 2014. The application of soil washing for treatment of polycyclic aromatic hydrocarbons contaminated soil: A case study in a petrochemical complex. *Environmental Progress and Sustainable Energy* 33(1): 107–13.

Mann, M. J. n.d. *Full-Scale Soil Washing at the King of Prussia (NJ) Technical Corporation Superfund Site*. Alternative Remedial Technologies.

Meers, E., F. M. G. Tack, and M. G. Verloo. 2008. Degradability of ethylenediaminedisuccinic acid (EDDS) in metal contaminated soils: Implications for its use in soil remediation. *Chemosphere* 70(3): 358–63.

Moon, D. H., J.-R. Lee, M. Wazne, and J.-H. Park. 2012. Assessment of soil washing for Zn contaminated soils using various washing solutions. *Journal of Industrial and Engineering Chemistry* 18(2): 822–25.

Moutsatsou, A., M. Gregou, D. Matsas, and V. Protonotarios. 2006. Washing as a remediation technology applicable in soils heavily polluted by mining-metallurgical activities. *Chemosphere* 63(10): 1632–40.

Mulligan, C. N. 2005. Environmental applications for biosurfactants. *Environmental Pollution* 133(2): 183–98.

Mulligan, C. N., and S. Wang. 2006. Remediation of a heavy metal-contaminated soil by a rhamnolipid foam. *Engineering Geology* 85(1–2): 75–81.

Mulligan, C., and F. Eftekhari. 2003. Remediation with surfactant foam of PCP-contaminated soil. *Engineering Geology* 70(3–4): 269–79.

Mulligan, C. N., R. N. Yong, and B. F. Gibbs. 2001. Surfactant-enhanced remediation of contaminated soil: A review. *Engineering Geology* 11(1–4): 371–80.

Nörtemann, B. 1999. Biodegradation of EDTA. *Applied Microbiology and Biotechnology* 51(6): 751–59.

Nowack, B., and J. M. VanBriesen, eds. 2005. *Biogeochemistry of Chelating Agents*. ACS Symposium Series, Vol. 910. Washington, DC: American Chemical Society.

Pichtel, J., and T. M. Pichtel. 1997. Comparison of solvents for ex-situ removal of Cr and Pb from contaminated soil. *Environmental Engineering Science* 14(2): 97–103.

Pickering, W. F. 1983. Extraction of copper, lead, zinc, or cadmium ions sorbed on calcium carbonate. *Water, Air, and Soil Pollution* 20(3): 299–309.

Pociecha, M., and D. Lestan. 2012. Recycling of EDTA solution after soil washing of Pb, Zn, Cd, and As contaminated soil. *Chemosphere* 86(8): 843–46.

Rajput, V. S., S. Pilapitiya, M. E. Singley, and A. J. Higgins. 1989. Detoxification of hazardous waste contamination soils and residues by washing and biodegradation. In: Yu, Y. C. (Ed.). *Physiochemical and Biological Detoxification of Hazardous Wastes*. Lancaster, PA: Technomic.

Reed, B. E., R. E. Moore, and S. R. Cline. 1995. Soil flushing of a sandy loam contaminated with Pb(II), $PbSO_4$, $PbCO_3$, or Pb-naphthalene: Column results. *Journal of Soil Contamination* 4(3): 243–67.

Ringbom, A. 1963. *Complexation in Analytical Chemistry*. New York: Wiley Interscience.

Rodriguez-Escales, P., T. Sayara, T. Vicent, and A. Folch. 2012. Influence of soil granulometry on pyrene desorption in groundwater using surfactants. *Water, Air, and Soil Pollution* 223(1): 125–33.

Sauer, N., E. Ehler, and B. Duran. 2004. Lead extraction from contaminated soil using water-soluble polymers. *Journal of Environmental Engineering* 130(5): 585–88.

Shirk, J. E., and C. W. Farrel. 1985. Approach to in-situ management of metals. In: *Proceedings of the 8th Madison Waste Conference*, September 18–19, Madison, WI.

Sillanpaa, M., and A. Oikari. 1996. Assessing the impact of complexation by EDTA and DTPA on heavy metal toxicity using Microtox bioassay. *Chemosphere* 32(8): 1485e1497.

Silva, A., C. Delerue-Matos, and A. Fiuza. 2005. Use of solvent extraction to remediate soils contaminated with hydrocarbons. *Journal of Hazardous Materials* B124(1–3): 224–29.

Slavek, J., and W. F. Pickering. 1986. Extraction of metal ions sorbed on hydrous oxides of iron(III). *Water, Air, and Soil Pollution* 28: 151–62.

Smith, L. A., J. L. Means, A. Chen, B. Alleman, C. C. Chapman, J. S. Tixier, S. E. Brauning, A. R. Gavaskar, and M. D. Royer. 1995. *Remedial Options for Metals-Contaminated Soils*. Boca Raton, FL: CRC Press.

Sposito, G., L. J. Lund, and A. C. Chang. 1982. Trace metal chemistry in arid-zone field soils amended with sewage sludge. I. Fractionation of Ni, Cu, Zn, Cd and Pd in solid phases. *Soil Science Society of America Journal* 46: 260–64.

Steele, M. C., and J. Pichtel. 1998. Ex-situ remediation of a metal-contaminated Superfund soil using selective extractants. *Journal of Environmental Engineering* 124(7): 639–45.

Sung, M., C. Y. Lee, and S. Z. Lee. 2011. Combined mild soil washing and compost-assisted phytoremediation in treatment of silt loams contaminated with copper, nickel, and chromium. *Journal of Hazardous Materials* 190(1–3): 744–54.

Tandy, S., A. Ammann, R. Schulin, and B. Nowack. 2006. Biodegradation and speciation of residual SS-ethylenediaminedisuccinic acid (EDDS) in soil solution left after soil washing. *Environmental Pollution* 40: 2753–58.

Tandy, S., K. Bossart, R. Mueller, J. Ritschel, L. Hauser, R. Schulin, and B. Nowack. 2004. Extraction of heavy metals from soils using biodegradable chelating agents. *Environmental Science and Technology* 38(3): 937–44.

Tessier, A., P. G. C. Campbell, and M. Bisson. 1979. Sequential extraction procedure for the speciation of particulate trace metals. *Analytical Chemistry* 51(7): 844–51.

Thirumalai Nivas, B., D. A. Sabatini, B. Shiau, and J. H. Harwell. 1996. Surfactant enhanced remediation of subsurface chromium contamination. *Water Research* 30(3): 511.

Tokunaga, S., S. W. Park, and M. Ulmanu. 2005. Extraction behaviour of metallic contaminants and soil constituents from contaminated soils. *Environmental Technology* 26(6): 673–82.

Torres, L. G., R. B. Lopez, and M. Beltran. 2012. Removal of As, Cd, Cu, Ni, Pb, and Zn from a highly contaminated industrial soil using surfactant enhanced soil washing. *Physics and Chemistry of the Earth*. doi: 10.1016/j.pce.2011.02.003.

Tsang, D. C. W., I. M. C. Lo, and K. L. Chan. 2007a. Modeling the transport of metals with rate-limited EDTA-promoted extraction and dissolution during EDTA-flushing of copper-contaminated soils. *Environmental Science and Technology* 41(10): 3660–67.

Tsang, D. C. W., W. H. Zhang, and I. M. C. Lo. 2007b. Copper extraction effectiveness and soil dissolution issues of EDTA-flushing of artificially contaminated soils. *Chemosphere* 68(2): 234–43.

Tuin, B. J. W., and M. Tels. 1990. Removing heavy metals from contaminated clay soils by extraction with hydrochloric acid, EDTA or hypochlorite solutions. *Environmental Technology Letters* 11(11): 1039–52.

Tunay, O., and N. I. Kabdasli. 1994. Hydroxide precipitation of complexed metals. *Water Research* 28(10): 2117–24.

Urum, K., S. McMenamy, S. Grigson, and T. Pekdemir. 2006. A comparison of the efficiency of different surfactants for removal of crude oil from contaminated soils. *Chemosphere* 62(9): 1403–10.

U.S. Environmental Protection Agency. 1990. *Superfund Record of Decision, King of Prussia*. New Jersey, September.

———. 1991. *Guide for Conducting Treatability Studies Under CERCLA: Soil Washing Interim Guidance*. EPA/540/2-91/020A. Washington, DC.

———. 1993. *Biogenesis Soil Washing Technology: Innovative Technology Evaluation Report*. EPA/540/R-93/510. Office of Research and Development.

———. 1994a. *Acid Extraction Treatment System for Treatment of Metal Contaminated Soils*. EPA/540/SR-094/513. Cincinnati: Superfund Innovative Technology Evaluation.

———. 1994b. *Remedial Action Report: Soil Washing Remediation*. New Jersey: King of Prussia Technical Corporation Site, Camden County, July.

———. 1995. *Soil Washing at the King of Prussia Technical Corporation Superfund Site, Winslow Township, New Jersey. Cost and Performance Report*. Office of Solid Waste and Emergency Response, Technology Innovation Office.

———. 1997. *Treatment Technology Performance and Cost Data for Remediation of Wood Preserving Sites*. EPA/625/R-97/009. Office of Research and Development.

———. 2018. *Superfund Site: King of Prussia Winslow Township, NJ*. Cleanup Activities. https://cumulis.epa.gov/supercpad/SiteProfiles/index.cfm?fuseaction=second.Cleanup&id=0200551#bkground.

Vandevivere, P. C., H. Saveyn, W. Verstraete, T. C. J. Feijtel, and D. R. Schowanek. 2001. Biodegradation of metal-[S.S]-EDDS complexes. *Environmental Science and Technology* 35(9): 1765–70.

Wang, G., G. F. Koopmans, J. Song, E. J. M. Temminghoff, Y. Luo, Q. Zhao, and J. Japenga. 2007. Mobilization of heavy metals from contaminated paddy soil by EDDS, EDTA, and elemental sulfur. *Environmental Geochemistry and Health* 29(3): 221–35.

Zhang, W., D. C. W. Tsang, H. Chen, and L. Huang. 2013. Remediation of an electroplating contaminated soil by EDTA flushing: Chromium release and soil dissolution. *Journal of Soils and Sediments*. doi: 10.1007/s11368-012-0616-8.

Zeng, Q., W. Hendershot, S. Sauve, and H. Allen. 2005. Recycling EDTA solutions used to remediate metal-polluted soils. *Environmental Pollution* 133(2): 225–31.

Chapter 8

Solidification/Stabilization

Crystals grew inside rock like arithmetic flowers. They lengthened and spread, added plane to plane in an awed and perfect obedience to an absolute geometry. . . .

—Annie Dillard

8.1 INTRODUCTION

Solidification and stabilization (S/S) technologies are employed in situations where large quantities of toxic and/or relatively immobile contaminants occur, large debris (e.g., drums, battery casings) are scattered extensively, and/or soil Ks is not suitable for flushing (i.e., massive soil structure). An example of such a scenario may be a sprawling waste site laden to variable depths with lead battery casings and free metals of varying chemistries and toxicities.

S/S technology is conducted by mixing contaminated soil, either in situ or ex situ, with a binding agent to form a crystalline or polymeric matrix that incorporates the contaminated material. Inorganic binders include cement, cement kiln dust, fly ash, and blast furnace slag. Certain organic wastes can be immobilized by organic binders such as bitumen (asphalt), polyethylene, or urea formaldehyde. During solidification, contaminants are immobilized within a solid matrix in the form of a monolithic block. Stabilization converts contaminants to a less- or nonreactive form, typically by chemical processes. It follows that the S/S process does not remove contaminants from the affected soil; rather, it serves to physically sorb, encapsulate, or alter the physical or chemical form of the contaminants, producing a less leachable material. S/S can: (1) improve the handling and other physical characteristics of the contaminated soil by converting a liquid or sludge to a solid; (2) decrease the toxicity of the soil by altering contaminant physical and chemical properties; (3) decrease the exposed surface area of contaminants, thereby reducing their solubility; and (4) limit the contact of transport fluids (groundwater, infiltrating rainwater) with contaminants. Many wastes containing metal or inorganic contaminants are suitable for treatment with S/S technologies; however, potential interferences and incompatibilities must be addressed. The primary technologies are grouped as inorganic (cement-based) S/S and organic polymerization (using thermoplastic- and thermoset-based polymers).

8.2 CEMENT-BASED TECHNOLOGIES

In cement-based S/S, contaminated materials are mixed with a suitable ratio of cement or similar binder. The primary binder categories are cement and pozzolans. Examples of common pozzolans are fly ash, lime kiln dusts, pumice, and blast furnace slag. Cement-based S/S processes have been used for site remediation and hazardous waste treatment more than other S/S technologies. Cement-based S/S has been applied to plating wastes containing Cd, Cr, Cu, Pb, Ni, and Zn. Studies on contaminated soil treatment showed cement-based S/S effective for immobilization of As, Pb, Cu, Zn, Cd, and Ni. Cement has also been successful with organic-contaminated materials including PCBs, oily sludges, vinyl chloride, plastics, and sulfides. The composition of the cement or pozzolan, the type(s) of contaminants present, and the amount of water and aggregate determine set time, cure time, pouring characteristics, and engineering properties (e.g., compressive strength) of the treated soil. Cement and pozzolan compositions, including those used in S/S, are classified according to ASTM standards (ASTM 1989, 1992, 1996).

Portland cements typically are composed of calcium silicates, aluminates, aluminoferrites, and sulfates. Cementation of the waste-binder mixture begins when water is added, either separately or from the soil. Once the cement contacts water, hydration of tricalcium aluminate occurs, causing rapid setting. The water hydrates calcium silicates and aluminates to form calcium silicate-hydrate. Thin silicate fibers grow out from the cement grains and interweave, hardening the mixture. Hydration of tricalcium and dicalcium silicates results in the formation of crystalline calcium hydroxide and other minerals, which provide for strength development after setting (figure 8.1) (Thomas and Jennings 2008).

Figure 8.1 Electron micrograph showing plates of calcium hydroxide and ettringite crystals during cement curing. *Source*: US Department of Transportation.

Pozzolanic reaction is a relatively inexpensive treatment for contaminated soil. In general, however, pozzolan-solidified soils are not considered as durable as Portland cement-treated soils (Qian et al. 2006; Dermatas and Meng 2003; Kamon et al. 2000). With pozzolanic S/S, silica or silica alumina solids are used that have minimal cementation properties but can react with lime or cement to produce a cement-like product. The primary mechanism involves physical entrapment of the contaminant within the pozzolan matrix. In contrast to lime-based materials, pozzolans contain significant quantities of silicates. The final product varies from a soft, fine-grained material to a hard, cohesive solid similar to cement, depending on the amount of reagent added and the types and amounts of wastes treated (Walker and Pavía 2011). Pozzolanic reactions are generally slower than those for cement. Waste materials that have been stabilized/solidified with pozzolans include oily sludges, plating sludges containing various metals (Al, Ni, Cu, Pb, Cr, and As), waste acids, and creosote.

Numerous hydrolysis reactions affect the solubility of metals in S/S-treated soil and waste. Hydrolyzed metals form various hydroxides, oxides, carbonates, and sulfates, among other species. The log K_{sp} values for certain metal-hydroxides are Cd, 14.3; Cr(III), 30.2; Pb, 19.9; and Hg(II), 25.5; all of which indicate the formation of highly insoluble products. Published solubility data is useful in assessing which form of a hazardous metal is most stable, and to what degree stabilization is possible given site conditions, available materials, and technology. Actual concentrations of precipitated species depend on several solution properties including pH, redox potential, and composition of salts. In an early study of effectiveness of S/S (U.S. EPA 1990a), $Cd(OH)_2$ and $Pb(OH)_2$ were found to have comparable stabilities; however, the degree of leaching differed for the two metals. Leaching of Pb was substantially higher than that of Cd due to the differing precipitation reactions for each metal hydroxide.

Additives may be included in cement-based S/S treatment to promote immobilization of specific contaminants or to improve physical characteristics. Activated carbon, organophilic clay, silica compounds, and other sorbents are added to aid in the immobilization of organics (Gong and Bishop 2003; Rho et al. 2001). Soluble silicate additives are used to speed setting and reduce free water. Silicates can also precipitate certain metal forms. Sulfide molecules promote reducing reactions which can form very low-solubility metal sulfides.

Efficacy of cement-based S/S can be improved by modifying cement phase compositions, controlling temperature, water/solid ratios, particle size, and other factors that affect setting and strength development and long-term durability of solidified waste forms (Chen et al. 2009).

Advantages of cement-based processes include the local availability of materials, the low cost both of materials and mixing equipment, binder applicability to a wide range of waste types, knowledge of hardening and setting reactions, ease of use in the field, use of naturally occurring materials as the binder, and flexibility for different applications.

8.3 ORGANIC STABILIZATION/SOLIDIFICATION

Organic stabilization/solidification using thermoplastic binders has been applied to certain hazardous wastes; however, it is used to a lesser degree than S/S with cements and pozzolans.

Thermoplastic S/S is often a *microencapsulation* process in which the contaminated soil does not react chemically with the encapsulating material. A thermoplastic material such as asphalt (bitumen) or polyethylene coats the waste constituents into a stabilized/solidified mass. The most common thermoplastic material for S/S is asphalt. The asphalt may be heated before being mixed with dry soil, or it may be applied at ambient temperature. In the latter case, compaction removes additional water from the thermoplastic/soil mixture. Asphalt may be used for stabilizing/solidifying oil- and gasoline-contaminated soils. The final consistency will vary depending on the density of the hydrocarbon mixed into the asphalt and the amount of aggregate added to the mixture. Thermoplastic encapsulation has also been applied to electroplating sludges, painting and refinery sludges containing metals and organics, incinerator ash, baghouse dust, and radioactive waste (Zhao 2017; Spence and Shi 2005; Nagy et al. 1991).

Organic-based S/S can also involve incorporation of contaminated soil within a polymer in order to immobilize contaminants. The technique involves dispersing soil into a hot plastic mass which subsequently cools, incorporating the soil within a solid block. The soil/thermoplastic mix is commonly extruded into a container, for example a metal drum, to provide a convenient form for transport and disposal. Polyethylene, polypropylene, urea formaldehyde, or paraffin can be employed for thermoplastic S/S of specific wastes (ASTM 1985). Organic polymerization has been used to treat radioactive debris and has been applied to organic chlorides, phenols, paint sludges, cyanides, and arsenic. Polymerization can also be applied to flue gas desulfurization sludge, electroplating sludges, Ni-Cd battery wastes, kepone-contaminated sludge, and certain chlorine product wastes (U.S. EPA 1989a; Kyles et al. 1987).

8.4 FIELD TECHNOLOGIES: EX SITU

S/S can be accomplished either in situ or ex situ. Vigorous mixing is required to disperse the binder within the soil matrix; therefore, S/S processes are often applied to excavated materials. Ex situ processing involves removing the contaminated material from the affected site, dry or wet screening to remove large debris and produce a well-graded size distribution, mixing soil with binder and water, removal and treatment of off-gases (if dusts or volatile gases are produced), and transfer of the treated materials to a disposal area (figure 8.2).

Ex situ S/S treatment can be accomplished in a fixed facility or in a mobile treatment unit which is transported to the site. Screening and crushing of soil may be needed prior to blending operations to handle oversize materials such as rocks and other debris. Portable S/S plants have been developed that include bulk chemical feed and other blending equipment. Processing rates for large portable plants range from 500 to over 1,000 tons per day (AFCEE 1992). Pilot-scale plants typically can process 100 tons per day and may be transported on one trailer.

Common approaches to mixing the binder include in-drum mixing, in-plant mixing (e.g., rotary drum), and area mixing. In-drum mixing typically is used for highly toxic materials or small volumes of soil; binder and soil are combined and mixed in a 55-gal drum. In-plant mixing can be either continuous or batch. Continuous processes often employ a pug mill mixer equipped with paddles on a rotating shaft to mix the contents. (figure 8.3) Batch operations typically use some form of drum mixer. For example, a rotary drum mixer is an inclined container that includes internal baffles and rotates to mix

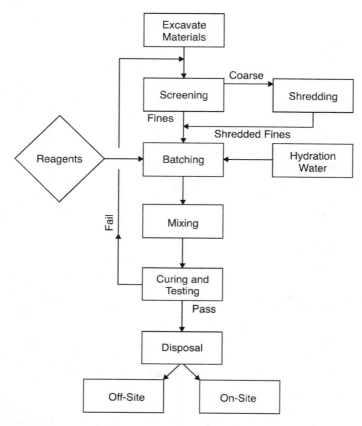

Figure 8.2 Treatment train for ex situ S/S of contaminated soil.

the contents. Area mixing involves excavating soil, transferring to a large contained area and then mixing with binder using a backhoe or other earth-moving equipment.

The loading and mixing of waste and binder inevitably results in particulate air emissions. If the contaminated material contains volatile organics, mixing and heating due to binder hydration will release organic vapors. Control of dust is often needed, and control of organic vapors is necessary in some situations.

8.5 IN SITU S/S TREATMENT

During in situ S/S treatment the binder material is introduced directly to the site by surface application or use of augers. A major advantage of in situ S/S

Figure 8.3 A self-loading cement mixer is useful for continuous S/S processes.

is that it eliminates the labor, energy, and costs involved with soil excavation, transport, and replacement or disposal. Another advantage is that it is applicable at space-limited sites such as alongside buildings or near pipes, tanks, and other obstructions. Furthermore, dust generation is greatly reduced. A critical challenge in utilizing in situ S/S, however, is complete and uniform mixing of binder with contaminated soil. Other disadvantages of auger methods are that they are not feasible in the presence of bedrock or boulders and are impeded in the presence of clay, oily sands, and cohesive soil. Poor progress under these circumstances may require ex situ processes.

In situ chemical treatment reagents must not create an environmental or health hazard. Treatment systems can potentially introduce oxidizing, reducing, or neutralizing chemicals into the groundwater system; therefore, adequate knowledge of the subsurface environment is essential to avoid contamination. Also, injection of treatment chemicals may trigger the requirements for land disposal as per federal regulations. In such cases, the selection of reagents for chemical treatment will be limited by the Land Disposal Restrictions (LDRs) on introducing chemicals into the soil (40 CFR 2016).

The chemical forms of soil contaminants must be considered during the treatment process. For example, Cr(VI), which is both toxic and mobile in soil, should first be reduced to the less hazardous Cr(III) species prior to in situ treatment. A common approach to Cr reduction involves acidification followed by chemical reduction and neutralization. The Cr(VI) ion is a strong oxidizing agent under acidic conditions and in many cases readily converts to Cr(III). Acidification is accomplished using mineral acids. With the pH adjusted to < 3.0, ferrous sulfate is added to convert Cr(VI) to Cr(III). After chemical reduction, solution pH is increased to > 7.0 to coprecipitate Cr(III) with ferric and ferrous iron (Qin et al. 2005). Cr(III) is also readily precipitated by addition of calcium hydroxide. Other in situ Cr reducing agents include FeS, leaf litter, and acid compost. Treatments are also available for Cr reduction in the neutral pH range. Reagents include sodium metabisulfite, sodium bisulfite, and ferrous ammonium sulfate (Brandhuber et al. 2004; Jacobs 1992). Reduction at neutral pH generates less sludge, so the total waste volume is reduced.

In situ treatment processing involves reagent preparation, mixing of binder and contaminated soil, and off-gas treatment. The basic approaches for mixing binder with soil include in-place mixing, vertical auger mixing, and injection grouting. *In-place mixing* is applicable only to shallow or surface contamination. The process involves spreading and mixing binder reagents with contaminated soil by conventional earth-moving equipment (e.g., draglines, backhoes, dozers) and allowing the mixture to set and cure. *Vertical auger application* is adapted from the construction industry. A set of large diameter vertical crane-mounted augers injects and mixes binder into the soil (figures 8.4 and 8.5). A 10-ft-diameter auger mounted on a crawler crane is applicable. Dry reagents and water are pneumatically dispersed into the soil

Figure 8.4 An engineer directs the positioning of multiple augers. Three augers overlap. Successful fixation of contaminants requires that the soil be thoroughly mixed with the cementing mixtures. *Source*: Reproduced with kind permission of Boulanger and Duncan, University of California, Davis. Boulanger and Duncan 2018.

Figure 8.5 Close-up showing that the middle auger is recessed relative to the outer two augers. The grout, which is a mixture of cementing agents and water, is injected through ports near the ends of the augers. *Source*: Reproduced with kind permission of Boulanger and Duncan, University of California, Davis. Boulanger and Duncan 2018.

as the auger creates a pattern of overlapping 10-ft-diameter columns. Mixing can be accomplished to depths up to 40 ft and can process 500–1,000 yd^3 of soil per day. Deep drilling is also possible, reaching depths of up to 150 ft. A cluster of two to four augers, each up to 3 ft in diameter, is assembled to loosen subsoil and mix in the binder (figures 8.6 and 8.7). During *injection grouting*, the binder containing dissolved or suspended treatment agents is forced into the contamination zone under pressure and allowed to permeate the soil. Grout injection can be applied to contaminated formations deep below the surface. The injected grout cures in place to form an in situ treated mass.

After reagents are applied the site is allowed adequate time for binder setting and curing. Once the materials have cured, site development can take

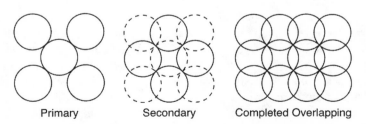

Figure 8.6 Overlapping column arrangement in S/S-treated soil. *Source*: U.S. Environmental Protection Agency.

Figure 8.7 Close-up of soil auger. *Source*: Reproduced with kind permission of Boulanger and Duncan, University of California, Davis. Boulanger and Duncan 2018.

place. If the site is located in an industrial area it may be possible to build upon the surface, using the solidified matrix as a base. Alternatively, a layer of soil can be distributed over the surface which is subsequently prepared for planting of selected vegetation.

8.6 TECHNICAL CONSIDERATIONS

S/S treatment is usually not applicable to wastes containing volatile organic compounds or high levels of semivolatile organics unless off-gas removal and treatment is employed. Otherwise, the waste can be prepared for S/S by steam stripping (Smith et al. 1995). For a site contaminated with both metals and organics, pretreatment may be needed to transform the solids to forms more suitable for S/S. Addition of silicates or modified clays to the binder system may improve S/S performance with organics. Other pretreatment steps may include screening for debris removal, size reduction, neutralization, Cr(VI) reduction, and removal or destruction of organics. Posttreatment steps include disposal of treated residuals, control of dust, and collection and treatment of off-gas.

Success of S/S is evaluated depending on the desired final use(s) of the solidified mass. If the operator's goal is simple soil detoxification, the TCLP should be conducted. This test provides a practical measure of the potential leachability of toxic elements within the treated material. It may also be necessary to test the permeability of the solidified mass. If the site is to be used for construction or similar engineered purposes, the durability of the cured materials must be determined, and other cement industry requirements (e.g., unconfined compressive strength, wet/dry, freeze/thaw) must be addressed. In some cases it may be necessary to inspect the material on a microstructural basis (Vespa et al. 2006; Dermatas and Meng 2003). Tests may include X-ray diffraction for crystalline structure, optical and scanning microscopy, and EDX (energy dispersive X-ray spectrometry) for elemental analysis of crystal structures.

8.7 OVERVIEW OF S/S TECHNOLOGIES

S/S technology is a widely accepted technology for treatment of metal-, inorganic-, and certain organic-contaminated wastes. The technology can be conducted either in situ or ex situ depending on practical considerations (e.g., chemical and physical state of contaminants), available equipment, subsurface characteristics (presence of utilities and other structures), and cost. The overall process is relatively simple and inexpensive, and uses common chemicals and construction equipment. Familiar materials, including certain nonhazardous industrial wastes (as in the case of pozzolans), can be employed as binding agents.

The technology can treat simultaneously a wide variety of contaminants. This is especially true for inorganics of similar chemistry (e.g., di- and trivalent metals). Some types of chemical treatment such as oxidation, reduction, and pH neutralization are simple and inexpensive to implement, particularly for single-contaminant wastes. S/S improves the handling and physical characteristics of the contaminated material; in other words, sludges are transformed to solids. The S/S technique decreases waste surface area and reduces solubility of the treated material, generally by chemical processes, thereby reducing leaching of contaminants. The permeability of the treated zone is significantly reduced. The stabilized material can be engineered to produce a subbase or slab for subsequent commercial/industrial use at the site. If a site is to be revegetated, however, placement of a soil cover of sufficient depth will be necessary.

Although S/S technology is relatively simple in principle, there are disadvantages to its use. No single binder is effective for stabilization of all metals. In some situations one metal will be stabilized while a second concurrently becomes more soluble (e.g., cement immobilizes Pb but may increase mobility of As). The potential for interferences and incompatibilities must be considered in pilot tests. General types of interferences include inhibition of bonding between waste and binder materials, slow or incomplete setting, reduced stability of the matrix resulting in leaching of contaminants, and insufficient physical strength of the final product. The operator must be aware of possible chemical incompatibilities; for example, the oxidation of soil containing reduced Cr, Hg, and Pb can convert these metals to more toxic or mobile forms; the neutralization of alkaline waste can increase metal mobilities by redissolving hydroxide precipitates; and the oxidation of certain organic wastes can increase toxicity due to formation of unwanted by-products. Also, arsenic wastes are difficult to treat due to the complex chemistry of arsenic. Soils with unacceptable physical characteristics such as being too solid or too viscous to mix will cause difficulties for S/S. Lastly, the technology does not reduce total contaminant content in the soil, and volume expansion is an inevitable result of S/S treatment.

Complications for cement-based S/S are posed by materials containing organics (Maijala et al. 2009). An oil and grease content > 1 percent interferes with contact between reactant and soil. Nonvolatile and semivolatile organics can be difficult to treat. Many contaminants interfere with bonding between soil and binder (therefore increasing time of setting and decreasing durability); these include cyanides, arsenates, sulfates, sulfides, metal salts, and metal complexes (Shi 2004; Stegemann 2004; Means et al. 1995). Pretreatment may therefore be necessary. Wastes that contain hydrocarbons as the primary contaminant will also cause difficulties. Some may be removed by volatilization processes; others may or may not react with the binder. The S/S process is very waste-specific, and site-specific treatability studies should be conducted.

Atmospheric emissions of dust and volatile hydrocarbons are a concern with S/S. Adding dry waste and a binder to a mixer can generate significant dust emissions. Volatilization and emission of hydrocarbons may pose a hazard during mixing, and emissions control may be required. Heat from exothermic binder hydration reactions can cause a rapid release of both dust and volatile organics.

There are some specific concerns regarding the use of in situ S/S. For example, delivering reagents to the subsurface and achieving uniform mixing and treatment in situ may be difficult. Further, in situ S/S has relatively slow treatment throughput compared to ex situ S/S.

8.8 CASE HISTORY

Chemical treatment of lumber had taken place at the Selma Pressure Treating (SPT) facility since 1942. The site is located southeast of Fresno, California, in the San Joaquin River Valley. The entire SPT site covers 18 acres; however, the actual wood-treatment area measures 3 to 4 acres. The property is located less than a quarter-mile from homes and businesses. Groundwater resources near the site are classified as a beneficial use, sole-source aquifer.

This aquifer provides the domestic water supply for surrounding communities and rural homes, and surface irrigation systems are also supplemented by this water (U.S. EPA 1992, 2005).

The original wood preserving process involved dipping lumber into a mixture of PCP and oil, and allowing the excess fluid to drip as the wood dried on open storage racks. In 1965, site operators converted to a pressure-treating process that involved conditioning the lumber to reduce moisture content and increase permeability, followed by impregnating the wood with chemical preservatives. From 1942 to 1971, wastes from the treatment plant were disposed via:

- runoff into drainage ditches and a percolation ditch;
- drainage into dry wells;
- spillage on open ground;
- placement into an unlined pond and a sludge pit;
- disposal in an adjacent vineyard.

The plant ceased operations in 1981.

Chemical preservatives used at the site included fluor-chromium arsenate-phenol, Woodtox 140 RTU, heavy oil penta solution, LST concentrate, copper-8-quinolinoate, PCP, and chromated-copper-arsenate (CCA). Contaminated groundwater was found to be migrating from the property, and soil was contaminated beneath the site. The primary metal contaminants were As, Cr, and Cu. PCP was also detected along with associated degradation and impurity products including polychlorinated dibenzo-*p*-dioxins (PCDD), polychlorinated dibenzofurans (PCDF), and chlorinated phenols. Other hydrocarbons were detected including benzene, toluene, and xylene, and PAHs such as naphthalene and pyrene. The highest levels of contamination occur in the first 5 ft of the soil profile.

Federal and state agencies have been jointly involved in regulatory actions at the site since the 1970s. EPA scored the site at 48.83 using the Hazard Ranking System (HRS) (U.S. EPA 2006), and the site was consequently placed on the Superfund NPL in 1983. Following a remedial investigation/feasibility study (RI/FS), a Record of Decision (ROD) was signed in 1988.

The Silicate Technology Corporation (STC) immobilization technology is a solidification/stabilization process that uses a proprietary organophilic material (FMS silicate) to adsorb organic compounds up to twenty times its weight. When combined with a cementitious binder material, the reagents selectively adsorb organic and inorganic contaminants and produce a high-strength monolith. The resulting solid materials have passed federal and state regulatory threshold levels for TCLP leachate tests. Leachability has also been found to decrease with age.

A backhoe/front-end loader collected contaminated soil from the unlined disposal pond. An approximately 300 ft^2 area was excavated to a depth of 3 ft. The excavation was lined with a layer of 20-mil HDPE and backfilled with 1 ft of sand overlaid by 1 ft of crushed stone (1-in diameter) and clean soil. The contaminated soil was transported directly to a processing area; storage piles were covered by a 10-mil HDPE prior to processing. Each batch was mixed in a batch mixer prior to addition of reagents. A schematic of the process appears in figure 8.8.

Materials and equipment at the site included:

- A 15 × 50 ft area lined with a 20-mil HDPE membrane liner to store the solidified waste. Treated waste was discharged into cardboard concrete forms mounted on pallets and placed in the storage area. The storage area was graded so that rainwater runoff from the solidified waste was collected.
- Electric generators to supply power for process equipment, the support trailer, equipment, and other needs.
- Process and wash water for the treatment unit and decontamination, obtained from the facility potable water supply. About 220 gal of water were required per treated batch.
- Scale for weighing reagents and raw wastes.
- Heavy equipment decontamination area for cleaning large equipment, bermed and lined with 20-mil plastic.
- A personnel decontamination station.
- Three-thousand gal tank to store decontamination water.
- Gasoline-powered high-pressure sprayer to clean the STC process equipment and other heavy equipment.

Arsenic was stabilized by STC treatment under neutral leaching conditions. Using the TCLP, reduction in As ranged from 35 to 92 percent. Some of the As may have been reduced from the (V) to the (III) species during the raw waste mixing process, thereby rendering the treated contaminants more mobile and easily leached under the acidic conditions of the TCLP. The STC process was not effective in converting arsenite to arsenate or a species that could be chemically immobilized; minor amounts of both arsenite and arsenate were detected in some of the treated waste.

Pentachlorophenol (PCP) was the main organic contaminant of concern at the SPT site. Based on data from a treatability study, samples were also analyzed for semivolatile organics such as tetrachlorophenol (TCP), phenanthrene,

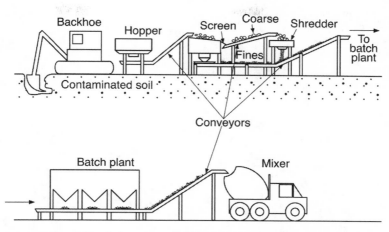

Figure 8.8 Schematic of the STC solidification/stabilization process. *Source*: U.S. Environmental Protection Agency.

naphthalene, and phenol. All were detected in negligible concentrations. PCP concentrations in the soil extracts varied after the STC stabilization process (table 8.1). All raw and treated waste concentrations were, however, below TCLP federal regulatory threshold levels of 100 mg/L for PCP. The percent reductions of oil and grease in the treated waste ranged from 32 to 52 percent. Although the treatment process was not highly effective in reducing the quantity of extractable oil and grease, the presence of small quantities (< 2%) of oil and grease did not adversely affect the solidification process.

Raw waste samples had pH values ranging from 6.3 to 7.1. Treated wastes were very basic, with pH values at approximately 12.5. The sand and water additives had pH values of 8.6 and 8.0, respectively, and the reagent mixture had a pH of 12.5.

Moisture content of the raw waste ranged from 3.9 to 5.8 percent, and from 1.9 to 9.7 percent in the treated waste (table 8.2). Average bulk densities ranged from 1.42 to 1.54 g/cm^3 for the raw waste, and treated waste had bulk densities ranging from 1.55 to 1.62 g/cm^3. Volume of the treated waste ranged from 59 to 75 percent greater than that of raw waste. Permeability of treated waste ranged from 0.8×10^{-7} to 1.7×10^{-7} cm/sec. Unconfined compressive strength (UCS), defined as the load per unit area, ranged from 259 to 347 psi for treated waste. These values are well below the ASTM and American Concrete Institute (ACI) minimum required UCS of 3,000 psi for the construction of sidewalks (ASTM 1991). However, values are well above the EPA minimum guideline of at least 50 psi for hazardous waste solidification (U.S. EPA 1988).

Table 8.1 TCLP results for selected analytes in the raw and treated waste, Selma Pressure Treating Superfund Site

Test	Batch	Raw Waste mg/L	Treated Waste mg/L
PCP-TCLP	1	1.50	3.42
	2	2.27	< 0.25
	3	1.75	5.52
	4	2.28	0.90
PCP-TCLP Distilled H_2O	1	34.7	3.98
	2	40.0	0.58
	3	40.5	3.05
	4	79.9	120
As-TCLP	1	1.82	0.086
	3	1.06	0.101
	4	2.40	0.875
	5	3.33	0.548
As-TCLP Distilled H_2O	1	0.80	< 0.01
	3	0.73	< 0.01
	4	1.25	< 0.01
	5	1.07	0.011
As(V)	3	60.5	< 2.0
	4	19.5	21
	5	260	< 2.0
Cr-TCLP	1	0.13	0.245
	2	< 0.05	0.187
	3	0.10	0.287
	4	0.27	0.320
Cr-TCLP	1	0.19	< 0.05
	2	0.17	< 0.05
	3	0.07	0.056
	4	0.11	0.079
Cu-TCLP	1	3.42	0.09
	2	1.38	0.75
	3	6.53	0.10
	4	9.43	0.06
Cu-TCLP	1	0.45	0.031
	2	0.37	< 0.030
	3	0.99	0.054
	4	0.56	0.032

Source: U.S. EPA 1992.

Certain operational problems were encountered during the solidification/stabilization activity:

1. Certain contaminated soils (PCP-contaminated, in particular) were not well mixed after treatment; the treated waste contained large (up to 2-in

Table 8.2 Selected physical and engineering characteristics of the raw and treated waste, Selma Pressure Treating Superfund Site

Batch	Moisture Content Raw Treated %		Bulk Density Raw Treated g/cm³		Permeability Treated cm/sec × 10^{-7}	UCS Treated psi
1	5.8	2.6	1.42	1.57	1.7	301
3	5.7	1.9	1.54	1.55	1.5	278
4	4.2	9.7	1.54	1.58	0.9	259
5	3.9	8.8	1.54	1.62	0.8	347

Source: U.S. EPA 1992.

diameter) aggregates of untreated waste. The problem was solved in subsequent batches by forcing the raw waste through a set of screens prior to treatment, reducing the raw waste aggregate size to about 0.04 to 0.08 in (1–2 mm) diameter.
2. Generation of large quantities of contaminated dust from the movement of equipment, supplies, and site personnel. The dust caused fouling of the photoionization device used for air monitoring. The dust problem was remedied via application of water to the site from a water truck.
3. Generation of a dust cloud upon initial mixing of the dry reagents in the mixer. A tarp was secured over the top of the mixer to limit release of dust.

QUESTIONS

1. Describe the purpose of the TCLP. How is it applied to solidified/stabilized soil material?
2. Locate ASTM standards for concrete testing. How would you expect solidified soil to differ from Portland cement in terms of UCS and other properties?
3. What, if any, is the effect of the presence of oily wastes on S/S of metal-contaminated soil?
4. Ideally, Cr(VI) should be reduced to Cr(III) prior to S/S. Explain why this is so.
5. Volume expansion inevitably results from S/S treatment. Identify a case study of S/S treatment of contaminated soil and provide data which demonstrates volume expansion.

REFERENCES

Air Force Center for Environmental Excellence. 1992. *Remedial Technology Design, Performance and Cost Study*. Brooks Air Force Base, TX: Environmental Services Office.

American Society for Testing and Materials (ASTM). 1985. *Temperature Effects on Concrete*. Philadelphia: ASTM.

———. 1989. *Environmental Aspects of Stabilization and Solidification of Hazardous and Radioactive Waste*. STP 1033. Authors: Cote, P., and M. Gilliam. Conshohocken, PA: ASTM.

———. 1991. *Annual Book of ASTM Standards*. Conshohocken, PA: ASTM.

———. 1992. *Stabilization and Solidification of Hazardous, Radioactive, and Mixed Wastes*, 2nd Volume. STP 1123. Authors: Gilliam, T. M., and C. C. Wiles. Conshohocken, PA: ASTM.

———. 1996. *Stabilization and Solidification of Hazardous, Radioactive, and Mixed Wastes*, 3rd Volume. STP 1123. Authors: Gilliam, T. M., and C. C. Wiles. Conshohocken, PA: ASTM.

Andres, A., J. Viguri, A. Irabien, M. Fdex de Velasco, A. Coz, and C. Ruiz. 2002. Treatment of foundry sludges by stabilization/solidification with cement and siliceous binders. *Fresenius Environmental Bulletin* 11(10): 849–53.

Benge, O., and W. Webster. 1994. *Proceedings of the IADC/SPE Drilling Conference,* Dallas, 1994, 169–80.

Bettahar, M., J. Ducreax, G. Schafer, and F. Van Dorpe. 1999. Surfactant enhanced in situ remediation of LNAPL contaminated aquifers: Large scale studies on a controlled experimental site. *Transport in Porous Media* 37(3): 255–76.

Boulanger, R. W., and J. M. Duncan. 2018. *Deep Soil Mixing for Contaminant Fixation*. Davis: University of California. cee.engr.ucdavis.edu/faculty/boulanger/geo_photo_album/index.html.

Brandhuber, P., M. Frey, M. J. McGuire, P. Chao, C. Seidel, G. Amy, J. Yoon, L. McNeill, and K. Banerjee. 2004. *Low-Level Hexavalent Chromium Treatment Options: Bench-Scale Evaluation*. Denver, CO: AWWA Research Foundation.

Carmalin, S., and K. Swaminatian. 2005. Leaching of metals on stabilization of metal sludge using cement based materials. *Journal of Environmental Sciences* 17(1): 115–18.

40 CFR. 2016. *Part 268. Land Disposal Restrictions*. Washington, DC: U.S. Government Printing Office.

Chen, Q. Y., M. Tyrer, C. D. Hills, X. M. Yang, and P. Carey. Immobilisation of heavy metal in cement-based solidification/stabilisation: A review. *Waste Management* 29(1): 390–403.

Conner, J. R. 1990. *Chemical Fixation and Solidification of Hazardous Wastes*. New York: Van Nostrand Reinhold.

Dermatas, D., and X. Meng. 2003. Utilization of fly ash for stabilization/solidification of heavy metal contaminated soils. *Engineering Geology* 70(3–4): 377–94.

Donnelly, J., and W. Webster. 1996. From sediment to solid. *Civil Engineering: ASCE* 66(5): 41–3.

Fitch, J., and C. Cheeseman. 2003. Characterisation of environmentally exposed cementbased stabilised/solidified industrial waste. *Journal of Hazardous Materials* 101(3): 239–55.

Gong, P., and P. Bishop. 2003. Evaluation of organics leaching from solidified/stabilized hazardous wastes using a powder reactivated carbon additive. *Environmental Technology* 24(4): 445–55.

Jacobs, J. H. 1992. Treatment and stabilization of a hexavalent chromium containing waste material. *Environmental Progress* 11(2): 123–26.

Jones, L. 1989. *Interference Mechanisms in Waste Stabilization/Solidification Processes, Project Summary.* EPA 600/S2-89/067. Cincinnati: Risk Reduction Engineering Laboratory.

Kamon, M., T. Katsumi, and Y. Sano. 2000. MSW fly ash stabilized with coal ash for geotechnical application. *Journal of Hazardous Materials* 76(2–3): 265–83.

Kyles, J. H., K. C. Malinowski, and T. F. Stanczyk. 1987. Solidification/stabilization of hazardous waste—A comparison of conventional and novel techniques, toxic and hazardous wastes. In: *Proceedings of the 19th Mid-Atlantic Industrial Waste Conference, June 21–23*, edited by J. C. Evans. Lewisburg, PA: Bucknell University.

Kyu-Hong, A., and N. Jagga. 1989. Stabilization of heavy metal bearing sludges using cementitious binders. In: Wu, Y. C. (Ed.) *Physiochemical and Biochemical Detoxification of Hazardous Wastes.* Lancaster, PA: Technomic.

Maijala, A., J. Forsman, P. Lahtinen, M. Leppaenen, A. Helland, A.-O. H. Roger, and M. Konieczny. 2009. *Cement Stabilization and Solidification (STSO): Review of Techniques and Methods.* Oslo, Norway: Ramboll Norge AS. Rap001-Id01, 57.

McDaniel, E. W., R. D. Spence, and O. K. Tallent. 1990. *Research Needs in Cement-based Waste Forms.* Paper presented at the XIVth International Symposium on the Scientific Basis for Nuclear Waste Management, Meeting of the Materials Research Society, Boston.

Means, J. L., L. A. Smith, K. W. Nehring, S. E. Brauning, A. R. Gavaskar, B. M. Sass, C. W. Wiles, and C. I. Mashni. 1995. *The Application of Stabilization/Solidification to Waste Materials.* Boca Raton, FL: CRC Press.

Montgomery, D., J. Sollars, and R. Perry. 1988. Cement-based solidification for the safe disposal of heavy metal contaminated sewage sludge. *Waste Management and Research* 6(3): 217–26.

Nagy, B., F. Gauthier-Lafaye, P. Holliger, D. W. Davis, D. J. Mossman, J. S. Leventhal, M. J. Rigali, and J. Parnell. 1991. Organic matter and containment of uranium and fissiogenic isotopes at the Oklo natural reactors. *Nature* 354(6353): 472–75.

Qian, G., J. Tay, Y. Cao, and P. Chui. 2006. Utilization of MSWI fly ash for stabilization/solidification of industrial waste sludge. *Journal of Hazardous Materials* 129(1–3): 274–81.

Qiao, X., C. Poon, and C. Cheeseman. 2006. Use of flue gas desulphurisation (FGD) waste and rejected fly ash in waste stabilization/solidification systems. *Waste Management* 26(2): 141–49.

Qin, G., M. J. Mcguire, N. K. Blute, C. Seidel, and L. Fong. 2005. Hexavalent chromium removal by reduction with ferrous sulfate, coagulation, and filtration: A pilot-scale study. *Environmental Science and Technology* 39(16): 6321–27.

Rho, H., H. A. Arafat, B. Kountz, R. C. Buchanan, N. G. Pinto, and P. Bishop. 2001. Decomposition of hazardous organic materials in the solidification/stabilization process using catalytic-activated carbon. *Waste Management* 21(4): 343–56.

Shi, C. 2004. Hydraulic cement systems for stabilization/solidification. In: *Stabilization and Solidification of Hazardous, Radioactive, and Mixed Wastes,* edited by R. D. Spence and C. Shi. Boca Raton, FL: CRC Press.

Smith, L. A., J. L. Means, A. Chen, B. Alleman, C. C. Chapman, J. S. Tixier, S. E. Brauning, A. R. Gavaskar, and M. D. Royer. 1995. *Remedial Options for Metals-Contaminated Soils.* Boca Raton, FL: CRC Press.

Spence, R. D., and C. Shi. 2005. *Stabilization and Solidification of Hazardous, Radioactive, and Mixed Wastes.* Boca Raton FL: CRC Press.

Spencer, R. W., R. H. Reifsnyder, and J. C. Falcone. 1982. Applications of soluble silicates and derivative materials in the management of hazardous wastes. In: *National Conference on the Management of Uncontrolled Hazardous Waste Sites,* edited by R. Sims and K. Wagner. Silver Springs, MD: HMCRI.

Stegemann, J. 2004. Interactions between wastes and binders. In: *Stabilization and Solidification of Hazardous, Radioactive, and Mixed Wastes,* edited by R. D. Spence and C. Shi. Boca Raton, FL: CRC Press.

Thomas, J., and H. Jennings. 2008. *The Science of Concrete.* Monograph 3.5. Chicago, IL: Northwestern University.

Tittlebaum, M. E., R. D. Seals, F. K. Cartledge, S. Engles, and H. R. Fahren. 1985. State of the art on stabilization of hazardous organic liquid wastes and sludges. *Critical Reviews in Environmental Control* 15(2): 179–211.

U.S. EPA. 1988. *Technology Screening Guide for Treatment of CERCLA Soils and Sludges.* EPA/540-2-88/04. Washington, DC: Office of Emergency and Remedial Response.

———. 1989a. *HAZCON Solidification Process, Douglassville, PA: Applications Analysis Report.* EPA/540/A5-89/001. Office of Research and Development.

———. 1989b. *Stabilization/Solidification of CERCLA and RECRA Wastes: Physical Tests, Chemical Testing Procedures, Technology Screening, and Field Activities.* EPA/625/6-89/022. Center for Environmental Research Information.

———. 1990a. *Handbook on In Situ Treatment of Hazardous Waste-Contaminated Soils.* EPA/540/2-90/002. Cincinnati: Risk Reduction Engineering Laboratory.

———. 1990b. *Morphology and Microchemistry of Solidified/Stabilized Hazardous Waste Systems.* EPA/600/02-89/056. NTIS No. PB90-134156/AS. Cincinnati: Risk Reduction Engineering Laboratory.

———. 1992. *Silicate Technology Corporation's Solidification/Stabilization Technology for Organic and Inorganic Contaminants in Soils: Applications Analysis Report.* EPA/540/AR-92/010. Office of Research and Development.

———. 1995. *Contaminants and Remedial Options at Selected Metal-Contaminated Sites.* EPA/540/R-95/512.

———. 1997. *Recent Developments for In Situ Treatment of Metal Contaminated Soils.* EPA-542-R-97-004. Office of Solid Waste and Emergency Response.

———. 2005. *Region 9: Superfund.* Selma Treating Co. See: yosemite.epa.gov/r9/sfund/r9sfdocw.nsf/4b229bb0820cb8b888256f0000092946/f67ccbca080c9ad288257007005e9435!OpenDocument.

———. 2006. *Introduction to the HRS.* See: www.epa.gov/superfund/programs/npl_hrs/hrsint.htm.

Vandercasteele, C., G. Wauters, V. Dutre, and D. Geysen. 2002. Solidification/stabilization of arsenic bearing fly ash from the metallurgical industry. Immobilisation mechanism of arsenic. *Waste Management* 22(2): 143–46.

Vespa, M., M. Harfouche, E. Wieland, A. Scheidegger, R. Dahn, and D. Grolimund. 2006. Speciation of heavy metals in cement-stabilized waste forms: A microspectroscopic study. *Journal of Geochemical Exploration* 88(1–3): 77–80.

Walker and Pavía. 2011. Physical properties and reactivity of pozzolans, and their influence on the properties of lime–pozzolan pastes. *Materials and Structures* 44(6): 1139–50.

Wareham, D., and J. Mackenchnie. 2006. Solidification of New Zealand harbor sediments using cementitious materials. *Journal of Materials in Civil Engineering* 18(2): 311–15.

Yilmaz, O., K. Unlu, and E. Cokca. 2003. Solidification/stabilization of hazardous wastes containing metals and organic contaminants. *Journal of Environmental Engineering* 129(4): 366–76.

Zhao, Y. 2017. *Pollution Control and Resource Recovery: Municipal Solid Wastes Incineration: Bottom Ash and Fly Ash.* New York: Elsevier.

Chapter 9

Soil Vapor Extraction

For what is your life? It is even a vapor, that appeareth for a little time, and then vanisheth away.

—James 4:14

9.1 INTRODUCTION

Soil vapor extraction (SVE, in situ volatilization, soil venting) involves the withdrawal of volatile hydrocarbon compounds from the subsurface by drawing air currents through a network of wells. This treatment applies to the vadose zone, specifically the interstitial spaces, where hydrocarbon vapors can become entrained in the flow of extracted air and removed from pores. Within the saturated zone, a modification of the technology termed *air sparging* is used to treat groundwater.

In the vadose zone, SVE recovers the vapor-phase components of gasoline, aviation gasoline, jet fuel, and certain solvents (e.g., naphthol spirits). The low molecular weight, volatile contaminants are extracted more readily than are the heavier ones. Lighter alkanes include butane, pentane, and hexane; the lighter aromatics include benzene, toluene, xylenes, ethylbenzene, and other alkylbenzenes. Diesel fuel, crude petroleum, heating oils, used oil, and fuel oil are poorly suited to vapor extraction.

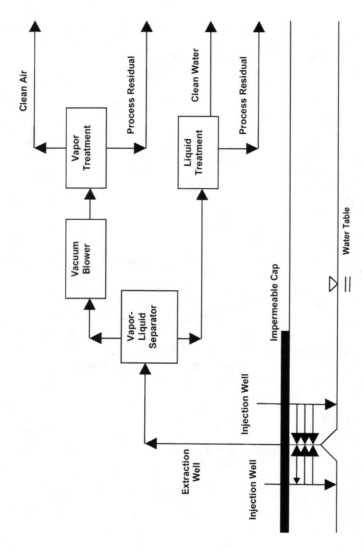

Figure 9.1 Schematic view of an SVE system. *Source*: U.S. Environmental Protection Agency.

A schematic of an SVE system is shown in figure 9.1. Unit operations include:

1. An air heater established upgradient of the injection port, to warm influent air. Warmed air raises subsurface temperatures and serves two purposes: (a) it increases the volatilization rate of hydrocarbons directly, and (b) it stimulates indigenous microbiological activity, which can further decompose organics. In cold climates air heaters additionally serve as freeze protection for the system (EPRI 1988).
2. A grid network of slotted pipe allows air to flow through the system.
3. Induced draft fans establish the air flow through the vadose zone.
4. A vapor treatment unit recovers and treats volatilized hydrocarbons; this unit operation minimizes emissions to the atmosphere. A number of techniques are available to oxidize or otherwise remove hydrocarbon emissions from the recovered air and will be discussed later in this chapter. The quality of gaseous emissions from the SVE unit must comply with air pollution standards.
5. Flow meters, flow control valves, and sampling ports are included in the design to support air flow and assess overall system operation (EPRI 1988).

9.2 SITE CHARACTERIZATION

Site investigation should begin with geophysical assessment (e.g., using an electromagnetic survey or ground-penetrating radar—see chapter 5) to determine the location of nonaqueous phase liquids, followed by soil gas monitoring to locate zones of highest concentrations, and conclude with soil sampling to determine the full extent of contamination and establish cleanup levels. Using a cone penetrometer equipped with sensing devices can reduce costs associated with sampling soil. Bench- and field-scale studies may be needed to determine treatability of the hydrocarbons.

When a shallow water table is present, it is important to investigate the potential for a rising water table, as such an event can result in the removal of less vapor and more water and adversely affect the overall efficiency of vapor extraction.

It is essential to identify geologic formations that occur between the surface and the base of the contaminant plume. Certain strata and components (i.e., clay lenses, large rocks and boulders, large cavities) can significantly impede vapor extraction. The most reliable method to identify them is by evaluating soil boring logs, either existing or conducted as part of the evaluation. Blow counts recorded from drilling operations can indicate densely compacted layers that might impede vapor extraction. Geophysical surveys such as electrical resistivity can also be conducted to delineate the locations of geologic formations. After the hydrocarbon plume and geologic formations have been identified they must be mapped. Such delineation will assist in determining where to locate the SVE system (i.e., where the contaminants are of highest concentration) and if any geologic structures will interfere. To evaluate this relationship, both plan view and cross-sectional maps should be generated; a 3-D computer-generated map serves as an excellent means to visualize the entire affected zone (figure 9.2).

Soil and groundwater are commonly analyzed for VOCs; base, neutral, and acid extractables (BNAs); and total petroleum hydrocarbons (TPH). For complex mixtures such as gasoline, diesel fuel, and solvent mixtures, it is more cost-effective to measure indicator compounds such as BTEX, or trichloroethylene rather than each compound present. The presence of vapor-phase hydrocarbons in the subsurface can be determined by sampling, either by soil gas recovery techniques or by monitoring wells. If the release has not aged significantly, free gasoline will contain a significant vapor phase. As a general rule, benzene levels (indicated by BTEX analysis) of approximately 200 ppm or greater indicate free gasoline. Biodegradation products should be considered as possible target compounds because they are sometimes more toxic than the parent compound (e.g., TCE may be converted to vinyl chloride).

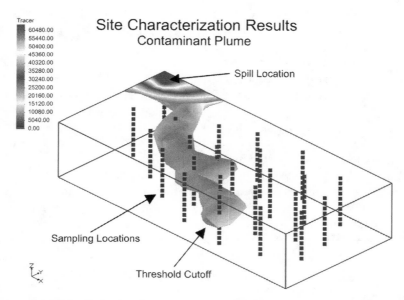

Figure 9.2 Three-dimensional computer view of a field site showing migration of contaminant plume. *Source*: Reproduced with kind permission of Aquaveo.com / GMS 10.4.

Because SVE may not remove all contaminants, soils should be analyzed for less volatile or nonvolatile contaminants (BNAs and TPH) to assess the need to remediate by other methods (biotreatment, soil washing, etc.).

9.3 FIELD DESIGN CONSIDERATIONS

Successful volatilization requires the incorporation of extraction wells and, frequently, injection wells (figure 9.3). The area from which a well extracts volatile compounds is termed the *zone of influence*. This area is rarely circular; however, such zones have often been assigned a *radius of influence*. The zone of influence of each well varies from 30 to 150 ft depending on soil type—for example, sandy versus clayey—and pumping rate.

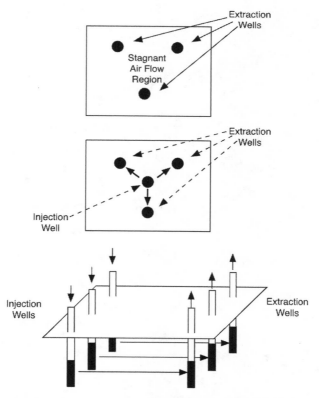

Figure 9.3 Arrangements of extraction and injection wells for SVE. *Source*: From Johnson et al. 1990.

Preheated air improves recovery, especially in colder weather, but this increases costs of operation. Steam injection has been used successfully in field tests but the hazard exists that increased subsurface pressure will cause the contamination to diffuse and spread.

Vapor extraction wells are similar to groundwater monitoring wells in terms of construction. A typical well is constructed from slotted plastic pipe, usually PVC. Wells should be slotted only through the zone of contamination. The slot size and number of slots per inch should be chosen to maximize the open area of the pipe. The slotted area is often encased in a nylon sock in order to prevent plugging by soil particles. A filter packing such as sand or gravel is placed between the borehole and pipe (figures 9.4 and 9.5). The filter packing should be as coarse as possible. Any dust carried by vapor flow can be removed by an aboveground filter. Bentonite pellets and a cement grout are placed above the filter packing. It is important that these be properly installed to prevent vapor flow from 'short-circuiting' (figures 9.4 and 9.6).

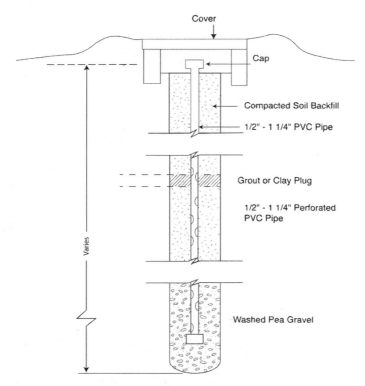

Figure 9.4 Vapor extraction well.

Soil Vapor Extraction

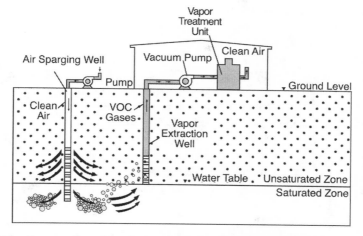

Figure 9.5 Cross section of an air sparging/vapor extraction well system.

Well locations should be chosen to ensure optimal vapor flow through the contaminated zone while minimizing vapor flow through other zones nearby. If one well is sufficient, it is placed in the geometric center of the contaminated zone and should be situated in order to maximize air flow through the entire

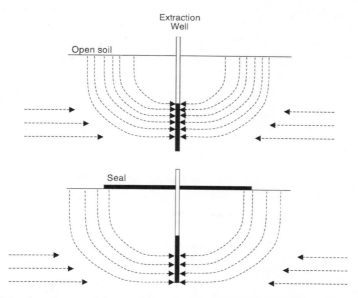

Figure 9.6 A surface seal in an SVE system, promoting greater lateral flow of soil vapors. *Source*: From Johnson et al. 1990.

contaminated zone. When multiple wells are used it is important to consider the effect that each well has on vapor flow to all other wells. For example, if three wells are required at a given site, a stagnant region may form in the central zone between the wells. This problem can be alleviated by insertion of either a passive well or a forced injection well within the zone of influence of the three wells. A passive well is one that is open to the atmosphere. Groundwater monitoring wells can be used for such applications. Forced injection wells are vapor wells into which air is pumped rather than removed. It is important that the locations of forced injection wells be carefully chosen so that vapors are captured by the extraction wells and not forced off-site.

A variation to conventional SVE technology involves the use of horizontal extraction wells. This method has shown great promise in areas where vertical drilling units may not be safe or feasible—for example, in soil directly beneath an active facility. A drill bit is guided at a selected angle into the subsurface, eventually leveling and penetrating the contaminated zone. The drill is subsequently directed to return to the surface on the other side of the structure. Screened PVC tubing is then drawn through the drill holes (figure 9.7). For shallow contamination zones (< 12 ft BGS) vapor extraction trenches combined with surface seals may be more effective than vertical or horizontal wells. Trenches are usually limited to shallow soil zones.

Surface seals are sometimes used to control vapor flow paths. For a shallow treatment zone (< 15 ft or 5 m) the surface seal imparts a significant effect on vapor flow paths. For wells screened below 25 ft (about 8 m), the influence of surface seals becomes less significant (Johnson et al. 1990). Depending on site characteristics, a variety of materials can be used to seal the surface. A geomembrane liner can be rolled over the site and removed when the remediation program is complete. Geomembranes are readily available in a variety of materials, of which high-density polyethylene

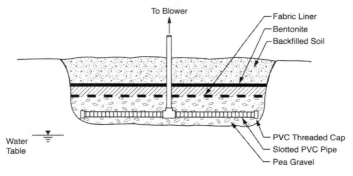

Figure 9.7 Horizontal drilling system for the installation of SVE wells. *Source*: U.S. Environmental Protection Agency.

Figure 9.8 Soil vapor extraction system. The paved surface is effective for gas migration control.

(HDPE) is most common. Alternatives to a synthetic membrane are clay or bentonite, which can be applied in any thickness. Clay liners are not easily removed, and both clay and geomembranes are susceptible to damage from personnel and equipment. A third alternative—the most common at commercial or industrial sites—is the use of a concrete or asphalt cap (figure 9.8). This alternative works well at sites that have been paved or will be paved (e.g., an industrial park or gasoline station).

In most areas an air permit or exemption must be obtained and exhaust air must be treated prior to release to the atmosphere. Common techniques for treating hydrocarbon-contaminated air include:

9.3.1 Condensation

Condensation can separate effluent VOCs from the carrier air and is usually accomplished by refrigeration. The efficiency of this technique is determined in part by the effect of temperature on the vapor pressure of the VOCs present. Condensation is more efficient for high vapor concentrations; the technology becomes less efficient as the cleanup progresses and vapor concentrations decrease. The method may be ineffective during the last stages of cleanup. Because vapors are not completely condensed, additional treatment steps may be required.

9.3.2 Vapor Combustion Units

Many vapors are incinerated with destruction efficiencies typically > 95 percent. A supplemental fuel such as natural gas is added before combustion, unless extraction well vapor concentrations are a few percent by volume. This method becomes less economical as vapor concentrations decrease below 10,000 ppm$_v$. If concentrations of extracted vapors are sufficiently high, they can be burned as a fuel supplement in an internal combustion engine which powers the pump system, thereby providing a free energy source for the system.

9.3.3 Catalytic Oxidation Units

Vapor streams are heated and then passed over a catalyst bed. Catalyzed metal powders, similar to those used in automotive catalytic converter devices, are commonly employed. These units remove hydrocarbons by oxidation to CO_2 and H_2O. Destruction efficiencies are typically > 95 percent. Catalytic oxidation units are used for vapor concentrations > 8,000 ppm$_v$ (Johnson et al. 1990). Catalytic systems are expensive to operate due mostly to high maintenance costs.

9.3.4 Carbon Beds

Carbon can treat almost any vapor stream, but it is economical only for very low emission rates (< 100 g/d). Granular activated charcoal filters clean exhaust air by chemical and physical adsorption of vapors. Such filters are too expensive to serve as the primary hydrocarbon removal method. Weekly maintenance is required to replace saturated filters. Granular activated charcoal filters require approximately 100 lb (45 kg) carbon for each 15 lb (6.8 kg) of hydrocarbon filtered (Cole 1994).

9.3.5 Chemical Oxidation

Collected vapors can be exposed to low concentrations of strong oxidants. A metallic catalyst (e.g., platinum, palladium) can be included that will allow for hydrocarbon oxidation at lower temperatures. Common oxidants include ozone, hydrogen peroxide, and UV radiation. If chlorinated compounds occur in contaminant vapors, there is concern that oxidation reactions will lead to the formation of chlorinated dioxins and/or furans, which may be toxic in the ppt range. It is therefore recommended that treatability studies be conducted prior to implementation of chemical oxidation. Continuous emission monitoring is also recommended.

9.3.6 Biofilters

A filter material such as sphagnum moss is packed into a large chamber and the packing material is inoculated with bacterial cultures and inorganic nutrients. Contaminated vapors are pumped through the column and are adsorbed onto the high surface area of the filter material and/or degraded by microbial action.

9.3.7 Diffuser Stacks

These do not treat vapors per se, but are a low-cost solution for locations in which they are permitted. Diffuser stacks must be designed to minimize health risks and maximize safety.

9.4 TECHNICAL CONSIDERATIONS

Successful removal of hydrocarbons from contaminated soil depends upon a number of site-specific factors; in other words, the controlling mechanisms for vapor and chemical diffusion are unique to each site. The primary factors include contaminant properties, site characteristics, and site management practices.

9.4.1 Contaminant-Specific

The chemical properties of hydrocarbon contaminants significantly influence the manner in which they interact with the soil matrix. As discussed in chapter 3, petroleum products are complex mixtures of hundreds of hydrocarbon compounds. The various components possess unique chemical and physical properties; as a result, some will be more amenable than others to removal by SVE. Properties such as vapor pressure, solubility, concentration, viscosity, and octanol-water partitioning coefficient affect the susceptibility of hydrocarbons to SVE. For example, compounds having high vapor pressure and low solubility in water are more efficiently removed from soil. The Henry's law constant, a measure of the equilibrium distribution between air and solution, is often used as an indicator of volatilization and thus ease of removal. The more volatile compounds such as benzene have a constant of approximately 0.25 or higher; less volatile compounds such as naphthalene have a constant closer to 0.02. By virtue of this property and the higher content of volatile organic compounds in gasoline, proportionally more gasoline will be removed by in situ volatilization than will kerosene or heavy heating oils. Experimental results indicate that gasoline can be almost completely removed within about 100 days from porous soil; however, fuel oil persisted at high levels even after 120 days (EPRI 1988).

Petroleum hydrocarbons chemically weather with time with a resultant partitioning into a number of phases. The most common phases are free product (NAPL), dissolved in water, sorbed to soil particles, and vapor (see chapter 5). The equilibrium distribution of phases in soil depends partly on the extent to which contaminants have aged. For example, fresh leaks in which gasoline is still predominantly in the bulk phase are more readily volatilized than older, aged releases containing higher percentages of heavier hydrocarbons.

The relative biodegradability of the hydrocarbon mixture also affects phase partitioning and therefore amenability to SVE. Hydrocarbons that are relatively biodegradable, for example MTBE, tend to partition into a number of phases including the vapor phase rather rapidly, thus becoming more available for volatilization.

9.4.2 Site-Specific

SVE is limited to the vadose zone. Soil water content influences volatilization rate by affecting the speed at which hydrocarbons diffuse through the vadose zone. Volatilization rates are generally greater in dry soil than wet soil. The SVE method works slowly, or not at all, if contaminants occur primarily in groundwater or if groundwater levels fluctuate over short periods, as for example during seasonal fluctuations.

The rate at which hydrocarbon compounds volatilize and migrate to the surface is a function of both travel distance and cross-sectional area available for flow. Porosity and permeability are important to the performance of the SVE system. Diffusion distance increases and cross-sectional flow area decreases with decreasing porosity. Vapor (and air) flow rates correspond, in general, to saturated hydraulic conductivity (K_s) which might range from a permeable 10^2 cm/sec (gravel) to 10^{-7} cm/sec (impermeable, homogeneous clay).

High soil clay content lowers permeability significantly and therefore inhibits volatilization. In porous soil, extraction of volatiles can be rapid; in heavy clays the method has little utility. The concentration of sorptive surfaces in the mineral and organic fractions of soil also affects volatilization. A high level of adsorption sites on clay surfaces, organic matter and other colloids will immobilize hydrocarbons; they are not available for volatilization until desorbed (EPRI 1988).

Volatilization of hydrocarbon compounds increases with increasing soil temperature. Additionally, increasing wind speed decreases the boundary layer of relatively stagnant air at the ground/air interface. Depending on hydrocarbon characteristics, this will assist volatilization. Water evaporation at the soil surface strongly affects the upward flow of water through the unsaturated zone. Hydrocarbon compounds that are dissolved in water can thus be transported to the surface, enhancing volatilization by evaporation.

Finally, the extent to which the liquid hydrocarbon phase has been smeared by fluctuating groundwater tables will affect partitioning in the subsurface.

9.4.3 Management-Specific

Those management practices that alter soil conditions may influence volatilization. Soil management techniques that decrease leaching, increase soil surface contaminant concentrations, or maximize soil aeration will enhance volatilization. Pumping rate, well positions, and well diameters all influence the degree of vapor recovery from a site.

9.5 CASE HISTORY 1

At the Hanford Site in Washington State, three disposal areas (the 216-Z-9 Trench, 216-Z-1A Tile Field, and 216-Z-18 Crib) were used from 1955 to 1973 for disposal of liquid wastes containing carbon tetrachloride (CCl_4). These liquids originated from plutonium finishing operations (U.S. DOE 2013).

Carbon tetrachloride in liquid waste volatilizes in the vadose zone, thus making SVE a preferred remediation technology. Between February 1992 and October 2012, SVE was implemented in the vicinity of the three disposal sites to recover CCl_4 from the vadose zone. Soil vapor was extracted from more than fifty wells.

SVE was first implemented as an interim action in 1992 (Smith and Stanley 1992). Later, the Record of Decision selected SVE as the final remedial action for carbon tetrachloride contamination in the vadose zone (U.S. EPA 2011).

SVE operations used a vacuum system to extract vapor-phase CCl_4 from either above and/or below a low-permeability caliche layer. Most wells are screened above the underlying water table, and most well screens/casing diameters are between 4 to 8 in (10.2 and 20.3 cm). The well screens are positioned in the relatively permeable sands and gravels above and/or below the lower permeability Cold Creek unit, which is located approximately 125 to 148 ft (38 to 45 m) BGS.

The extracted vapor passes sequentially through a flow meter, an air/water separator, a heat exchanger, and finally GAC canisters. Any vapor-phase CCl_4 that is present is removed by the GAC before the treated air is released to the atmosphere. Air samples are routinely collected for laboratory analysis to ensure that air discharge release limits are met. During SVE operations, GAC canisters typically require change-out every few weeks; several times per year a shipment of spent GAC canisters are sent off-site for regeneration.

Between 1992 and 1997, three SVE systems at 500, 1,000, and 1,500 ft³/min (14.2, 28.3, and 42.5 m³/min) were operated continuously to recover CCl_4 from the vadose zone. During the first years of operations, 165,018 lb (74,851 kg) of CCl_4 were removed. The SVE systems were not operated in 1997, so a rebound study could be conducted to determine the increase in CCl_4 vapor concentrations following temporary shutdown of the systems (figure 9.9).

Carbon tetrachloride concentrations in soil vapor extracted from the 216-Z-9 wells declined from approximately 30,000 ppm$_v$ at startup in 1993 to a maximum of 14 ppm$_v$ in 2012. Carbon tetrachloride concentrations in soil vapor extracted from the 216-Z-1A/216-Z-18 wells declined from approximately 1,500 ppm$_v$ at startup in 1992 to a maximum of 11 ppm$_v$ in 2012.

With continued extraction, the rebound of CCl_4 vapor concentrations gradually declined. In 2001 concentrations in the vadose zone near the 216-Z-9 Trench rebounded from 28 ppm$_v$ in fall 1999 to 224 ppm$_v$ in spring 2001 (an eightfold increase). In contrast, in 2012 the concentrations near the 216-Z-9 Trench rebounded from 8 ppm$_v$ in fall 2011 to 14 ppmv in spring 2012.

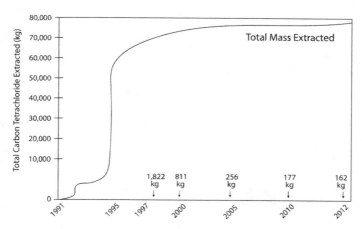

Figure 9.9 Annual carbon tetrachloride mass extracted by soil vapor extraction.
Source: U.S. DOE, 2013.

In 2011, a Record of Decision was issued that identifies 100 ppm$_v$ as the final cleanup level for CCl_4 in soil vapor within the disposal area (U.S. EPA 2011). CCl_4 concentrations in all SVE wells are well below this cleanup level. For this reason, SVE operations were not performed in 2013 to allow CCl_4 concentrations to rebound for the entire year. Carbon tetrachloride concentrations were re-evaluated in spring 2014.

9.6 CASE HISTORY 2

The Fairchild Superfund site, located in southern San Jose, California, houses a former semiconductor manufacturing facility. An underground organic solvent storage tank failed, and a mixture of solvents contaminated local soil and groundwater. An estimated 60,000 gal (227,000 liters) of solvents were released. Solvents detected in local soils included TCA, DCE, IPA, xylenes, acetone, freon-113, and PCE. Trichloroethane was measured at concentrations up to 3,530 mg/kg and xylenes up to 941 mg/kg. The maximum concentration of total solvents in soil (including TCA, 1,1-DCE, IPA, xylenes, acetone, Freon-113, and PCE) was 4,500 mg/kg.

The affected area ranges from flat to gently sloping and is underlain by several hundred feet of unconsolidated alluvial deposits over bedrock. The alluvium formation consists of layers of water-bearing sand and gravel alternating with more dense silt and silty clay layers. Four aquifer systems occur in the alluvium, the shallowest at a depth ranging from 10 to 40 ft BGS, and the next from 50 to 70 ft BGS. The alternating sand and gravel layers range in thickness from several feet to 140 feet in thickness.

During interim remedial cleanup activities the company removed the damaged UST and excavated and disposed 3,400 yd^3 of soil in a permitted hazardous waste facility. A groundwater extraction and treatment system was installed to prevent further migration of contaminants and to collect contaminated groundwater. A bentonite slurry wall was constructed around the site perimeter to contain contaminated groundwater on-site within the shallower aquifers.

SVE was selected as the remedial alternative for soil at the Fairchild site based on treatability study results.

The California Regional Quality Control Board established cleanup goals for the SVE remedial action for both individual vapor extraction wells and the entire SVE system. The board required air extraction from individual wells until the contaminant removal rate decreased to 10 percent or less of the initial removal rate, the contaminant removal rate declined at less than 1 percent per day for ten consecutive days, or SVE system operation achieved a total contaminant removal rate of less than 10 lb/day.

The SVE system installed at the Fairchild site contained thirty-nine extraction wells. Air inlet wells were includ to provide additional air into the zone of contamination. The slurry wall and groundwater extraction system were used to control groundwater flow and to prevent contaminant migration. Groundwater was extracted from recovery wells within the slurry wall enclosures to lower the water elevation inside the wall. These activities also contained soil vapors for the SVE system.

The extraction wells were connected to a vapor extraction and treatment system consisting of vacuum pumps, a dehumidifier, and GAC tanks. Two vacuum pumps with a capacity of approximately 4,500 ft^3 per minute (cfm) at 20-in Hg were used to remove soil vapors. Five GAC adsorption units captured the organic compounds extracted with the soil vapors. The vapors were first directed to two 3,000-lb GAC beds operating in parallel followed by a second set of GAC beds in parallel, and then to a final, single 3,000 lb GAC bed.

Each extraction well was equipped with a submersible pump to remove accumulated groundwater. The pumps in the vapor extraction wells were connected by underground piping to a groundwater treatment system that consisted of air stripping and discharge to surface water.

The SVE system was designed to operate continuously five days per week. At any one time, the system operated a maximum of twenty-five of the thirty-nine extraction wells.

The Fairchild site SVE system removed approximately 16,000 lb of solvents from the soil during 16 months of operation (427 days totaling 9,800 hours of operation). The extraction rate was maximized at 130 lb/day early in the program (figure 9.10). The rate of contaminant extraction increased rapidly during the initial stages of system operation (two months) and then decreased to a more modest rate. The cleanup goal, that is, < 10-lb/day contaminant removal rate, was achieved after eight months. The system was operated for another eight months after the 10 lb/day goal was achieved to remove additional contaminants (i.e., to the point where the soil was considered to no longer leach contaminants to groundwater). Less than 4 lb/day was extracted after sixteen months. Cumulative mass of contaminants removed over time is shown in figure 9.11. A test designed to evaluate potential rebound in the extraction wells revealed that shutting them off for two to six weeks did not cause soil vapor concentrations to increase.

The concentrations of many contaminants in soil borings decreased after seven months of operation (table 9.1); however, concentrations of several contaminants increased during this period, including acetone, TCA, xylenes, IPA, and TCE. The variation in contaminant concentrations in soil may be attributed to variation in contamination across the areas where soil borings were collected.

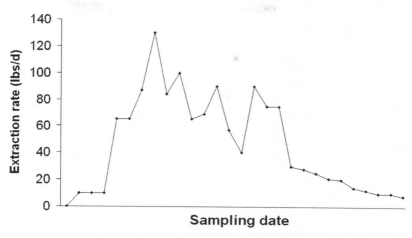

Figure 9.10 Contaminant removal rate from the Fairchild Superfund Site. *Source*: U.S. Environmental Protection Agency 1995.

Capital costs for the SVE program were $2,100,000 (not including costs for construction of the slurry wall or for aquifer pumping), and actual operation and maintenance costs totaled $1,800,000 for sixteen months of operation. This corresponds to $240 per pound of contaminants removed and $93 per yd^3 of soil treated. The actual costs for this project were 7 percent less than the projected costs because the time required for remediation was less than originally estimated (U.S. EPA 1989, 1992, 1995; Canonie 1989, 1993).

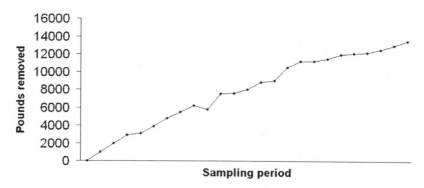

Figure 9.11 Cumulative mass of contaminants removed over time from the Fairchild Superfund Site. *Source*: U.S. Environmental Protection Agency 1995.

Table 9.1 Analysis of soil contaminant concentrations at two intervals during the SVE process

Boring	TCA t_1 mg/kg	TCA t_2	DCE t_1 mg/kg	DCE t_2	Xylenes t_1 mg/kg	Xylenes t_2	Acetone t_1 mg/kg	Acetone t_2	IPA t_1 mg/kg	IPA t_2
1	3,530	416	16.6	2.2	941	462	18	281	nd	134
2	40.6	79	3.4	2.5	19.2	156	nd	1.5	nd	0.9
3	266	37.3	12.5	1.5	189	85.6	7.7	3.5	0.02	1.8
4	12.2	7.8	1.6	0.3	4.8	5.5	7.6	1.9	nd	nd
5	6.4	5.5	0.5	1.5	nd	1.2	nd	2.9	nd	0.4
6	1.1	0.1	.05	.01	nd	nd	nd	nd	nd	nd

t_1 = before remediation; t_2 = after eighteen months; nd = not detected.
Source: U.S. EPA 1995.

QUESTIONS

1. How does soil temperature influence success in SVE?
2. Soil pH will affect SVE success. True or false? Explain.
3. Explain the effect of soil texture on establishing the radius of influence.
4. Explain why a cap is useful to control flow of vapors during SVE.
5. How does air sparging differ from SVE?
6. Explain how hydrocarbon vapors can be treated once they are extracted from the subsurface.

REFERENCES

Byrnes, M., V. Rohay, S. Simmons, J. Morse, E. Laija, and G. Sinton. 2014. *Success with Soil Vapor Extraction, 200-PW-1 Operable Unit, Hanford Site, Richland, Washington – 14017.* WM2014 Conference, March 2–6, 2014, Phoenix, AZ.

Canonie Environmental. 1989. *Interim Design Report, In-Situ Soil Aeration System.* Project 82-012, March.

———. 1993. *Five-Year Status Report and Effectiveness Evaluation.* December.

Cole, G. M. 1994. *Assessment and Remediation of Petroleum Contaminated Sites.* Boca Raton, FL: CRC Press.

Conger, R. M., and K. Trichel. 1994. *Drilling on Your Side.* As cited in Underground Tank Technology Update. Madison, WI: University of Wisconsin-Madison, College of Engineering 8: 5–7.

Diks, A., and R. Ottengrat. 1991. *In Situ and On-Site Bioreclamation, The Second International Symposium,* April 5–8, 1993, San Diego.

Electric Power Research Institute. 1988. *Remedial Technologies for Leaking Underground Storage Tanks.* Chelsea, MI: Lewis.

Friesen, K. A. 1990. *Chemical Changes of Biodegraded Gasoline in a Laboratory Simulation of the Vadose Zone.* MS thesis, Colorado School of Mines, Golden, CO.

Johnson, P. C., C. C. Stanley, M. W. Kemblowski, D. L. Byers, and J. D. Colthart. 1990. A practical approach to the design, operation, and monitoring of in situ soil-venting systems. *Ground Water Monitoring Review* 10 (2): 159–78.

Jury, W. A. 1986. *Volatilization from Soil, and S. C. Hern and S. M. Melancon, Guidelines for Field Testing Soil Fate and Transport Models.* Final Report, Appendix B. EPA 600/4-86-020. Las Vegas, NV: Environmental Monitoring Systems Laboratory.

Noonan, D. C., and T. J. Curtis. 1990. *Groundwater Remediation and Petroleum, A Guide for Underground Storage Tanks.* Chelsea, MI: Lewis.

Smith, R. F., and R. Stanley. 1992. *Action Memorandum: Expedited Response Action Proposal for 200 West Area Carbon Tetrachloride Plume. Letter No. 9200423 (to R. D. Izatt), U.S. Department of Energy, Richland Operations Office, Washington State.* Lacey, WA: Department of Ecology.

Turkall, R. M., G. A. Skowronski, and M. S. Abdel-Rahman. 1992. The effect of soil type of absorption of toluene and its bioavailability. In: *Petroleum Contaminated Soils*, vol. 3, edited by P. T. Kostecki and E. J. Calabrese. Chelsea, MI: Lewis.

Turner, D.A., J. Pichtel, Y. Rodenas, J. McKillip, and J. V. Goodpaster. 2013. Microbial degradation of ignitable liquids in soil: Comparison by soil type. *Journal of Bioremediation and Biodegradation* 5 (2): 216–21.

U.S. Department of Energy. 2012. *DOE/RL-2013-18, 2013, Hanford Site Environmental Report for CY 2012*, Rev. 0. Richland, WA: Richland Operations Office.

———. 2013. *Hanford Site Groundwater Monitoring*. 200-ZP. DOE/RL-2013-22, Rev. 0. August 2013. www.hanford.gov/c.cfm/sgrp/GWRep12/html/gw12_14_20 0ZP.pdf.

U.S. EPA (Environmental Protection Agency). 1989. *Superfund Record of Decision*. S. San Jose, California: Fairchild Semiconductor, March.

———. 1991. *Guide for Conducting Treatability Studies under CERCLA: Soil Vapor Extraction. Interim Guidance*. EPA/540/2-91-019A. Washington, DC: Office of Emergency and Remedial Response.

———. 1992. *Superfund Interim Site Close Out Report*. San Jose, California: Fairchild. U.S. EPA Region IX, March 25.

———. 1995. *Soil Vapor Extraction at the Fairchild Semiconductor Corporation Superfund Site, San Jose, California Cost and Performance Report*. Washington, DC: Office of Solid Waste and Emergency Response, Technology Innovation Office.

———. 1996. *Engineering Forum Issue Paper: Soil Vapor Extraction Implementation Experiences*. Publication 9200.5-223FS EPA 540/F-95/030 PB95-963315. Washington, DC: Office of Solid Waste and Emergency Response.

———. 2011. *Record of Decision, Hanford 200 Area Superfund Site 200-CW-5 and 200-PW-1, 200-PW-3, and 200-PW-6 Operable Units*. U.S. Department of Energy, Richland Operations Office, Richland, WA; and U.S. Environmental Protection Agency, Olympia, WA.

Chapter 10

Permeable Reactive Barriers

The meeting of two personalities is like the contact of two chemical substances; if there is any reaction, both are transformed.

—C.G. Jung

10.1 INTRODUCTION

Recovery of contaminated groundwater followed by chemical and/or biological treatment (i.e., 'pump-and-treat' system) has found widespread application and become a standard technology for cleanup of contaminated groundwater (see chapter 6). Several drawbacks are noteworthy to this approach, however. A pump-and-treat system requires an external energy source that becomes costly when operated over long periods. Also, because much of the water initially extracted is uncontaminated, pump-and-treat may waste groundwater resources.

Even when properly operated, pump-and-treat systems possess other inherent limitations (Voudrias 2001; Keely 1989): they may not work well with complex geologic materials or heterogeneous aquifers; they often cease to reduce contamination long before reaching intended cleanup levels; and, in some situations, they make sites more difficult to remediate by smearing contamination across the subsurface. To address such concerns, the installation of PRBs downgradient of a contaminant plume may permit low-cost, long-term treatment (figure 10.1). These barriers allow the passage of groundwater while promoting degradation or removal of contaminants by specific

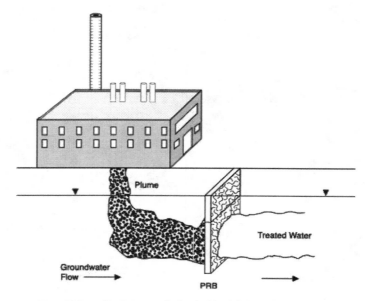

Figure 10.1   How PRB technology works in an ideal field situation.

chemical and physical mechanisms, most commonly reduction, precipitation, and sorption. PRBs have been used to successfully treat or remove metallic, radioactive, nonmetallic (e.g., nitrates), and hydrocarbon contaminants from groundwater.

10.2 INSTALLATION AND CONFIGURATION OF PRBS

The PRB is placed across the flow path of a contaminant plume (figure 10.2). The reactive media is installed in order to be in intimate contact with the surrounding aquifer material. The reactive treatment zone either decomposes the contaminant or restricts its movement via employing reactants and agents including zerovalent iron (ZVI) or other reduced metals, zeolites, humic materials, chelating agents, sorbents, active microbial cells, and others. Several variations of PRB configurations, including a *funnel-and-gate* system (figure 10.2), are available and combine a permeable gate and impermeable funnel sections to capture wide contaminant plumes. The reactive cell (gate) portion is supplied with the necessary treatment media.

Continuous reactive barriers are useful when the plume is not wide and/or contaminant concentrations are low. Vertical sections of pea

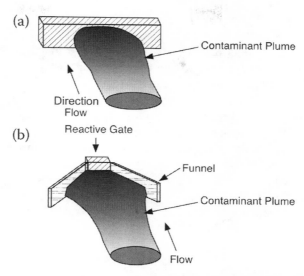

Figure 10.2 Two PRB configurations: (a) continuous reactive barrier and (b) funnel-and-gate system.

gravel can be installed upgradient of the reactive zone to promote uniform flow of affected groundwater through the reactive zone. Some PRBs are installed as permanent or long-term units across the flow path of the contaminant plume. Others are placed as in situ reactors that are readily accessible for removal and replacement of reactive media. The wall may be used to provide permanent containment for relatively nonhazardous residues.

Various construction methods are used to install PRBs, such as trenching, caisson deployment, clamshell digging, soil mixing, vertical hydraulic fracturing, or high-pressure injection.

Design of the PRB system begins with collecting contaminated groundwater from the site and performing a bench-scale treatability study to determine flow rates, reaction intermediates, and products. The required residence time of the contaminants and reactive cell dimensions are determined from the results of the treatability study, along with aquifer properties (flow velocity, etc.) and computer modeling.

A thorough site characterization is important to ensure proper design and installation of a reactive barrier. Characterization should include an evaluation of surface features, structures, and buried utilities to ascertain whether the site is suitable for PRB installation. If the site is amenable, the characterization will help determine what types of PRB emplacement technologies are feasible.

The aerial and vertical extent of the plume, contaminant chemical types and concentrations, and flow direction and velocity of groundwater must be accurately determined to achieve the required treatment levels. Detailed information on subsurface permeability, fracturing, and aqueous geochemistry is needed for PRB design. The plume must not pass over, under, or around the PRB, and the reactive zone must adequately treat the contaminant without rapidly clogging with precipitates or becoming passivated. (*Passivation* is defined as the changing of the chemically active surface of the PRB media to a less reactive state. This phenomenon is discussed further in section 10.6.)

The key components of site characterization that should be evaluated before implementing a PRB include:

- contaminant types and concentrations
- site hydrogeology
- site geochemistry
- site microbiological properties

10.3 TREATMENT MECHANISMS WITHIN PRBS FOR SPECIFIC CONTAMINANTS

Many reactive media or their combinations have been employed for use in PRBs, and numerous other media are under investigation. Iron metal, designated as Fe° or zero-valent iron (ZVI), is the most common reactive media in the majority of field-scale and commercial PRB projects. ZVI creates a low oxidation potential in groundwater, thus causing reduction of oxidized inorganic species and precipitation of low-solubility minerals. It can also destroy certain halogenated hydrocarbons via reductive dehalogenation (Morrison et al. 2002a). Scrap iron is an inexpensive material that is readily obtained in granular form in the large quantities necessary in PRBs. Contaminants that have been successfully treated by ZVI and other media are listed in table 10.1.

A range of biological and chemical mechanisms have been employed in PRBs to remove hydrocarbon and inorganic contaminants from groundwater, including abiotic reduction, chemical precipitation, sorption, ion exchange, and biotic reduction (Thiruvenkatachari et al. 2008; Morrison et al. 2002a). Their mechanisms of action on various groundwater contaminants are described below.

Table 10.1 Contaminants that can be treated by PRBs

Organic Compounds		Inorganic Compounds	
Methanes	Tetrachloromethane	Trace metals	Antimony
	Trichloromethane		Chromium
	Dichloromethane		Nickel
Ethanes	Hexachloroethane		Lead
	1,1,1-trichloroethane		Uranium
	1,1,2,-trichloromethane		Technetium
	1,1-dichloroethane		Iron
Ethenes	Tetrachloroethane		Manganese
	Trichloroethene		Selenium
	cis-1,2-dichloroethene		Copper
	trans-1,2-dichloroethene		Cobalt
	1,1-dichloroethene		Cadmium
	vinyl chloride		Zinc
Propanes	1,2,3-trichloropropane	Anion Contaminants	Sulfate
	1,2-dichloropropane		Nitrate
Aromatics	Benzene		Phosphate
	Toluene		Arsenic
	Ethylbenzene		
Other	Hexachlorobutadiene		
	1,2-dibromoethane		
	Freon 113		
	N-nitrosodimethylamine		

Source: IRTC 2011; Thiruvenkatachari et al. 2008; Naftz et al. 2002; U.S. EPA 1998.

10.4 CONTAMINANTS SUITABLE TO PRB TREATMENT

10.4.1 Reductive Treatment of Halogenated Organic Compounds

A substantial body of research has been conducted on the degradation of chlorinated solvents, for example chloroform (CCl_4), TCE and PCE, by reaction with surfaces of Fe° particles. The suggested degradation mechanism is abiotic reductive dehalogenation with consequent oxidation of the Fe° by the chlorinated hydrocarbon.

A combination of anodic and cathodic reactions occurring at the Fe° particle surface (eqs. 10.1 and 10.2, respectively) results in the net reductive reaction (eq. 10.3). Dechlorinated hydrocarbon products are generated as a result of the combined reactions (U.S. EPA 1998):

$$\text{Anode} \quad Fe^\circ \rightarrow Fe^{2+} + 2e^- \quad (10.1)$$

Cathode $\quad RCl + 2e^- + H^+ \rightarrow RH + Cl^-$ (10.2)

Net $\quad Fe^\circ + RCl + H^+ \rightarrow Fe^{2+} + RH + Cl^-$ (10.3)

Sequential hydrogenolysis is considered the primary degradative pathway during the reductive dechlorination of chlorinated methanes (e.g., CCl_4). The majority of the carbon tetrachloride is converted to chloroform, which is eventually converted to methylene chloride (CH_2Cl_2) (eq. 10.4) (U.S. EPA 1998; Roberts et al. 1996). No further reaction of methylene chloride has been detected in unamended ZVI (U.S. EPA 1998).

$$CCl_4 \rightarrow CHCl_3 + Cl^- \rightarrow CH_2Cl_2 + 2Cl^- \quad (10.4)$$

Accurate mass balances have been determined for the transformation of several chlorinated ethenes (Orth and Gillham 1996). Using Fe°, TCE is reduced primarily to ethene and ethane via two interconnected degradation pathways, that is, hydrogenolysis (A) and reductive β-elimination (B) (eq. 10.5) (Tobiszewski and Namieśnik 2012).

(10.5)

The intermediate products, *cis*-dichloroethene (cDCE) and vinyl chloride, are produced in the sequential hydrogenolysis pathway and are slower to degrade than is TCE. In contrast, the chloroacetylene produced via the β-elimination pathway is a short-lived intermediate and is rapidly reduced to ethene. The β-elimination pathway accounts for the rapid conversion of TCE to ethene and ethane, with relatively minor formation of intermediate products (Tobiszewski and Namieśnik 2012; U.S. EPA 1998).

Zero-valent iron is a mild reductant, and dehalogenation rates vary for different chlorinated solvents. Published data for degradation rates of halogenated hydrocarbons indicate that the primary variables responsible for affecting degradation rate are the surface area of iron per unit volume of pore water, particle size, and metal storage and pretreatment conditions (Tobiszewski and Namieśnik 2012; Cheng and Wu 2000; U.S. EPA 1998; Johnson et al. 1996).

10.4.1.1 Other Considerations with Reductive Dechlorination in PRBs

Incomplete dechlorination of a highly chlorinated hydrocarbon, for example tetrachloroethene, could result in the production of undesirable intermediates such as vinyl chloride, which is more hazardous and more persistent than the parent compound. Generation of even very low concentrations of undesirable intermediates in the PRB effluent must therefore be prevented. Comprehensive knowledge of contaminant types as well as subsurface conditions must be determined in advance. Bench-scale testing of the PRB reactive media with the contaminants is essential prior to field application.

Typically, PRBs are designed to provide adequate residence time in the treatment zone for degradation of the parent hydrocarbon compound and all toxic intermediates generated. At sites where groundwater is contaminated with a mixture of chlorinated hydrocarbons, PRB design is usually determined by the least reactive constituent. Highly halogenated hydrocarbons tend to be reduced more rapidly than the less halogenated congeners; likewise, dechlorination is more rapid on saturated carbons (e.g., CCl_4) than on unsaturated carbons (e.g., TCE or vinyl chloride) (U.S. EPA 1998).

Under aerobic conditions, dissolved oxygen is an effective terminal electron acceptor. Chlorinated hydrocarbons such as PCE and CCl_4, however, have oxidizing potentials very similar to that of O_2. Should the affected groundwater have a relatively high Eh, oxygen could potentially compete with the chlorinated hydrocarbons as the oxidant (U.S. EPA 1998; Archer and Harter 1978; see equation 10.6).

$$2Fe^\circ + O_2 + 2H_2O \rightarrow 2Fe^{2+} + 4OH^- \qquad (10.6)$$

10.4.2 Reductive Treatment of Inorganic Anions

Anions and oxyanions of chromium, arsenic, selenium, technetium, and antimony are significant groundwater contaminants. Such anionic species are typically repulsed by negatively charged colloidal surfaces that tend to predominate in soil and aquifers under neutral pH conditions (see chapter 4). The resultant high solubility of these anions results in a potential hazard to water supplies. PRB technology is often appropriate for their treatment and removal.

10.4.2.1 Chromate

As discussed in chapter 2, chromium typically occurs in soil, water, and sediments in two oxidation states, Cr(III) and Cr(VI). Trivalent Cr is relatively nontoxic and also serves as a micronutrient. It forms hydroxide precipitates under certain conditions and is also readily adsorbed by various minerals. Hexavalent Cr, on the other hand, is quite toxic and forms relatively soluble oxyanions, resulting in its mobility in contaminated aquifers (Marsh et al. 2000).

Under typical groundwater pH and Eh conditions, Cr(VI) usually speciates as chromate, CrO_4^{2-}. Reduction of chromate to trivalent Cr species, followed by precipitation of insoluble Cr(III) hydroxide precipitates, has been studied extensively in PRBs. Several solid phases containing reduced iron including Fe° promote the reduction and precipitation of Cr(VI) (Stanin and Pirnie 2004; U.S. EPA 1998). Iron-bearing oxyhydroxides and iron-bearing aluminosilicate minerals have also been successful in Cr(VI) reduction (Tang 2003). Rates of Cr(VI) reduction are dependent on the nature of the Fe° (i.e., how it was manufactured, levels of impurities) and whether certain aluminosilicate-containing aquifer materials are present and in contact with the iron (ITRC 2011; Powell et al. 1995a; 1995b; Powell and Puls 1997). The overall reactions for Cr(VI) reduction by Fe° and the subsequent precipitation of Cr(III) and Fe(III) oxyhydroxides are shown in equations 10.7 and 10.8.

$$CrO_4^{2-} + Fe^\circ + 8H^+ \rightarrow Fe^{3+} + Cr^{3+} + 4H_2O \tag{10.7}$$

$$(1-x)Fe^{3+} + (x)Cr^{3+} + 2H_2O \rightarrow Fe_{(1-x)}Cr_xOOH_{(s)} + 3H^+ \tag{10.8}$$

Cr(III) and Fe(III) hydroxide phases are known to precipitate on the surface of reacted Fe, indicating coprecipitation and the likelihood of formation of a solid phase having the general formula $(Cr_x, Fe_{1-x})(OH)_3$ (Powell et al. 1994; Powell et al. 1995a). Other studies indicate that the dominant mineral product generated is a mixed Fe-Cr oxyhydroxide phase with the mineral structure of goethite (FeOOH) as shown in equation 10.8.

Reduction and precipitation of soluble metals within a plume can also be accomplished by addition of dissolved dithionite, $S_2O_4^{2-}$, into groundwater.

The dithionite ion reduces solid phase ferric iron (eq. 10.9) (Morrison et al. 2002a; Ammonette et al. 1994; Scott et al. 1998). Dithionite oxidizes to sulfite, $S_2O_3^{2-}$ as Fe^{3+} is reduced to Fe^{2+}. As was the case with Fe°, ferrous iron (Fe^{2+}) will reduce Cr(VI) to Cr(III) which precipitates as a solid solution of Cr(III) and Fe(III) hydroxide (eq. 10.10) (Morrison et al. 2002a).

$$S_2O_4^{2-} + 2Fe^{3+}_{(s)} + 2H_2O \rightarrow 2SO_3^{2-} + 2Fe^{2+}_{(s)} + 4H^+ \qquad (10.9)$$

$$CrO_4^{2-} + 3Fe^{2+}_{(s)} + 5H^+ \rightarrow Cr(OH)_{3(s)} + 3Fe^{3+}_{(s)} + H_2O \qquad (10.10)$$

10.4.2.2 Arsenate

Arsenic commonly occurs as a dissolved species in two oxidation states, As(V) and As(III), and less commonly in other oxidation states including As(0), As(-I), and As(-II) (see chapter 2). The As(V) species forms the mineral H_3AsO_4; As(III) forms H_3AsO_3. Arsenic reduction to the As° species followed by precipitation or incorporation into a secondary arsenic sulfide has been proposed as a treatment technique for soluble arsenic in groundwater (McRae et al. 1997). In batch tests conducted using Fe°, McRae and others (1997) measured rapid reduction of As(V), from concentrations of 1000 μg/L to < 3 μg/L. Similar experiments using As(III) and mixtures of As(III) and As(V) indicated equally rapid removal rates.

10.4.3 Reductive Treatment of Heavy Metals

Biotic treatment in PRBs has been developed for the reductive precipitation of heavy metals. A common approach is to convert soluble metals to insoluble metal sulfides. Biologically-mediated sulfate reduction in natural and constructed wetlands has been used for decades to remove metal cations generated from acid mine drainage. This mechanism has also been applied to treatment in PRBs.

Biotic reduction is carried out by supplying an electron donor and nutrients to support microorganisms present in the PRB. Possible electron donors include leaf mulch, sawdust, wheat straw, and other plant wastes (ITRC 2011; Shen and Wilson 2006). MSW and MSW compost, and composted wastewater solids are rich microbial nutrient sources (Naftz et al. 2002; Benner et al. 1997). Dissolved sulfate serves as an electron acceptor. Biologically mediated reduction of sulfate to sulfide, with the concurrent formation of metal sulfides, occurs via equations 10.11 and 10.12.

$$2CH_2O_x + SO_4^{2-} + 2H^+ \rightarrow 2CO_2 + 2H_2O + H_2S^- \quad (10.11)$$

$$Me^{2+} + H_2S^- \rightarrow MeS_{(s)} + H^+ \quad (10.12)$$

where CH_2O represents an organic compound and Me^{2+} represents a divalent soluble metal cation.

Reducing conditions generated within the PRB result in precipitation of metals and other redox-sensitive inorganic contaminants. Groundwater conditions (e.g., temperature) and the chemical milieu (e.g., pH, nutrient status, salinity) will affect the populations and activities of microorganisms and the consequent rate of sulfate reduction.

10.4.4 Chemical Precipitation of Metals

Groundwater contaminated with strong acids will contain significant concentrations of dissolved metals. For example, acid mine drainage (AMD) may achieve pH values as low as 2.0 and often contains high concentrations of soluble Fe, Al, and Mn.

Limestone, $CaCO_3$, and apatite have been used successfully in PRBs to precipitate soluble metals in groundwater (Golab et al. 2006; Skinner and Schutte 2006; Amos and Younger 2003). Limestone PRBs have been used extensively for treating AMD by both neutralizing acidity and precipitating dissolved metals, especially Fe.

Limestone dissolves in the acidic water, increases the alkalinity, and raises solution pH (equation 10.13). Metal contaminants precipitate as hydroxides (equation 10.14) or carbonates (equation 10.15) if pH has increased sufficiently.

$$CaCO_3 + H^+ \rightarrow HCO_3^- + Ca^{2+} \quad (10.13)$$

$$Me^{2+} + 2(OH)^- \rightarrow Me(OH)_2 \quad (10.14)$$

$$Me^{2+} + HCO_3^- \rightarrow MeCO_{3(s)} + H^+ \quad (10.15)$$

where Me^{2+} = metallic cation.

Phosphate from apatite stabilizes soluble Pb by forming hydroxypyromorphite—$Pb_{10}(PO_4)_6(OH)_2$—a low-solubility mineral (Ren et al. 2018). Apatite also retains Cd and Zn (Sneddon et al. 2006; et al. 2004), but not as effectively as it does Pb (Morrison et al. 2002a; Chen et al. 1997).

10.4.4.1 Uranium

In carbonate-dominated groundwater having pH values around neutral, reductive precipitation of uranium proceeds as shown in equation 10.16 (Morrison et al. 2002a).

$$Fe^\circ + UO_2(CO_3)_2^{2-} + 2H+ \rightarrow UO_{2(s)} + 2HCO_3^- + Fe^{2+} \qquad (10.16)$$

U^{6+} is additionally reduced by ferrous iron (Morrison et al. 2002a).

Apatite has been used in PRBs to promote chemical precipitation and to accumulate uranium (Fuller et al. 2002). Commercial sources of apatite include mined phosphate rock deposits and bone material. Uranium precipitates occurring on the surface of phosphate particles were identified as meta-autunite—$Ca(UO_2)_2(PO_4)_2 \cdot 6.5H_2O]_{14}$—which suggests mineral precipitation as the mechanism of U immobilization on apatite (Morrison et al. 2002a).

10.4.5 Adsorptive Treatment of Metals

Inorganics that are not susceptible to reductive or precipitation processes must be removed from solution by other means. These contaminants may be sorbed onto mineral surfaces. Most adsorption reactions are reversible and occur at relatively rapid rates. Some reactions are selective, with attachment occurring at specific sites. Other adsorption reactions are less specific and therefore ions compete for all attachment sites (Morrison et al. 2002a).

Several adsorbents are useful in PRBs to remove inorganic contaminants from groundwater. Amorphous ferric oxides have a high affinity for adsorption of U and metal contaminants by virtue of a high surface area per unit mass (Duff et al. 2002; Waite et al. 1994; His and Langmuirm 1985). Simon and others (2003) found that hydroxyapatite could sorb more than 2,900 mg/kg uranium. Zeolites have also been used in PRBs to treat inorganic contaminants. Zeolites are a group of hydrated sodium aluminosilicates, either natural or synthetic, having significant ion exchange properties (figure 10.3).

The use of sphagnum peat, limestone, and hydrated lime to remove U, As, Mo, and Se in laboratory tests was described by Conca et al. (2002), Thomson et al. (1991), Longmire et al. (1991), and others. Several researchers have proposed the use of industrial by-products as reactants to remove U, As, and Mo via precipitation or sorption processes in PRBs (Chung et al. 2007; Morrison et al. 2002a). Lee and others (2004), using waste greensands from the iron foundry industry, measured high removal capacities for Zn. The mechanism is attributed to the action of clay, organic carbon, and iron particles. Furthermore, high pH values in

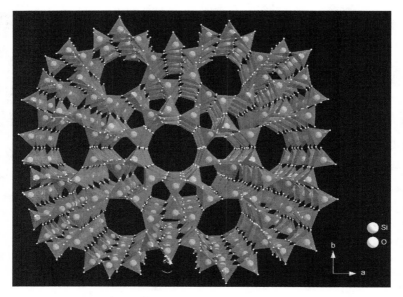

Figure 10.3 Structure of a zeolite.

the presence of clay and iron particles apparently enhance sorption and precipitation of Zn.

10.5 DESIRED CHARACTERISTICS OF PRB REACTIVE MEDIA

For optimal long-term treatment with minimal effects on the local environment, the media employed in PRBs must be compatible with local soils, strata, and groundwater. In other words, it should cause no adverse chemical reactions nor generate any undesirable by-products when reacting with constituents in the contaminant plume. Likewise, the media itself must not become a source of contamination. It cannot pose a hazard to site workers. The PRB media must, therefore, be well understood and characterized.

To control treatment costs using PRBs, the reactive material should be readily available at moderate cost and persist for long periods (i.e., it should not be highly soluble nor become depleted due to excessive reactivity). The PRB media must not restrict groundwater flow. This requires that particle size not be excessively small; similarly, it should not contain a wide range of particle sizes that might result in blocked intergranular spaces. Ideally, then, the material is of uniform grain size.

10.6 A NOTE ABOUT Fe⁰ LONGEVITY

The lifespan of effective ZVI-based PRBs is not clearly established. Cost effectiveness is directly linked to longevity of the PRB reactant media.

Longevity of ZVI is potentially reduced by three phenomena (Morrison et al. 2002a): (1) dissolution of the iron; (2) mineral precipitation leading to significant reduction in permeability; and (3) passivation of the ZVI (i.e., the changing of the chemically active surface of the iron to a much less reactive state) resulting from alteration of Fe⁰ particle surfaces.

ZVI dissolves in the presence of oxygenated groundwater (eq. 10.17). When sufficient oxygen is present the Fe^{2+} generated in equation 10.17 further oxidizes to Fe^{3+} (equation 10.19) and can precipitate as ferric hydroxide or (oxy)hydroxides (equation 10.20).

$$2Fe^{\circ} + 4H^{+} + O_2 \rightarrow 2H_2O + Fe^{2+} \tag{10.17}$$

$$2Fe^{\circ} + O_2 + 2H_2O \rightarrow 2Fe^{2+} + 4OH^{-} \tag{10.18}$$

$$4Fe^{2+} + 4H^{+} + O_2 \rightarrow 4Fe^{3+} + 2H_2O \tag{10.19}$$

$$Fe^{3+} + 3OH^{-} \rightarrow Fe(OH)_{3(s)} \tag{10.20}$$

Iron oxidation can generate large quantities of iron oxides and (oxy)hydroxide precipitates, resulting in significant chemical and physical impacts within the reactive system. Rapid consumption of dissolved O_2 at the entrance to an Fe barrier results in precipitate formation that might restrict system hydraulic conductivity upgradient of the reactive media (IRTC 2011). This mineralized zone resulted in low permeability in laboratory column experiments by Mackenzie and others (1995), who suggested that placing a high-porosity zone of ZVI mixed with gravel upgradient from the PRB can extend the life of the reactive barrier. Upgradient gravel zones are commonly used in PRBs for this purpose (Morrison et al. 2002b; Naftz et al. 2002). Sulfide minerals may also precipitate in ZVI-based PRBs (see equations 10.11–10.12) and further restrict permeability.

ZVI will also experience corrosion in the absence of O_2. Anaerobic corrosion of Fe by water (equation 10.21) is a slow process, however. Reaction 10.21 may increase solution pH, yielding ferrous (oxy)hydroxides (equation 10.22). Aqueous corrosion of iron is mediated by the layer of oxides, hydroxides, and oxyhydroxides present at the iron-water interface. The formation of these precipitates might coat iron surfaces and limit their reduction-oxidation capability.

$$Fe^{\circ} + 2H_2O \rightarrow Fe^{2+} + H_2 + 2OH^- \qquad (10.21)$$

$$Fe^{2+} + 2OH^- \rightarrow Fe(OH)_{2(s)} \qquad (10.22)$$

Little is known about the phenomena of surface passivation. Minerals tend to precipitate on the ZVI surface, leading to surface deposit formation that is sufficiently thick to prevent electron transfer (ITRC 2011).

10.7 SUMMARY OF PRB TECHNOLOGY

As with any remedial technology, it is essential to fully understand all variables that will influence successful implementation and remediation. Comprehensive knowledge of local hydrogeology and plume boundaries is required prior to PRB installation. The rate of groundwater flow through the reactive zone of the PRB must be determined. This is necessary to establish the groundwater/contaminant residence time per unit thickness of reactive media which, when combined with the contaminant transformation rate as it passes through the media, determines the required thickness of reactive media (U.S. EPA 1998).

The stability, mobility, and toxicity of the transformation products passing through the PRB must also be assessed. If these products are regulated compounds, they must not exit the reactive zone of the PRB without themselves being immobilized or transformed to innocuous compounds.

10.8 ADVANTAGES OF PRB TECHNOLOGY

PRB technology has shown much promise in field treatment of contaminated aquifers. Some significant advantages include (NFESC n.d.):

- passive in situ detoxification of groundwater
- no external energy source
- potential to treat contaminants to very low levels
- the land surface can be maintained in its original state and function
- long-term unattended operation

- more cost-effective than pump-and-treat systems

10.9 DISADVANTAGES OF PRB TECHNOLOGY

The technology has some drawbacks worth noting, however:

- unknown long-term effects from chemical and/or biological precipitate formation;
- construction complications from subsurface utilities and/or aboveground structures;
- limited to depths of less than 50 ft using current construction technologies.

QUESTIONS

1. Explain the process of decomposition of chlorinated hydrocarbons (e.g., PCE) within a PRB. Are certain reactions promoted by aerobic versus anaerobic conditions? Explain.
2. How can PRBs be constructed or maintained so that occlusion (clogging) on the upgradient side is limited?
3. Explain passivation. How can this process be overcome or at least limited in a ZVI-reactive barrier?
4. What industrial wastes and by-products might be suitable for use in a PRB? Check the scientific literature and explain how such materials may chemically reduce chlorinated hydrocarbons and/or certain oxidized metals.

REFERENCES

Ammonette, J. E., J. E. Szecsody, H. T. Schaef, J. C. Templeton, Y. A. Gorby, and J. S. Fruchter. 1994. Abiotic reduction of aquifer materials by dithionite: A promising in-situ remediation technology. In: *Thirty-Third Hanford Symposium on Health and the Environment, In-Situ Remediation: Scientific Basis for Current and Future Technologies*, edited by G. W. Gee and N. R. Wing, 851–81. Columbus, OH: Battelle.

Amos, P. W., and P. L. Younger. 2003. Substrate characterization for a subsurface reactive barrier to treat colliery spoil leachate. *Water Research* 37(1): 108–20.

Archer, W. L., and M. K. Harter. 1978. Reactivity of carbon tetrachloride with a series of metals. *Corrosion-National Association of Corrosion Engineers (NACE)* 34(5): 159–62.

Benner, S. G., D. W. Blowes, and C. J. Ptacek. 1997. A full-scale porous reactive wall for prevention of acid mine drainage. *Ground Water Monitoring and Remediation* 17(4): 99–107.

Blowes, D. W., and C. J. Ptacek. 1992. *Geochemical Remediation of Groundwater by Permeable Reactive Walls: Removal of Chromate by Reaction with Iron-Bearing Solids*. Dallas: Subsurface Restoration Conference, U.S. EPA, Kerr Laboratory.

Blowes, D. W., C. J. Ptacek, C. J. Hanton-Fong, and J. L. Jambor. 1995. *In Situ Remediation of Chromium Contaminated Groundwater Using Zero-valent Iron*. Paper Presented at the American Chemical Society, 209th National Meeting, Anaheim, California, April 2–7.

Blowes, D. W., R. W. Puls, T. A. Bennett, R. W. Gillham, C. J. Hanton-Fong, and C. J. Ptacek. 1997. *In-situ Porous Reactive Wall for Treatment of Cr(VI) and Trichloroethylene in Groundwater*. Paper presented at the International Containment Technology Conference and Exhibition, St. Petersburg, FL, February 9–12.

Bostick, W. D., R. J. Stevenson, R. J. Jarabek, and J. L. Conca. 1999. Use of apatite and bone char for the removal of soluble radionuclides in authentic and simulated DOE groundwater. *Advances in Environmental Research* 3: 488–98.

Bowers, A. R., C. A. Ortiz, and R. J. Cardozo. 1986. Iron process for treatment of Cr(VI) wastewaters. *Metal Finishing* 84: 37.

Chen, X., J. V. Wright, J. L. Conca, and L. M. Peurrung. 1997. Evaluation of heavy metal remediation using mineral apatite. *Water, Air, and Soil Pollution* 98(1–2): 57–78.

Cheng, S.-F., and S.-C. Wu. 2000. The enhancement methods for the degradation of TCE by zero-valent metals. *Chemosphere* 41(8): 1263–70.

Chung, H. I., S. K. Kim, Y. S. Lee, and J. Yu. 2007. Permeable reactive barrier using atomized slag material for treatment of contaminants from landfills. *Geosciences Journal* 11(2): 137–45.

Conca, J., E. Strietelmeier, N. Lu, S. D. Ware, T. P. Taylor, and J. Wright. 2002. Treatability study of reactive materials to remediate ground water contaminated with radionuclides, metals and nitrates in a four-component permeable reactive barrier. In: *Groundwater Remediation of Trace Metals, Radionuclides, and Nutrients, with Permeable Reactive Barriers*, edited by S. Naftz, J. A. Morrison, and C. C. Fuller, 221–52. San Diego, CA: Academic Press.

Duff, M. C., J. Urbanik Coughlin, and D. B. Hunter. 2002. Uranium co-precipitation with iron oxide minerals. *Geochimica et Cosmochimica Acta* 66(20): 3533–47.

Eary, L. E., and D. Rai. 1989. Kinetics of chromate reduction by ferrous ions derived from hematite and biotite at 25°C. *American Journal of Science* 289(2): 180–213.

Fuller, C. C., J. R. Bargar, J. A. Davis, and M. J. Piana. 2002. Mechanisms of uranium interactions with hydroyapatite: Implications for groundwater remediation. *Environmental Science and Technology* 36(2): 158–65.

Golab, A. N., M. A. Peterson, and B. Indraratna. 2006. Selection of potential reactive materials for a permeable reactive barrier for remediating acidic groundwater in acid sulphate soil terrains. *Quarterly Journal of Engineering Geology and Hydrogeology* 39(2): 209–23.

Gould, J. P. 1982. The kinetics of hexavalent chromium reduction by metallic iron. *Water Research* 16(6): 871–77.

His, C. D., and D. Langmuirm. 1985. Adsorption of uranyl onto ferrick oxyhydroxide: Application of the surface complexation site-binding model. *Geochimica et Cosmochimica Acta* 49(9): 1931–41.

ITRC (Interstate Technology and Research Council). 2011. *Permeable Reactive Barrier: Technology Update*. Technical/Regulatory Guidance. Washington, DC.

Johnson, T. L., M. M. Scherer, and P. G. Tratnyek. 1996. Kinetics of halogenated organic compound degradation by iron metal. *Environmental Science and Technology* 30(8): 2634–40.

Keely, J. F. 1989. *Performance Evaluations of Pump-and-Treat Remediation*. U.S. Environmental Protection Agency, Superfund Ground Water Issue. EPA/540/4-89/005. Robert S. Kerr Environmental Research Laboratory.

Kent, D. B., J. A. Davis, L. C. D. Anderson, B. A. Rea, and T. D. White. 1994. Transport of chromium and selenium in the suboxic zone of a shallow aquifer: Influence of redox and adsorption reactions. *Water Resources Research* 30(4): 1099–114.

Lee, T., J.-W. Park, and J.-H. Lee. 2004. Waste green sands as reactive media for the removal of zinc from water. *Chemosphere* 56(6): 571–81.

Longmire, P. A., D. G. Brookins, B. M. Thomson, and P. G. Eller. 1991. Application of sphagnum peat, calcium carbonate, and hydrated lime for immobilizing U tailings leachate. In: *Scientific Basis for Nuclear Waste Management XIV*, vol. 212, edited by T. Abrajano Jr. and L. H. Johnson, 623–31. Pittsburgh: Materials Research Society.

Ma, Q. Y., T. J. Logan, and S. J. Traina. 1995. Lead immobilization from aqueous solutions and contaminated soils using phosphate rocks. *Environmental Science and Technology* 29(4): 1118–26.

MacKenzie, P. D., S. S. Baghel, G. R. Eykholt, D. P. Horney, J. J. Salvo, and T. M. Sivavec. 1995. *Pilot Scale Demonstration of Chlorinated Ethene Reduction by Iron Metal: Factors Affecting Iron Lifetime*. Paper Presented at Emerging Technologies in Hazardous Waste Management VII, Special Symposium of the American Chemical Society, American Chemical Society, Atlanta, September 17–20.

Mackenzie, P. D., T. M. Sivavec, and D. P. Horney. 1997. Extending hydraulic lifetime of iron walls. In: *Proceedings of the International Containment Technology Conference and Exhibition*, St. Petersburg, FL, February 9–12, 781–87.

Marsh, T. L., N. M. Leon, and M. J. McInerney. 2000. Physiochemical factors affecting chromate reduction by aquifer materials. *Geomicrobiology Journal* 17(4): 291–303.

Matheson, L. J., and P. G. Tratnyek. 1993. *Processes Affecting Reductive Dechlorination of Chlorinated Solvents by Zero-valent Iron*. Paper presented at the 205th ACS National Meeting, American Chemical Society, Denver, March 28–April 2.

McRae, C. W., D. W. Blowes, and C. Ptacek. 1997. *Laboratory-scale Investigation of Remediation of As and Se Using Iron Oxides*. Paper presented at the Sixth Symposium and Exhibition on Groundwater and Soil Remediation, Montreal, March 18–21.

Morrison, S. J., D. L. Naftz, J. A. Davis, and C. C. Fuller. 2002a. Introduction to groundwater remediation of metals, radionuclides, and nutrients with permeable reactive barriers. In: *Handbook of Groundwater Remediation Using Permeable Reactive Barriers*, edited by D. L. Naftz, S. J. Morrison, C. C. Fuller, and J. A. Davis. New York: Academic Press.

Morrison, S. J., C. E. Carpenter, D. R. Metzler, T. R. Bartlett, and S. A. Morris. 2002b. Design and performance of a permeable reactive barrier for containment of uranium, arsenic, selenium, vanadium, molybdenum, and nitrate at Monticello, Utah. In: *Handbook of Groundwater Remediation Using Permeable Reactive Barriers*, edited by D. L. Naftz, S. J. Morrison, C. C. Fuller, and J. A. Davis. New York: Academic Press.

———. 1993. Chemical barriers for controlling groundwater contamination. *Environmental Progress* 12(3): 175–81.

Morrison, S. J., and R. R. Spangler. 1992. Extraction of uranium and molybdenum from aqueous solutions: A survey of industrial materials for use in chemical barriers for uranium mill tailings remediation. *Environmental Science and Technology* 26(10): 1922–31.

Naftz, D. L., S. J. Morrison, J. A. Davis, and C. C. Fuller, eds. 2002. *Handbook of Groundwater Remediation Using Permeable Reactive Barriers: Applications to Radionuclides, Trace Metals, and Nutrients*. New York: Academic Press.

Naval Facilities Engineering Service Center. n.d. *Permeable Reactive Barrier Remediation of Chlorinated Solvents in Groundwater*. enviro.nfesc.navy.mil/erb/erb_a/restoration/technologies/remed/phys_chem/PRBSuccess.pdf.

Orth, W. S., and R. W. Gillham. 1996. Dechlorination of trichloroethene in aqueous solution using Feo. *Environmental Science & Technology* 30: 66–71.

Palmer, C. D., and R. W. Puls. 1994. *Natural Attenuation of Hexavalent Chromium in Ground Water and Soils*. U.S. EPA, Ground Water Issue. EPA/540/S-94/505.

Peld, M., K. Tonsuaadu, and V. Bender. 2004. Sorption and desorption of Cd^{2+} and Zn^{2+} ions in apatite-aqueous systems. *Environmental Science and Technology* 38(21): 5626–31.

Powell, R. M., and R. W. Puls. 1997. *Permeable Reactive Subsurface Barriers for the Interception and Remediation of Chlorinated Hydrocarbon and Chromium (VI) Plumes in Ground Water*. U.S. EPA Remedial Technology Fact Sheet. EPA/600/F-97/008.

Powell, R. M., R. W. Puls, S. K. Hightower, and D. A. Sabatini. 1995a. Coupled iron corrosion and chromate reduction: Mechanisms for subsurface remediation. *Environmental Science and Technology* 29(8): 1913–22.

Powell, R. M., R. W. Puls, S. K. Hightower, and D. A. Clark. 1995b. *Corrosive and Geochemical Mechanisms Influencing In Situ Chromate Reduction by Metallic Iron*. Paper presented at the 209th ACS National Meeting, American Chemical Society, Anaheim, CA, April 2–7.

Powell, R. M., R. W. Puls, and C. J. Paul. 1994. Chromate reduction and remediation utilizing the thermodynamic instability of zero-valence state iron. In: *Proceedings of the Water Environment Federation, Innovative Solutions for Contaminated Site Management*. Water Environment Federation. Miami, March 6–9.

Ren, J., Z. Zhang, M. Wang, G. Guanlin, P. Du, and F. Li. 2018. Phosphate-induced differences in stabilization efficiency for soils contaminated with lead, zinc, and cadmium. *Frontiers of Environmental Science and Engineering* 12(2): 1–9.

Roberts, A. L., L. A. Totten, W. A. Arnold, D. R. Burris, and T. J. Campbell. 1996. Reductive elimination of chlorinated ethylenes by zero-valent metals. *Environmental Science and Technology* 30(8): 2654–59.

RTDF (Remediation Technologies Development Forum). 2001. *Permeable Reactive Barrier Installation Profiles.* https://rtdf.clu-in.org/public/permbarr/prbsumms/default.cfm.

Scott, M. J., F. B. Metting, J. S. Fruchter, and R. E. Wildung. 1998. *Research Investment Pays Off.* Soil Groundwater Cleanup, October 6–13.

Shen, H., and J. T. Wilson. 2006. *Plant Mulch to Treat TCE in Ground Water in a PRB.* National Risk Management Research Laboratory. Presented at 5th International Battelle Conference, Monterey, CA, May 22–25, 2006.

Simon, F.-G., C. Segebade, and M. Hedrich. 2003. Behaviour of uranium in iron-bearing permeable reactive barriers: Investigation with 237U as a radioindicator. *Science of the Total Environment* 307(1–3): 231–38.

Sivavec, T. M., D. P. Horney, and S. S. Baghel. 1995. *Reductive Dechlorination of Chlorinated Ethenes by Iron Metal and Iron Sulfide Minerals.* Paper Presented at Emerging Technologies in Hazardous Waste Management VII, Special Symposium of the American Chemical Society, Atlanta, September 17–20.

Skinner, S. J. W., and C. F. Schutte. 2006. The feasibility of a permeable reactive barrier to treat acidic sulphate- and nitrate-contaminated groundwater. *Water SA* 32(2): 129–35.

Sneddon, I. R., P. F. Schofield, E. Valsami-Jones, M. Orueetxebarria, and M. E. Hoodson. 2006. Use of bone meal amendments to immobilize Pb, Zn and Cd in soil: A leaching column study. *Environmental Pollution* 144(3): 816–25.

Stanin, F. T., and M. Pirnie. 2004. The transport and fate of Cr(VI) in the environment. Chapter 5. In: *Chromium (VI) Handbook*, edited by J. Guertin, J. A. Jacobs, and C. P. Avakian. Boca Raton, FL: CRC Press.

Tang, W. Z. 2003. *Physicochemical Treatment of Hazardous Wastes.* Boca Raton, FL: CRC Press.

Thiruvenkatachari, R., S. Vigneswaran, and R. Naidu. 2008. Permeable reactive barrier for groundwater remediation. *Journal of Industrial and Engineering Chemistry* 14: 145–56.

Thomson, B. M., S. P. Shelton, and E. Smith. 1991. Permeable barriers: A new alternative for treatment of contaminated groundwater. In: *45th Purdue Industrial Waste Conference Proceedings.* Chelsea, MI: Lewis.

Tobiszewski, M., and J. Namieśnik. 2012. Abiotic degradation of chlorinated ethanes and ethenes in water. *Environmental Science and Pollution Research International* 19(6): 1994–2006.

U.S. EPA. 1998. *Permeable Reactive Barrier Technologies for Contaminant Remediation.* EPA/600/R-98/125. Washington, DC: Office of Solid Waste and Emergency Response, Office of Research and Development.

Voudrias, E. A. 2001. Pump-and-treat remediation of groundwater contaminated by hazardous waste: Can it really be achieved? *Global nest. The International Journal* 3(1): 1–10.

Waite, T. D., J. A. Davis, T. E. Payne, G. A. Waychunas, and N. Xu. 1994. Uranium (VI) adsorption to ferrihydrite: Application of a surface complexation model. *Geochimica Cosmochima Acta* 58(24): 5465–78.

Waybrant, K. R., D. W. Blowes, and C. J. Ptacek. 1995. Selection of reactive mixtures for the prevention of acid mine drainage using in situ porous reactive walls. In: *Sudbury '95, Mining and the Environment.* Ottawa, ON: CANMET.

White, A. F., and M. L. Peterson. 1998. *The Reduction of Aqueous Metal Species on the Surfaces of Fe(II)-containing Oxides: The Role of Surface Passivation.* American Chemical Society Symposium 715: 323–41.

Chapter 11

Microbial Remediation

*I hear there's microbes in a kiss. This rumor is most rife.
Come, lady dear, and make of me—an invalid for life.*

—Anonymous

Support bacteria—they're the only culture some people have.

—Anonymous

11.1 INTRODUCTION

The process of bioremediation at a contaminated site involves the engineered use of active microbial biomass for the destruction, detoxification, and/or uptake of pollutants from soil, waste piles, groundwater, surface water, or other environmental media. Numerous technologies are applicable to solids, slurries, and liquids, and, as with many of the strategies presented thus far, bioremediation can be performed either in situ or ex situ. Processes range from the simple, for example application of inorganic nutrients and oxygen to the subsurface via wells, to complex, involving addition of specialized cells to a soil slurry in aboveground reactor vessels supplied with elaborate temperature, pH and climate controls, effluent polishing, and wastewater treatment.

Bioremediation refers primarily to treatment of organic (hydrocarbon) contamination, where microbial cells utilize these molecules as a carbon source, thereby extracting energy for respiration and carbon for biomass. There are reported cases, however, in which metals or inorganics have been treated via bioremediation (Green-Ruiz 2006; Umrania 2006; Baxter and Cummings 2006; Krishna and Philip 2005).

Biological processes can be applied to a broad range of organic contaminants. The technology is essentially a destruction process based on either oxidation or reduction reactions. Biological treatment is designed to accomplish: (1) transformation to smaller and less toxic molecules, and, where applicable, (2) dehalogenation. The overall process for the biologically-mediated destruction of a hazardous organic molecule may be represented as:

$$(HC)_x + (N, P, S) + O_2 \rightarrow (\text{acid intermediates}) \rightarrow$$
$$CO_{2(g)} + H_2O + NO_3^- + SO_4^{2-} + \text{energy} + \text{microbial biomass}$$

(11.1)

11.2 OVERVIEW OF RELEVANT MICROBIOLOGICAL PRINCIPLES

In most applications, bioremediation is carried out using communities of microbial species, rather than one or a limited number of species. This is because most soils are already enriched with stable and complex microbial communities that have adapted to the physical and chemical milieu specific to a site. Communities are also more efficient because diverse populations of organisms will survive under a range of conditions (i.e., dry vs. wet, cool vs. warm soils, etc.). Finally, no single group will be capable of acting on all contaminant types; a range of organisms, and hence a continuum of physiological processes, will provide for more complete contaminant removal.

Bacteria, actinomycetes, fungi, algae, and protozoa are among the most numerous and important microorganisms in soil from the standpoint of both nutrient cycling and organic matter transformation. Of the groups listed, the first three comprise the microorganisms most responsible for transformations of organic contaminants in soil. Most research in bioremediation has addressed the exploitation of bacteria, but fungi are also known to play an important role, especially with halogenated compounds (e.g., pentachlorophenol, a wood preservative, and polychlorinated biphenyls [PCBs]).

Relevant morphologic and physiologic characteristics of various soil organisms were discussed in chapter 4, "Subsurface Properties and Remediation."

11.3 MICROBIAL REQUIREMENTS FOR GROWTH AND REMEDIATION

The driving force of all microbial physiological reactions is the acquisition of energy for survival, including support of metabolic processes, reproduction, motility, and so on. Microorganisms are often classified by the methods in which energy is acquired. For example, those that gain energy from the oxidation of inorganic compounds are termed *chemotrophs*; those that acquire energy from sunlight are *phototrophs*. Microbes are also categorized based on carbon requirements. Organisms that use CO_2 as their main carbon source are labeled *autotrophs*. In contrast, those that feed on preformed organic materials are *heterotrophs*. There are, of course, myriad examples in which microorganisms exist under more than one classification.

Microorganisms must experience vigorous growth (i.e., optimization of metabolic processes as well as reproduction) in order for bioremediation to be successful. The microbial populations essential to bioremediation are heterotrophic. Ideally, the hydrocarbon components of the contaminant plume serve as the energy source. Additionally microbes will require appropriate electron acceptors for oxidation-reduction processes, other nutrients, and a suitable physical and chemical environment.

11.3.1 Oxygen

Most microorganisms active in bioremediation processes are aerobic—that is, they require free O_2 to carry out metabolic functions. The aerobic heterotrophic group offers the greatest potential in bioremediation systems, as hydrocarbon decomposition is most efficient (i.e., the greatest energy yield per mole of substrate is available). Some treatment trains make use of anaerobic microorganisms; however, such processes are used only infrequently in environmental cleanup, for example, in the reductive dechlorination of trichloroethylene (Young et al. 2006; Aulenta et al. 2006; Sulfita and Sewell 1991). Microorganisms established in aqueous reactors, aquifers, or soil may be supplied with oxygen by pumping air or oxygen-supplying compounds (e.g., hydrogen peroxide) into the reaction zone. Cells growing in surface soil horizons may be supplied with oxygen by conventional tillage to facilitate incorporation of air.

11.3.2 Moisture

Most microorganisms that are exploited for bioremediation exist in water, whether as free water (occurring in soil pores, aquifers or tank reactors), or as a film on the surface of a soil particle or oil droplet. Moisture affects microbial activity in several ways. Water is the major component of protoplasm; therefore, an adequate supply is essential for vegetative development. Second, in order for a substrate molecule to be ingested, both it and the cell must be in contact with water. This contact facilitates movement of the cell to substrate or vice versa. Additionally, exoenzymes, secreted by the cell in order to initiate substrate decomposition, require water for transport. Finally, when soil moisture becomes excessive, microbial growth is hindered. The effect is not due to a direct toxic effect of water, but its excess limits gaseous exchange and lowers the available O_2 supply, thereby creating anaerobic conditions (Arora 2015).

Microorganisms are strongly affected by the osmotic potential of the local environment. Osmotic potential affects the ability of the microbial cell to maintain an adequate balance of water. If the environment is too dry or if water in the local environment contains excessive concentrations of solutes, the cell is unable to maintain the proper balance of protoplasmic water. This imbalance interferes with bioremediation projects where, for example, contaminated soil contains high levels of dissolved salts. Increased osmotic potential inhibits microbial activity by upsetting normal metabolic processes and, in extreme cases, results in lysis, or disintegration of cell walls.

11.3.3 pH

The pH range within which bioremediation processes operate most efficiently is approximately 5.5 to 8. It is no coincidence that this is also the optimum pH range for many heterotrophic bacteria, the major microbial players in most bioremediation technologies. The most favorable pH for a particular situation, however, is site-specific. Soil pH is influenced by a complex relationship between organisms, contaminant chemistry, and physical and chemical properties of the local environment. Additionally, as biological processes proceed in the contaminated media, pH may shift and therefore must be monitored regularly. The pH can be adjusted to the desired range by addition of acid- or base-forming substances (i.e., $Al_2(SO_4)_3$ or limestone, respectively).

11.3.4 Nutrients

Heterotrophic bacteria, actinomycetes, and fungi possess fairly complex nutritional requirements. Nutrients serve three primary functions: providing the necessary building blocks for synthesis of protoplasmic components; supplying the needed energy for cell growth and biosynthetic reactions; and serving as electron acceptors for energy-related reactions in the cell (table 11.1). Energy sources for heterotrophs include sugars, starch, cellulose, hemicellulose, lignin, pectic substances, inulin, chitin, proteins, amino acids, and organic acids. Given the correct microbial populations and chemical and physical properties of an affected site, organic contaminant molecules will also serve as an energy source. The oxidation of these organics releases energy, a portion of which is used for synthesis of protoplasm.

Table 11.1 Nutrients required by microorganisms

Inorganics (minerals)	N, P, K, Ca, Mg, S, Fe, Mn, Cu, Zn, Co, Mo
Carbon sources	CO_2, HCO_3^-
Energy sources	Organic compounds (glucose, polysaccharides, cellulose, etc.)
	Inorganic compounds (Fe^{2+})
	Light
Electron acceptors	O_2
	NO_3^-, SO_4^{2-}, Fe^{3+}, CO_2
	Organic compounds
Growth factors	
a. Amino acids	Alanine, cysteine, histidine, serine, tyrosine, and so on.
b. Vitamins	Nicotinic acid, riboflavin, pantothenic acid, biotin, PABA, pyridoxine, thiamine, B_{12}, folic acid, etc.
c. Other	Purine bases, pyrimidine bases, peptides, and so on.

Carbon dioxide, a product of both aerobic and anaerobic metabolism, is important not only because it completes the carbon cycle but also because of its direct influence on growth. In the classic terminology of microbiology, both chemoautotrophic and photoautotrophic microorganisms require CO_2 because it is their sole carbonaceous nutrient. However, the gas is stimulatory to and often required by many heterotrophs, and growth of many species will not proceed in the absence of CO_2. A portion of the CO_2 present is incorporated into the cell structure, even for heterotrophs. The requirement for this gas rarely presents a problem in soil because of its continual evolution from decaying organic matter. Nitrogen, P, K, Mg, S, Fe, Ca, Mn, Zn, Cu, Co, and Mo are integral parts of the cell's protoplasmic structure. These nutrients plus C, H, and O are needed for the synthesis of the microbial cell. In a bioremediation program mineral nutrients are usually supplied as soluble salts in fertilizers. Carbon may be supplied in the form of animal manures (which will also supply many mineral nutrients), wood chips, or similar organic materials. A proper balance is needed between the various mineral nutrients (e.g., N and P) and carbon, or microorganisms cannot make optimum use of the carbon source. For most bioremediation situations, biodegradation is optimal at C:N ratios in the range of 10 to 30 to 1, and N:P ratios of about 10 to 1, weight basis. These ratios vary widely depending on the type of carbonaceous material present (Rojas-Avelizapa et al. 2007; Cookson 1995).

The availability of nutrients to microorganisms is strongly influenced by pH. In the circumneutral pH range, trace metals (e.g., Cu, Co, Ni, Zn) are typically available in micro quantities. This degree of availability is generally adequate for most biota, as excess quantities, for example under acid pH regimes, are known to be toxic and inhibitory. Also at neutral pH, P is maximally available. A soil pH of 5.5 to 8 is therefore generally recommended during bioremediation.

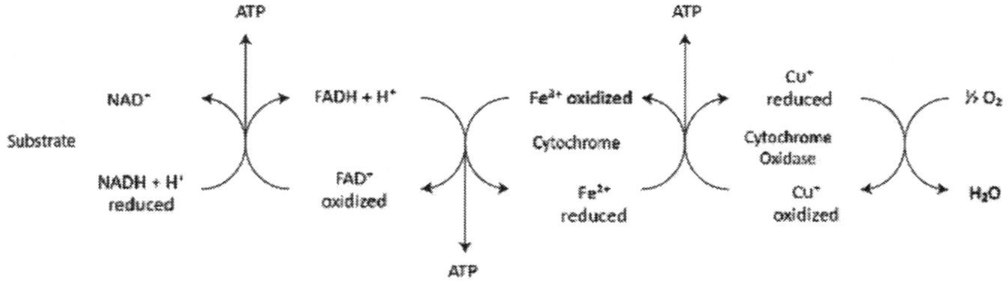

Figure 11.1 The electron transport chain.

Energy generation within a microbial cell is an oxidation process in which electrons are transported along the respiration pathway. This electron flow generates energy through the electron transport chain, which consists of molecules that undergo repeated oxidation and reduction while transferring the electron from one molecule to the next (figure 11.1). The electrons are transported in the system by several compounds, with one of the more significant being nicotinamide adenine dinucleotide phosphate, NADP (figure 11.2). The pyridine nucleotides and related compounds experience reversible oxidation and reduction which results in the storage of energy within the microbial cell as energy-rich chemical bonds (Cookson 1995).

In order to be utilized, the substrate must penetrate the cell. Often, the energy source enters with no difficulty, but microbial cells are impermeable to many complex molecules. These compounds must first be solubilized and simplified prior to serving within the cell as energy sources. Here again, exoenzymes are important in the initial dissociation of a complex substrate molecule, allowing for its eventual incorporation into the cell. Once across the membrane, substrates are catabolized (i.e., degraded) by microorganisms using one of three general pathways of metabolism:

1. In *aerobic respiration*, organic chemicals are oxidized to carbon dioxide and water or other end-products using molecular oxygen as the terminal electron acceptor. We would expect aerobic respiration to occur under highly oxygenated conditions. The following is a generalized reaction for aerobic respiration involving a glucose molecule:

$$C_6H_{12}O_6 + 6O_2 \rightarrow 6CO_2 + 6H_2O + \text{energy} \tag{11.2}$$

2. If the oxygen supply decreases, as may occur in a dense clay soil with active microbial biomass, *anaerobic respiration* may be initiated. In this mode, microorganisms metabolize hydrocarbons in the absence or near-absence of molecular oxygen and use inorganic substrates as terminal

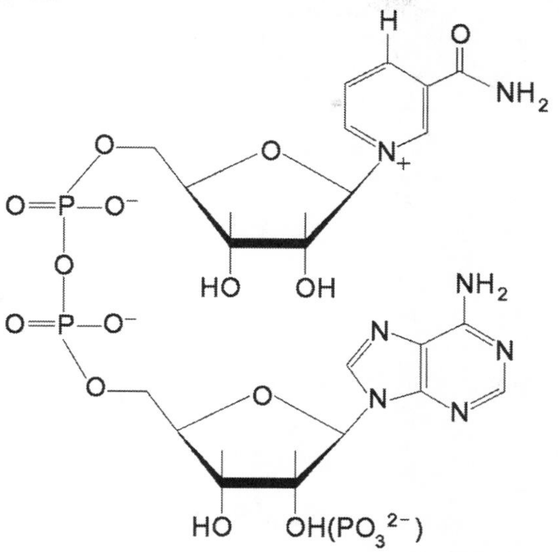

Figure 11.2 Structure of nicotinamide adenine dinucleotide phosphate (NADP), a coenzyme critical to the electron transport chain.

electron acceptors. Common electron acceptors include nitrate, NO_3^-, sulfate, SO_4^{2-}, and Fe^{3+}. In anaerobic respiration, nitrate is reduced to nitrogen (N_2) or ammonium (NH_4^+), sulfate to sulfide (S_2^-), ferrous iron (Fe^{2+}) to ferric iron (Fe^{3+}), and CO_2 to methane (CH_4).

3. Under highly reducing conditions and in the absence of inorganic electron acceptors, *fermentation* occurs. Hydrocarbons are degraded independent of oxygen, and organic compounds function as electron acceptors. Fermentation results in end-products including acetate, ethanol, propionate, and butyrate (table 11.2).

11.3.5 Contaminant Properties

Microbial acclimation and growth, and bioremediation success are affected in part by contaminant concentration. A microbial consortium will maximize decomposition reactions within a preferred range of contaminant concentrations. Below this range microbial action on the molecule may be minimal; above this range microbial activity will be inhibited. The concentrations at which microbial growth is supported or inhibited vary with the contaminant, properties of the media (soil versus slurry versus groundwater), and distribution of individual species. Given long-term exposure, microbes have been known to acclimate to very high contaminant concentrations and other stressful conditions.

Table 11.2 Metabolism modes of microbial flora

Type	Electron Acceptor
Aerobic respiration	Oxygen (O_2)
Anaerobic respiration:	
Denitrification	Nitrate (NO_3^-)
Nitrate reduction	Nitrate (NO_3^-)
Sulfate reduction	Sulfate (SO_4^{2-})
Ferric iron reduction	Iron (Fe^{3+})
Sulfur reduction	Sulfur (S)
Fermentation	Organic compound
Methane fermentation	Carbon dioxide (CO_2)

The bioavailability of a contaminant to microorganisms depends on several factors: (1) solubility in water; (2) tendency to sorb to soil solids; (3) potential as a source of energy or nutrients; and (4) potential toxicity. Long-chain aliphatics or condensed aromatics are unlikely to occur in the aqueous phase and will be difficult to biodegrade. Limited bioavailability can be due to adsorption to soil solids rather than insolubility. Hydrophobic contaminants partition from soil water and concentrate in organic matter, resulting in unavailability to microbes. Contaminant properties that affect sorption include molecular weight, structure, solubility, and polarity. Relevant soil properties include the types and amounts of organic matter and clays, and presence of metal hydrous oxides. Bioavailability is also a function of the utility of the molecule to the organism, that is, whether it acts as a substrate or co-substrate. When the compound cannot serve as a metabolic substrate but is oxidized in the presence of a substrate already present, the process is referred to as co-oxidation and the contaminant molecule is the co-substrate. Co-oxidation is important for biodegradation of high molecular weight PAHs and some chlorinated solvents such as trichloroethene (TCE).

11.4 PATHWAYS OF HYDROCARBON METABOLISM

Heterotrophic microbes are now recognized to decompose and detoxify a broad range of organic contaminants. Table 11.3 presents some common RCRA-regulated organic compounds that are susceptible to biodegradation. A commonly encountered soil and aquifer contaminant is gasoline, which is a mixture of hydrocarbons including alkanes, cycloalkanes, and aromatics (see chapter 3). Soil microorganisms initiate aerobic degradation of the alkane components through attack of either the terminus of the molecule (most common) or at some point along the chain. A typical degradative route involves oxidation of the terminal methyl group to an alcohol

Table 11.3 Biodegradable RCRA-regulated organic compounds

Substrate	Respiration		Fermentation	Oxidation	Co-oxidation
	Aerobic	Anaerobic			
Straight-chain alkanes	+	+	+	+	+
Branched alkanes	+	+	+	+	+
Saturated alkyl halides		+		+	+
Unsaturated alkyl halides		+		+	
Esters, glycols, epoxides	+	+	+	+	
Alcohols	+	+		+	
Aldehydes, ketones	+	+		+	
Carboxylic acids	+	+		+	
Amides	+	+			
Esters, glycols, epoxides	+	+			
Nitriles	+	+			
Amines	+	+			
Phthalate esters	+	+		+	
Nitrosamines		+			
Thiols					
Cyclic alkanes	+		+	+	+
Unhalogentated aromatics	+	+		+	
Halogenated aromatics	+	+		+	+
Aromatic nitro cmpds	+	+			
Phenols	+	+	+	+	+
Halogenated side chain Aromatics	+		+	+	
Nitrophenols		+			
Halophenols	+			+	
2- and 3-ring PAHs	+			+	
Biphenyls	+				
Chlorinated biphenyls	+				
4-ring PAHs	+				
Organophosphates	+	+			
Pesticides and herbicides	+	+			

(figure 11.3). This alcohol undergoes a series of dehydrogenation steps to form the corresponding aldehyde and eventually a fatty acid. The fatty acid is further metabolized by beta oxidation, a process in which two carbons are excised from the chain and released as two moles of CO_2. The chain, now less two carbons, receives a new carboxylic acid group at the terminus, which is again available for beta oxidation.

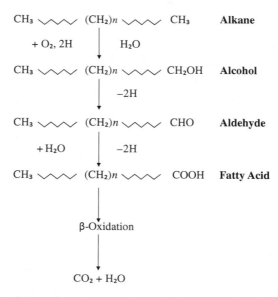

Figure 11.3 Oxidation of an alkane molecule by heterotrophic microorganisms.

Microorganisms also use cycloalkanes as a source of carbon and energy. For example, cyclohexane hydroxylation by a microbially produced monooxygenase leads to the formation of an alicyclic alcohol. Subsequent dehydrogenation of the alcohol forms a ketone, and further oxidation results in formation of a lactone ring structure. The lactone is a substrate suitable for ring opening and is eventually converted to a dicarboxylic acid which, in turn, is decomposed via beta oxidation (figure 11.4) (Neilson and Allard 2008; Cookson 1995).

Aromatic compounds are subject to microbial attack by a wide range of bacterial and fungal types. In the case of benzene, a bacterial dioxygenase enzyme incorporates both atoms of molecular oxygen to form *cis*-benzene dihydrodiol that is subsequently dehydrogenated, resulting in ring cleavage (figure 11.5) (Cookson 1995; U.S. EPA 1989).

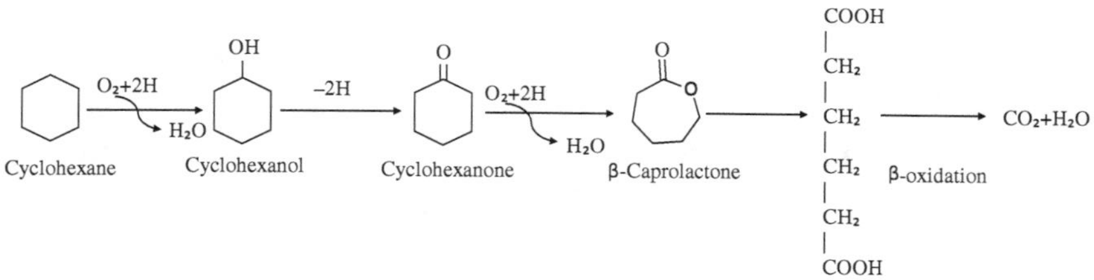

Figure 11.4 Oxidation of a cycloalkane molecule by heterotrophic microorganisms.

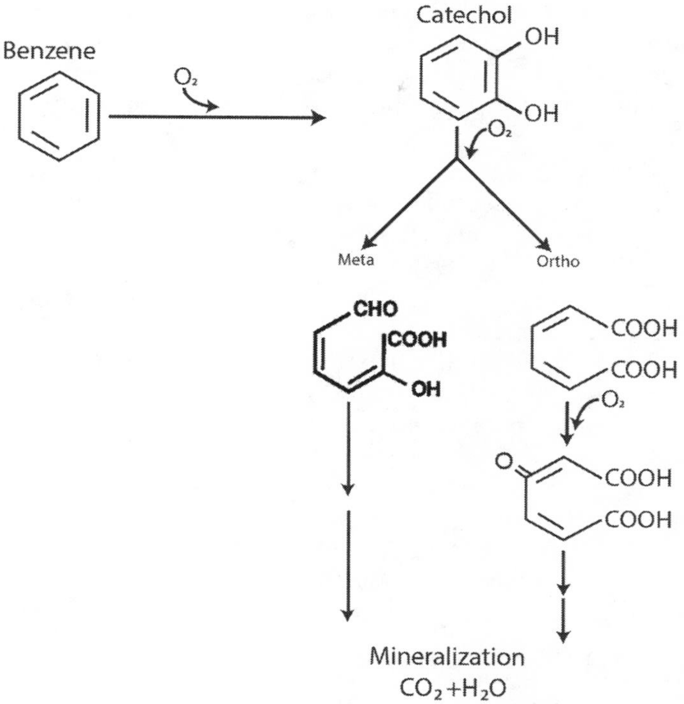

Figure 11.5 Oxidation of an aromatic molecule by heterotrophic microorganisms.

All of the metabolic pathways discussed above require oxygen as co-reactant. In addition, the organisms catalyzing these bioconversions use oxygen as a terminal electron acceptor. Hydrocarbon metabolism therefore places a significant demand on oxygen resources; as a result, designs of biorestoration systems should consider how this oxygen demand is to be supplied. Field technologies are discussed later in this chapter.

Some general rules of thumb regarding biodegradation of alkanes, alkylaromatics, and aromatics are as follows (Das and Chandran 2011; Cookson 1995; EPRI 1988; Bossert and Bartha 1984):

1. Gaseous n-alkanes (CH_4–C_4H_{10}) are lost by volatilization.
   Some are biodegradable but are used only by a narrow range of specialized hydrocarbon degraders such as methanotrophs.
2. The n-alkanes, alkylaromatic, and aromatic hydrocarbons in the C_5 to C_9 range are readily volatilized due to their low molecular weights. These compounds are, however, biodegradable at low concentrations.

3. The n-alkanes, n-alkylaromatic, and aromatic compounds in the C_{10} to C_{22} range are the most readily biodegradable and the least toxic to microbes.
4. Branched alkanes and cycloalkanes of the C_{10} to C_{22} range are less biodegradable than their n-alkane and aromatic counterparts. Branching interferes with beta oxidation.
5. Highly condensed aromatic and cycloparaffinic systems, with four or more condensed rings, and the partially oxygenated and condensed components of tar, bitumen, and asphalt degrade slowly.

Depending on intended use, petroleum products contain a wide range of additives such as detergents, flow improvers, pour depressants, oxidation and corrosion inhibitors, octane improvers, anti-icing additives, combustion aids, biocides, and dyes. These additives possess a wide array of chemical structures; however, soil microorganisms are capable of transforming many of them.

11.5 INDIGENOUS VERSUS EXOGENOUS PARTICIPANTS

As part of the environmental site assessment, it is recommended that the contaminated soil be assayed to determine whether the appropriate microbial consortia are present to carry out effective bioremediation. *Indigenous microorganisms* are defined as those that are native to the affected site. For remediation to be efficient and complete it may be necessary to stimulate the growth of these microbes; therefore, soil conditions such as pH, oxygen, and nutrient content may need to be adjusted. Once conditions are appropriate, indigenous microorganisms utilize the hydrocarbon contaminants in the soil as a substrate. Under ideal circumstances, these contaminants will ultimately be converted to nonhazardous substances such as CO_2 and H_2O.

Indigenous microorganisms are typically employed at sites undergoing bioremediation. The soil of interest should be sampled and brought to a laboratory where microbial types and their optimal growth conditions are determined. If the indigenous microorganisms successfully degrade the contaminant within a reasonable time frame, field operations should be relatively straightforward. If the desired reactions do not occur, it may be necessary to modify soil conditions. Furthermore, different microbial populations may be needed. If microbes required for contaminant decomposition do not occur in the affected soil, microbes collected from other areas (*exogenous microorganisms*), whose effectiveness has been documented for bioremediation, are incorporated into the soil. This is termed *bioaugmentation*. Once again, soil conditions may need to be modified to ensure that the exogenous cells will survive and carry out necessary reactions (Singh et al. 2011; U.S. EPA 1992a).

Exogenous microorganisms are harvested from soil collected from other locations and cultured in the laboratory. The cells are placed under optimal conditions for proliferation (i.e., ideal temperature range, soil pH, concentration of nutrients, etc.). Once their *titer* (total numbers) is sufficiently high, biomass is collected, treated and stored, transported to the site, and applied to the zone of operation. Once degradation of the contaminants is complete, most of the exogenous microorganisms will die because they have depleted their food source and because the local environment will no longer be modified to suit their growth needs. The dead microbial cells should pose no contamination risk (U.S. EPA 1992a).

Relying on indigenous microorganisms is appropriate if useful strains are present and concentrated in the area of contamination. If indigenous organisms are already surviving in the original soil conditions, the process of optimizing the soil environment is straightforward. Using indigenous microorganisms also tends to be less expensive than culturing and introducing exogenous cells into the soil. For all of these reasons, most bioremediation technologies make use of native microorganisms whenever possible.

11.6 FIELD TECHNOLOGIES

A bioremediation system must be designed in order to optimize and control all relevant microbial biochemical cycles. Numerous technologies have been developed to support the survival and proliferation of microbial communities in contaminated environments, and in carrying out degradative reactions. The specific technology employed is determined by the contaminants present, site conditions, types of microbes available, and cost.

11.6.1 In Situ Processes

Under the appropriate conditions, bioremediation can be successfully accomplished in situ; that is, soil microorganisms may be stimulated and/or added to treat low-to-moderate concentrations of organic contaminants in place, without excavating or otherwise disturbing contaminated soil. The reaction zone may be a soil horizon at or near the surface, or may occur in an aquifer many feet below the surface.

In some situations bioremediation occurs solely by the action of the indigenous microbial populations without application of supplemental materials. This process, labeled *intrinsic bioremediation* or *passive degradation*, is normally extremely slow. For in situ bioremediation to be successful and reasonably rapid, it is necessary to stimulate the growth of microbes that react with and degrade contaminant molecules. Oxygen, nutrients, and other amendments are introduced as required into the affected soil and groundwater. These additions correct many of the factors that may limit microbial activity and, thus, the rate and extent of contaminant degradation. Depending on cleanup goals, in situ biodegradation can be used as the sole treatment technology or as one component of a remediation system involving various chemical and physical technologies.

In situ systems commonly employ infiltration galleries together with groundwater pumping to introduce aerated, nutrient-enriched water into the contaminated zone through an array of injection wells, sprinklers, or trenches. Sufficient time is allowed for microbial communities to act upon the contaminants, and the treated water is eventually recovered downgradient (figure 11.6). The recovered water may be further treated (e.g., passage over GAC) and reintroduced to the affected soil. Otherwise it may be discharged to a municipal wastewater treatment plant or to surface water.

In situ biodegradation is effective at degrading a range of organic compounds (Sui et al. 2006; Dettmers et al. 2006; Goi et al. 2006; U.S. EPA 1994, 1990b; Norris et al. 1993). As discussed earlier, water-soluble organic substrates possess a relative advantage in terms of biodegradability. However, somewhat insoluble contaminants may be degraded if a suitable surfactant is supplied to the system, thereby increasing solubility. Compounds suitable for biodegradation include petroleum hydrocarbons (e.g., gasoline and diesel fuel), chlorinated and non-chlorinated pesticides, non-chlorinated solvents (e.g., ketones and alcohols), wood-treating wastes (e.g., creosote and pentachlorophenol), and some chlorinated aliphatics (e.g.,trichloroethene). In situ biodegradation is typically not used to treat inorganics (e.g., acids and metals); however, it has been successful in the treatment of water contaminated with nitrate and phosphate (U.S. EPA 1994).

Several bacterial genera are known to be well suited for biodegradation and include *Pseudomonas, Arthrobacter, Alcaligenes, Corynebacterium, Flavobacterium, Achromobacter, Acinetobacter, Micrococcus, Nocardia,* and *Mycobacterium* (Chikere et al. 2011; Flocco et al. 2009; Peng et al. 2008). Others are listed in table 11.4. The most suitable microbial consortia for a particular site is a function of contaminant chemistry and properties of the recipient soil.

Table 11.4 Bacteria and yeasts that oxidize aliphatic hydrocarbons

Bacteria	Yeasts
Achromobacter	Aspergillus
Acinetobacter	Candida
Actinomycetes	Cladosporium
Aeromonas	Cryptococcus
Alcaligenes	Cunninghamella
Arthrobacter	Debaryomyces
Bacillus	Endomyces
Beneckea	Eupenicillum
Brevibacterium	Hansenula
Corynebacterium	Mycotorula
Desulfatibacillum	Pichia
Flavobacterium	Rhodotorula
Methylobacter	Saccharomyces
Methylobacterium	Selenotila
Methylococcus	Sporidiobolus
Methylocystis	Sporobolomyces
Methylomonas	Torulopsis
Methylosinus	Trichosporon
Micromonospora	
Mycobacterium	
Nocardia	
Pseudomonas	
Spirillum	
Vibrio	

Source: Chikere et al. 2011; Flocco et al. 2009; Peng et al. 2008.

11.6.1.1 Removal of Free Product

Residual contamination, that is, hydrocarbons occurring as free product (NAPL) or adsorbed to soil colloids, commonly occurs with discovery of the release and at the initiation of a remediation activity. An excess of NAPL compounds may be directly toxic to native microbial cells. Furthermore, immiscible compounds are not available to the cell. Microbes are capable of degrading only those contaminants that come into intimate contact with the cell or its exoenzymes; in other words, compounds are bioavailable when they are at least partially dissolved in water. Free liquids (NAPL) should therefore be removed before the bioremediation gallery is established. Pumping of free product may be required.

11.6.1.2 Water Treatment

An initial step in an in situ biodegradation system (see figure 11.6) involves pretreating infiltration water, if necessary, to remove metals such as iron. Groundwater, municipal drinking water, or trucked water may be used as

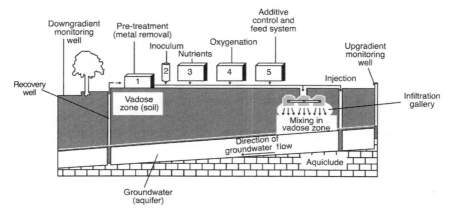

Figure 11.6   Schematic of an in situ bioremediation system.

infiltration water. Specific pretreatment steps are based upon site conditions (i.e., soil chemical properties, pH, etc.), water source, contaminant properties, chemistry of the metal(s) of concern, and treatment system used. The relevant chemical properties of the water must be determined in advance in order to avoid adverse reactions. At a minimum pH, salinity level, total organic carbon, and dissolved Fe concentration should be assessed.

If groundwater is used, excess dissolved Fe may bind applied phosphates which are required for microbial growth. Soluble Fe will also deplete hydrogen peroxide or other oxidizing agents, which are often used as oxygen sources. The Fe can be complexed early in the treatment system by applying excess phosphate to the infiltration water. Likewise, addition of $CaCO_3$ or CaO will increase pH and result in the formation of iron oxides, carbonates, and other precipitates. Clogging of the system by metal precipitates must be considered during this phase (see "Clogging and Its Control," later in this chapter). Another consideration in using groundwater or recycled water is that any toxic metals which were solubized and released from the contaminated soil may have to be removed from the recycled infiltration water so that the active microbial biomass is not inhibited.

Following pretreatment of the infiltration water, a microbial inoculum, containing either indigenous or exogenous microbial cells, can be added to augment the indigenous microbial community. An inoculum enriched from site samples may be injected at the site, or commercially available cultures which are recognized for degrading the contaminants is supplied. The ability of exogenous microbes to survive in a foreign and toxic environment, as well as the ability to metabolize the contaminants of concern, should first be assessed in pilot tests.

Table 11.5 Nutrients required for biological treatment

Major	C, H, O, N, P
Minor	Ca, Mg, S, Fe, K, Na, Cl
Micro	B, Co, Cu, Mn, Mo, Ni, Zn, V

Nutrients essential to the activity of both indigenous and exogenous organisms are often applied to the in situ treatment train. Optimum nutrient levels are specific to the site and are best determined via standard soil fertility analysis. Nitrogen and phosphorus are commonly applied and should be included at a minimum. Trace elements may be added, but they typically occur at sufficient levels in soil and groundwater. Macro- and micronutrients important for microbial establishment and growth are listed in table 11.5. Nitrogen application rates must be carefully determined to avoid excessive nitrate formation. Excessive quantities of nitrates in groundwater are linked with health problems in humans, especially infants. Similarly, P application rates must be calculated to prevent precipitation as calcium and iron phosphates.

Before water is introduced to the contaminated site, the pH can be adjusted using simple additives. A neutral pH is recommended for most systems as this favors the activity of many indigenous bacteria. Water can be acidified with $Al_2(SO_4)_3$ or simple inorganic acids; conversely, pH is increased with pulverized lime, $Ca(OH)_2$. Limestone ($CaCO_3$) may also be used in pulverized form; however, it is less soluble than lime. High Ca concentrations may result in precipitation of added phosphate, which will thus become unavailable to microbial populations.

11.6.1.3 Hydraulics of the In Situ System

A major design function of the in situ reaction system is to control water movement through and out of the contaminated zone. Such control is provided via installation of a series of injection wells, recovery wells, and barriers. The general principles and equipment for in situ bioremediation are relatively simple; however, achieving complete control is difficult in field applications due to the heterogeneity of a site (Song and Seagren 2008; Cookson 1995). For example, the presence of impermeable layers, buried USTs, or drums will affect flow patterns, the ultimate degree of reaction, and effectiveness of recovery of liquids.

The delivery system provides the necessary reagents to the contaminated zone in a manner that minimizes losses and clogging, and provides for uniform distribution of reagents within the zone. Gravity delivery methods involve application of treatment chemicals to the site surface for eventual downward percolation; forced delivery methods inject the necessary additives through a perforated pressurized pipe into the affected zone.

11.6.1.4 Surface Application

Surface application of reagents via gravity application include flooding, spray irrigation, recharge basins, and trenches. Surface gravity delivery methods are applicable for soils contaminated to shallow depths (i.e., within several feet of the surface), sites of high permeability such as gravel and sand beds, and localized contamination (Abdel-Moghny et al. 2012). Hydraulic conductivity, thickness of the vadose zone, and depth to water table all affect the rate of travel of applied water into the contaminated zone.

Hydraulic conductivity usually decreases after startup and then stabilizes. Continuous application will result in reduced hydraulic conductivity. Infiltration is maximized when a cycle of application and drying is used. One goal of the remedial program is to hold drying time to a minimum while maintaining hydraulic conductivity and optimizing microbial activity (Cookson 1995).

Although surface application systems are relatively low in cost, they limit the ability to maintain overall process control. Other drawbacks include potential for flooding and the consequent need for runoff controls and a collection system. Flooding is avoided by installation of berms, dikes, and drainage ditches, and by establishing cycling times between flooding and drying.

Surface application methods are not appropriate where impermeable layers occur between the surface and the contamination zone. Surface applications such as flooding and spray irrigation may not be appropriate in irregular terrain and on significant inclines. Slopes greater than 3 to 5 percent can result in runoff and reduced effectiveness of applied reagents (Cookson 1995). An additional consideration when evaluating the feasibility of gravity feed is soil organic matter content: a high organic concentration may immobilize nutrients and decompose or adsorb electron acceptors. As a result, the progress of the bioremediation program may be hindered.

11.6.1.5 Subsurface Application

Trenches and subsurface drains are common methods for introducing bioremediation reagents. Subsurface drains are generally limited to shallow depths; the infiltration gallery consists of a gravel-filled trench. The applied solution fills the pores in the gallery and is distributed to the surrounding soil vertically and horizontally. The system can also be constructed by excavating a trench and installing perforated drainage pipe, which will serve as a distribution medium. The trench is then backfilled with gravel and covered with soil. Infiltration trenches have been placed as deep as 15 ft. Drains are often protected from clogging by lining with filter fabric—perforated pipe can be wrapped and/or filter fabric is placed on the gravel backfill. Trench lengths have measured over 150 ft. The pit resulting from an excavated UST also can be used as an infiltration gallery. If the above delivery systems are not sufficient to support the required level of biological activity, forced injection may be necessary.

Forced injection involves the delivery of remediation reagents under pressure into the contaminated zone. A series of injection wells is installed at the site based on aerial extent of the plume and hydrogeologic factors. This system provides for flexibility in process control because rate and location of delivery are carefully managed. Forced injection systems are applicable to soils with hydraulic conductivities greater than 10^{-4} cm/s (e.g., fine sands, loams, silt loams) and high porosities (25%–50%). A maximum injection pressure must be set to prevent hydraulic fracturing and uplift in the subsurface, which would cause the pumped reagent liquids to migrate upward, rather than through the contaminated zone (NAVFAC 2013; Sims et al. 1993).

Forced injection usually involves the use of vertical wells, but horizontal wells have been employed in some situations. The advantage of the horizontal well is that reagents can be delivered to selected regions of the subsurface that cannot be reached by vertical wells because of the presence of utilities or surface structures.

Oxygen is a key limiting factor for in situ bioremediation success. The zone of influence is often established by the distance that oxygen can be carried to concentrations of at least 1 mg/L. This will determine the necessary well spacing and overall costs for the forced injection system. Ideally, injection wells and recovery wells should be situated at a distance that allows added nutrients to reach the zone of contamination within less than six weeks from time of injection. Travel time and reaction rates for injected chemicals are estimated from treatability studies and should be confirmed by field monitoring. A low oxygen supply rate relative to the contaminant load results in extended remediation times. Excessive oxygen inputs results in the potential for soil gas binding and high project costs (Vance 2008; Cookson 1995).

11.6.1.6 Oxygen Sources

In the majority of applications, bioremediation is an oxidation process. During oxidation of contaminants, microorganisms extract energy via electron transfer. Electrons are detached from the contaminant and transferred to a terminal electron acceptor which, during aerobic biodegradation, is oxygen. During decomposition of the organic substrate, oxygen concentrations in the subsurface may become depleted. Air, oxygen, or other oxygen sources (e.g., hydrogen peroxide, ozone) may need to be added to the infiltration water. If anaerobic degradation is being used, nitrate or sulfate may be added instead to serve as electron acceptors.

Oxygen can be added to the zone by pumping oxygen directly or by initiating a chemical reaction with the consequent release of O_2. Pure oxygen or air can be dissolved in water and pumped into the contaminated zone. The rate of oxygen supplied is a function of project goals, hydrocarbon concentration, injection rate, and initial dissolved oxygen concentration. This method, however, provides only a modest rate of oxygen delivery. Pumping of oxygenated water into the subsurface can result in biofouling, which will inhibit the flow of outgoing air into the well bore (Clu-in 2005; Cookson 1995).

Water can contain dissolved oxygen only up to about 20 mg/L. A supplemental chemical oxygen supply involves the addition of a compound such as hydrogen peroxide that can be converted to oxygen. With the addition of H_2O_2 to water, the O_2 content is greatly increased. Hydrogen peroxide is highly soluble and decomposes to water and oxygen:

$$H_2O_2 \rightarrow H_2O + 1/2\,O_2 \qquad (11.3)$$

H_2O_2 solutions from 100 to 500 mg/L have been beneficial to microorganisms involved in hydrocarbon degradation. At 100 mg/L or above, hydrogen peroxide keeps injection wells free of heavy biological growth. This prevents fouling, allowing more equal and more rapid transmission of oxygen to the treatment area. In a study in the Midwest, a solution of 1000 mg/L was found suitable to support indigenous microbial populations for the decomposition of diesel fuel in a contaminated aquifer under a fleet facility.

A saturated subsurface is needed if hydrogen peroxide is employed in order to prevent significant decomposition; in other words, H_2O_2 is not practical for bioremediation of the unsaturated zone. Forced injection is usually necessary since gravity feed will result in substantial losses due to reaction with surface soil constituents such as organic matter, ferrous iron (Fe^{2+}), and other reduced metals. Hydrogen peroxide decomposes readily because it is a nonselective oxidizing agent.

Adverse reactions sometimes occur with the use of hydrogen peroxide and ozone. For example, they may react violently with other compounds present in the soil, reduce the soil sorptive capacity (due primarily to organic matter decomposition), and produce oxygen gas bubbles that block soil pores. If the solution is too concentrated, it will oxidize bacterial cell membranes, thus killing the cells. Hydrogen peroxide at concentrations of approximately 1,500–2,000 mg/L has been found to be toxic to microorganisms.

Hydrogen peroxide is relatively expensive to use, and commercial formulations (25%–30%) damage skin and are highly reactive with organic debris. One method to alleviate some of the problems associated with oxygen delivery to the subsurface involves addition of a solid oxygen supply compound to the contamination zone (U.S. EPA 2018; White et al. 1998; Koenigsberg 1997; Brubaker 1995; Bianchi-Mosquera et al. 1994). For example, solid magnesium peroxide can be installed in a permeable reactive wall through which a contaminated plume passes. The material can also be packed into replaceable filter socks or can be injected directly in slurry form. Solid oxygen supplies are generally insoluble and release oxygen in a rate-controlled manner. These supplements can last from several months to a year depending on contaminant loading (Koenigsberg 1997).

Nitrate and sulfate have also been studied as electron acceptors in bioremediation projects. Microorganisms that use nitrate as an electron acceptor have been stimulated in situ by injecting acetate as a primary substrate and nitrate as the electron acceptor. Sulfate-reducing bacteria are also stimulated by acetate injections.

Situations with multiple contaminants, each having differing chemistries, may be treated by in situ technologies. Multiple metabolic modes, for example, a design that promotes a sequence of anaerobic and aerobic metabolic processes, can be devised for an in situ treatment program (Lee et al. 2002; Master et al. 2002). A two-zone in situ treatment approach was applied for the treatment of pentachlorophenol (Meade and D'Angelo 2005; Vira and Fogel 1991). Contaminants requiring anaerobic dehalogenation are first reduced in an anaerobic metabolic zone. The reduced products and other contaminants not treated by anaerobic activity are then transferred to an aerobic metabolism zone for additional treatment.

11.6.1.7 Soil Treatment

The affected soil may require pretreatment in order to render all process reagents effective. Soil pH adjustment using lime is dependent on several factors including texture, types and amounts of clay, and organic matter content.

Changes in soil pH will influence dissolution or precipitation of soil metals and may increase the mobility of certain hazardous compounds. Therefore, soil buffering capacity should be evaluated prior to application of amendments. A simple buffer test such as the SMP procedure may be used (McLean 1982). As with most recommendations addressing optimization of the microbial community, liming or acidification requirements should be determined on a site-specific basis.

11.6.1.8 Transformation versus Decomposition; Residuals

Bioremediation technologies are intended to mineralize organic contaminants into innocuous by-products such as carbon dioxide, water, and inorganic salts. In many cases, however, only partial degradation of a contaminant may occur with the consequent generation of intermediate products. It is important to determine the identity, potential toxicity, and mobility of these partially degraded compounds, since they may be nonbiodegradable in the affected soil. There are numerous examples of intermediate compounds being formed which are more toxic than the parent compound. Additional remedial actions may be necessary for their removal.

Gaseous emissions will also be generated during in situ biodegradation. Off-gases will vary in terms of contaminant composition (including potential toxicity), concentration, and volume. These emissions, which may consist of the original contaminant or any volatile degradation products, must be monitored and may require collection and treatment. Vapor treatment commonly includes passage over GAC, catalytic oxidation with metal powders, biofiltration, venting, or direct burning. Any by-products of emissions treatment will require appropriate disposal. It should be noted that activated charcoal is not well suited for removal of halogenated aliphatic compounds and short chain aliphatics. It does, however, have an affinity for a range of aliphatic compounds and aromatics such as BTEX.

11.6.1.9 Recovery Systems

Groundwater at an in situ treatment system will need to be intercepted and ultimately recovered for treatment and disposal. During passive (i.e., gravity) treatment, the process water is intercepted downgradient from the contamination zone, typically collected by gravity flow in open ditches or buried drains. Active suction via standard well pumps is occasionally used; however, this will increase maintenance costs.

11.6.1.10 On-Site Concerns

A number of variables must be regularly monitored and controlled in order for in situ bioremediation to successfully decompose the hydrocarbon plume. Microbial activity is reduced by deficiencies in nutrients, moisture and oxygen levels. Extremes in soil pH and soil temperature limit microbial diversity and activity. Spatial variation of soil moisture, oxygen, pH, and nutrients may result in inconsistent biodegradation due to variations in biological activity. Low hydraulic conductivity can restrict movement of water, nutrients, and electron acceptors (e.g., oxygen and nitrate) through the contamination zone. Low percolation rates may cause amendments to be assimilated by soils immediately surrounding application points, preventing them from reaching remote areas.

High metal concentrations can adversely affect the bioremediation of organics in soil or groundwater. Several metals can be transformed (i.e., oxidized, reduced, methylated, etc.) by organisms to produce new contaminants. The solubility, volatility, and sorption potential of the original soil contaminants can subsequently be altered, leading to potential toxicological effects. Examples include the methylation of Hg and As (see chapter 2).

11.6.1.11 Clogging and Its Control

A common operating problem during in situ bioremediation is clogging of the injection wells and aquifer. This is commonly a result of the presence of suspended solids, metal precipitates, and biological growth (biofouling). When suspended solids clog aquifers, hydraulic conductivity is significantly decreased. Suspended solids occur due to the presence of clays from the recovery well, floc from poor surface treatment of recovered water, or biological growth. Suspended solids should be reduced to levels less than 2 mg/L in the injected water (Vance 2008). Chemical precipitation results from the interaction of nutrients, substrate, oxygen, and biological activity with aquifer components. Changes in soil redox potential and the addition of nutrients may stimulate precipitation reactions. Groundwater in an aquifer typically has a low redox potential because natural processes will have exhausted most of the native oxygen. This frequently results in high levels of dissolved minerals, especially Fe. When groundwater is recovered and treated, oxygen will inevitably be introduced even if an oxygen source is not added. Reintroduction of this oxygenated water can result in precipitation of Fe and other metals. Phosphate has caused aquifer clogging due to the formation of orthophosphate (Jeong et al. 2017; Cookson 1995). Phosphate should not be injected into calcareous soil because much added phosphate will be adsorbed and precipitated. Alternate phosphate sources have been used for in situ bioremediation and include sodium tripolyphosphate and other polyphosphates (Cookson 1995; Norris et al. 1993).

Microbial clogging of injection wells is another significant problem with in situ bioremediation. The addition of oxygen and nutrients stimulates biological growth at the well openings and in adjacent soil. To reduce clogging at the injection well, a pulse injection procedure can be utilized. The electron acceptors and nutrients are injected separately at alternating intervals. The separate injection of agents results in suboptimal conditions at the injection point, thus discouraging biological growth (Cookson 1995; Norris et al. 1993).

A common form of biofouling results from growth of iron bacteria. Two species associated with well biofouling are *Gallionella ferruginea* and *Leptothrix* spp. (Fleming et al. 2014; Tuhela et al. 1993). *Gallionella* is a key organism responsible for biofouling and is ubiquitous in the biosphere. The organism derives carbon from CO_2 or from organic compounds and oxidizes ferrous iron to the ferric form:

$$4Fe^{2+} + O_2 + 4H^+ \rightarrow 4Fe^{3+} + 2H_2O \qquad (11.4)$$

Ferric oxides with low solubilities such as ferrihydrite ($Fe_2O_3 \cdot 0.5H_2O$) subsequently form.

$$5Fe^{3+} + 12H_2O \rightarrow Fe_2O_3 \cdot 0.5H_2O + 4H_2O + 15H^+ \qquad Ksp = 10^{-40}$$

(11.5)

Precipitates will form on well screens and in surrounding aquifer media. Such deposits can be controlled by periodically adding higher concentrations of H_2O_2. It may also be possible to use HCl to solubilize Fe precipitates.

11.6.1.12 Summary

In situ bioremediation, which avoids excavation and many emissions control costs, is generally cost-effective. This is attributed in part to low operation and maintenance requirements. During setup and operation, material handling requirements are minimal, resulting in limited worker exposure and reduced health impacts. Although in situ technologies are generally slow and at times difficult to control with precision, a large volume of soil may be treated.

Although in situ biodegradation may be successful in degrading and removing organic contaminants from one site, the identical treatment system may not be effective at another. Contaminant mixtures may be highly complex and create a hostile environment for biota. Elevated concentrations of metals, pesticides, chlorinated organics, and inorganic salts may occur in the soil which inhibit microbial activity and overall treatment performance. Additionally, variations in soil organic matter content, moisture regime, pH, and other factors will influence microbial activity. For these reasons, it is strongly recommended that treatability studies be conducted in order to assess the effectiveness of a particular in situ bioremediation strategy for a site.

11.6.2 Slurry Biodegradation

In slurry biodegradation, contaminated soil or sludge is removed from the affected area and mixed with water to produce a slurry. This suspension is transferred to a reactor vessel or a lined lagoon where it is continuously mixed and aerated. . Decomposition of organic contaminants takes place via aerobic processes. Slurry biodegradation has been shown capable of treating soils with contaminant concentrations up to 250,000 mg/kg. The technology has been used to treat a range of organic contaminants such as coal tars, wood-treating wastes, API separator sludge, pesticides, fuels, and some halogenated organic compounds (Thomas et al. 2006; Kim and Weber 2005; Collina et al. 2005). It is especially suited to the treatment of coal tars, refinery wastes, and organic and chlorinated organic sludges. The presence of heavy metals and other potential toxins may inhibit microbial metabolism

Table 11.6 RCRA-listed hazardous wastes treated by slurry biodegradation

K001	Wood-treating wastes
K048	Dissolved air-flotation (DAF) wastes
K049	Slop oil emulsion solids
K051	American Petroleum Institute (API) separator sludge

Source: U.S. EPA 1990c.

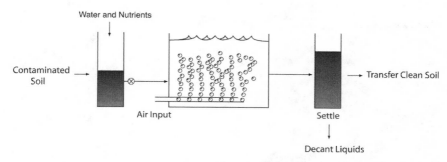

Figure 11.7 Schematic of a slurry bioremediation system. Source: Reproduced with kind permission of the McGraw-Hill Companies, Inc.

and require pretreatment. Listed Resource Conservation and Recovery Act (RCRA) wastes that have been treated are shown in table 11.6 (U.S. EPA 1989). A schematic of a slurry bioremediation system is shown in figure 11.7.

A significant benefit with use of slurry biodegradation is the accelerated rate of contaminant degradation, a direct result of enhanced contact between microorganisms and hazardous compounds. The agitation of contaminants in the water phase provides for a high degree of solubilization and greater homogeneity.

Slurry phase bioremediation is usually conducted as a batch process. Waste preparation includes excavation, screening to remove large debris, and transferring the soil material to the reactor area. Reactors include lagoons and open and closed systems (figure 11.8). The treatment train typically requires a settling tank or thickener and solids dewatering. Other preparation steps may include particle size reduction, water addition, nutrient and other microbial enhancer (e.g., surfactant) additions, and pH adjustment. Optimal feed characteristics for slurry biodegradation are (Robles-Gonzalez 2008; Cookson 1995; U.S. EPA 1990c; Richards 1965):

organic content	0.25%–25% (w/w)
temperature	15°–35°C
solids content	10%–30% (w/v)
pH	6.75–7.25
water	70%–90% (w/w)
solids particle size	< 1/4 in

Figure 11.8 Slurry bioremediation in an open lagoon. *Source*: Reprinted with kind permission of Slashbuster.com, Montesano, WA.

The soil is mixed with water, suspended in slurry form, and mixed in a vessel to maximize contact between contaminants and populations of biodegrading microorganisms. The slurry phase can be 70 to 90 percent water by weight. Aeration is provided by floating or submerged aerators, compressors and spargers, or by dredging (figure 11.9).

Figure 11.9 Aeration during slurry bioremediation via vigorous mixing in an open lagoon. *Source*: Reprinted with kind permission of Slashbuster.com, Montesano, WA.

Aeration and mechanical agitation promote mixing. Nutrients are supplied to support microbial proliferation and activity; likewise, pH is adjusted as needed. Other supplements such as surfactants can be used to increase substrate availability for biodegradation (Adrion 2016). Microorganisms may be added initially to seed the bioreactor or added continuously to maintain a preferred concentration of biomass.

It is necessary to monitor process control frequently in slurry reactors. The high loading rates of organic-contaminated soil plus the effects of mixing and microbial processes can cause rapid changes in pH, microbial populations, nutrient needs, and electron acceptors.

Once biodegradation of contaminants is complete, the treated slurry is dewatered. A clarifier for gravity separation, hydrocyclones, or any standard dewatering equipment can be used to separate the solid and aqueous phases of the slurry.

11.6.2.1 Reactors

The slurry reactor can be an engineered containment unit or a lagoon. Reactor designs are available that differ in mechanics of oxygenation and mixing of the solid suspension. Floating direct-drive mixers with draft tubes, turbine mixers, and spargers provide both aeration and mixing. Systems also vary significantly in terms of energy requirements. Because retention times are often long, reactors that use less energy are the most cost-effective. The use of open lagoons as holding ponds for hazardous and nonhazardous waste sludges had been common practice for decades. These may measure as large as several acres. A waste lagoon can be converted to a slurry bioremediation vessel. The relatively minimal materials handling reduces risks to workers, nearby populations, and the local environment. However, engineering lagoon bioremediation provides more difficult challenges than for surface bioreactors. Sludge lagoons can contain a broader chemical mixture as compared with pre-treated material placed in an engineered containment unit. In addition, the chemical nature of sludges at one end or depth of a lagoon may be substantially different at another location. Lagoon solids frequently require physical breaking and continuous dispersion to accomplish reasonable rates of biodegradation.

Mixing is an important process variable in slurry systems. Mixing must (Robles-Gonzalez 2008; Cookson 1995):

1. Maintain biomass, soil, and other solid particles in a suspended state
2. Lift soil from the reactor base
3. Break up larger, insoluble particles into smaller ones that remain suspended
4. Emulsify oils and other NAPLs

5. Vary the degree of mixing to prevent an increase in toxicity, while maximizing degradation rates

Air sparging alone has not been adequate in maintaining mixing in slurry reactors. Most soils and sludges are either too viscous or have settling velocities too great for suspension by sparging equipment. Air sparging also causes excessive volatiles emissions.

A key engineering decision is selection of the microbial metabolism mode for degradation of the hazardous compounds. Because of the possible wide range of chemical properties of a contaminated soil or sludge, it is unlikely that one system will achieve complete treatment. If an aerobic system is selected, stripping of volatile compounds becomes a concern to the operator. Volatile emissions have been a serious problem at some lagoon treatment facilities. Oxygen supply can be augmented via the addition of pure oxygen and hydrogen peroxide. These will serve to minimize volatile emissions. In contrast, because air contains 20.9 percent O_2, large volumes of air must be injected for delivery of a given amount of oxygen. This excess gas is released from the lagoon along with the transport of volatile emissions.

In the case of soil contaminated with chlorinated hydrocarbons, an anaerobic metabolism mode can be provided, followed by an aerobic phase. The anaerobic phase has the advantage of dehalogenating the chlorinated volatile compounds. In a subsequent phase, aeration is provided for aerobic degradation of the remaining hydrocarbons. Disadvantages of anaerobic treatment include production of odors and overall slower rate of degradation. Odor collection and treatment increases project costs. The application of an anaerobic metabolism mode followed by an aerobic phase has several advantages, however:

1. Anaerobic conditions can dehalogenate chlorinated volatile compounds.
2. There is greater detoxification of chlorinated aromatic compounds.
3. Release of chlorinated volatile compounds to the atmosphere will be limited.
4. After anaerobic dehalogenation of hydrocarbons, oxygen can be provided for aerobic degradation of the remaining compounds.
5. Operating cost is reduced because oxygen is not required during anaerobic treatment.

11.6.2.2 Residuals and Wastes

The primary waste streams generated in the slurry biodegradation system are treated solids (sludge or soil), process water, and off-gases. Solids are dewatered and may be further treated if they are still contaminated with organics. Air emissions (e.g., BTEX) are possible during system operation.

Air pollution control, for example passage of off-gases over a bed of activated carbon, may be necessary.

11.6.2.3 Summary

Slurry biodegradation is not a feasible treatment method for all sites. Treatability tests should be conducted in order to determine the biodegradability of the contaminants and the solids/liquid separation that occurs at the end of the process.

Advantages of slurry phase bioremediation include greater and more uniform process control compared to other biological treatment technologies. The continuous mixing provides for increased contact between microorganisms and contaminants; improved distribution of nutrients, electron acceptors, or primary substrates; breaking of soil-sludge agglomerations; and more rapid biodegradation rates as compared to solid (e.g., in situ) systems. Furthermore, addition of surfactants can enhance solubilities of contaminants.

Slurry phase systems have several disadvantages, which are related to additional process requirements and materials handling, resulting in higher costs. Significant quantities of wastewater can result from solids separation and dewatering after slurry remediation. Wastewater may require treatment before discharge. However, these disadvantages are often offset by the improved performance of slurry bioremediation over solid phase treatment.

11.6.3 Liquid Phase Bioremediation

There are situations in which it is advantageous to recover contaminated groundwater and treat it in an aboveground bioreactor. Microorganisms are encouraged to proliferate and are retained under optimized process conditions. Treated water is subsequently returned to the site or sent to a POTW. The original principles of design originate from municipal and industrial wastewater treatment facilities. Several technologies are well adapted to treatment of contaminated groundwater; however, some of the theory behind operations will differ. In the treatment of domestic wastewater, reactor designs are based on the destruction and removal of a predictable range of contaminants, for example, dissolved organic matter. In the case of site remediation, however, contaminants vary from site to site and often occur in highly complex mixtures; therefore, designs of bioreactors are typically site-specific. Once again, treatability studies are recommended before designing field reactors.

Bioreactors usually fall into one of two categories: suspended growth and fixed film. In suspended growth reactors, microbial populations are suspended as aggregates in the liquid to be treated (typically groundwater). The biomass is termed *activated sludge*. In fixed-film reactors, biomass attaches to a solid support having very high surface area. This support layer is provided with a regular supply of oxygen while at the same time contacting the contaminated groundwater as a microbial carbon source.

11.6.3.1 Activated Sludge

The activated sludge process promotes microbial activity within a suspended growth reactor, typically operated in a continuous mode. The major components of the process are an aeration tank and a sedimentation tank (figure 11.10). In the aeration tank biomass reacts with hydrocarbons in the influent stream. The solution is aerated and mixed thoroughly. Mixing is provided using physical mixers or air spargers fed by compressors. After the appropriate reaction time in the bioreactor, the biomass is transferred to a sedimentation tank. The suspended biomass accumulates quickly; therefore, sedimentation is necessary. The biomass is allowed to coagulate and settle. The clarified liquid is collected and is either discharged or treated further. The sludge is removed from the tank bottom. A portion of the biomass is recycled to the bioreactor, which allows for sustained microbial decomposition of the contaminants. The recycled biomass is already acclimated to the waste stream constituents; there will, therefore, be no lag time for the microbes to act on the newly added waste.

Activated sludge treatment systems are susceptible to shock loadings. Also, because of the difficulty in maintaining biomass, this technique is of only limited use for treating organic-contaminated groundwater (Langwaldt and Puhakka 2000; Cookson 1995). In limited cases, activated sludge treatment of groundwater has been coupled with industrial wastewater treatment.

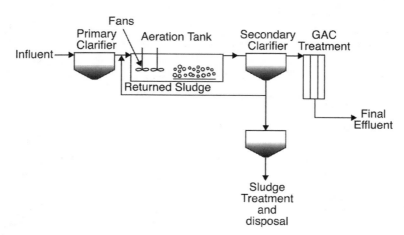

Figure 11.10 Schematic of an activated sludge process.

11.6.3.2 RBCs

Rotating biological contactors (RBCs) are an example of a fixed-film reactor for treatment of contaminants in aqueous waste streams. An RBC is composed of a series of closely spaced corrugated plastic disks mounted on a horizontal shaft. The disks rotate through the aqueous waste; approximately half of the RBC surface is immersed. The remainder is exposed to the atmosphere, which provides oxygen to attached microorganisms and promotes oxidation of the contaminants. Microbes proliferate on the surface of the disks and degrade the organics occurring in the water. The substantial microbial population provides a high degree of contaminant treatment within a short time.

A typical RBC unit consists of 12-ft (4-m) diameter plastic disks mounted along a 25-ft (7.6-m) horizontal shaft. The total disk surface area is normally 100,000 ft^2 (9290 m^2) for a standard unit and 150,000 ft^2 (1390 m^2) for a high-density unit (figure 11.11) (Ecologix 2018; U.S. EPA 1992b).

The overall RBC treatment process involves a sequence of steps (figure 11.12): aqueous liquids are transferred from a storage tank to a mixing tank where reagents are added for pH adjustment, metal precipitation, and nutrient addition. The waste stream then enters a clarifier where floating debris, grease, metals, and suspended solids are separated from the raw influent by gravity (i.e., *primary treatment*). Clarifiers or screens remove material that could settle in the RBC tank or plug the disks. The wastes collected from primary clarification may contain metallic and organic contaminants and may require additional treatment.

The effluent from the clarifier enters the RBC. The disks rotate through the liquid at a rate of about 1.5 rpm. A microbial slime forms on the disks,

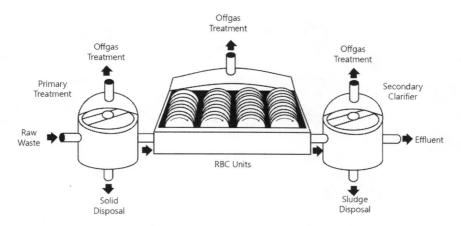

Figure 11.11 An RBC reaction chamber.

292   *Chapter 11*

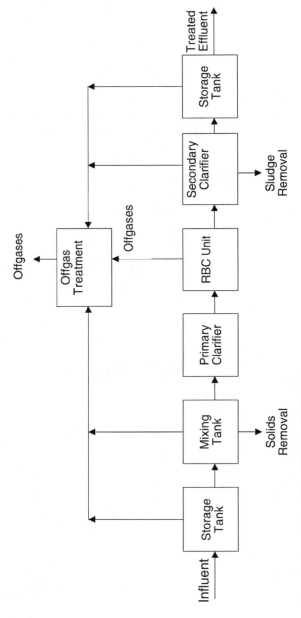

Figure 11.12 Schematic of the overall RBC system.

which degrade the organic and nitrogenous contaminants present in the waste stream. The rotating motion of the disks through the aqueous waste causes excess biomass to shear off regularly. After RBC treatment, the effluent enters a secondary clarification stage to separate suspended biomass solids. Additional treatment of solids and liquids may be required. Clarified secondary effluents may be discharged to a surface stream while residual solids must be disposed appropriately.

Biological systems can degrade only the soluble fraction of organic contamination; therefore, the applicability of RBC treatment is ultimately dependent on solubility of the contaminant. RBCs are generally applicable to influents containing organic concentrations of up to 1 percent organics, or between 40 and 10,000 mg/L of SBOD. RBCs can be designed to reduce influent BOD concentrations below 5 mg/L SBOD, and ammonia-nitrogen (NH_3-N) levels below 1 mg/L (Walker et al. 1988). RBCs are effective for treating solvents, halogenated organics, acetone, alcohols, phenols, cyanides, ammonia, and petroleum products.

Staging, which employs a number of RBCs in series, establishes biological cultures acclimated to successively decreasing organic loadings. As the waste stream passes from stage to stage, more aggressive levels of treatment occur (U.S. EPA 1992b; Envirex n.d.). Staging can also improve the system's ability to handle shock loads by absorbing the impact of a load in the initial stages. RBCs can be compartmentalized with baffles or with separate tanks to provide for distinct microbial communities (Cortez et al. 2013).

Removal efficiency of RBC systems is affected by the type and concentration of organics present, liquid residence time, rotational speed, media surface area exposed, and pre- and posttreatment. Issues to be addressed in RBC design include organic hydraulic loading rates, design of the disk system, rotational velocity, tank volume, media area exposed, retention time, primary treatment and secondary clarifier capacity, and sludge storage and treatment (U.S. EPA 1992b, 1987). Storage should be provided to hold the aqueous treated product until it has been tested to determine its acceptability for disposal or reuse.

Gases may be generated throughout the RBC treatment process as a result of aeration by mixing and sparging processes, and by volatilization of hydrocarbons. The gases must be collected for treatment.

As is the case of the bioremediation systems discussed thus far, RBCs are not effective at removing inorganics or nonbiodegradable organics. Water containing high concentrations of metals and certain pesticides, herbicides, or highly chlorinated organics resist RBC treatment by inhibiting microbial activity. In such cases, pretreatment of waste streams may be necessary to remove toxic compounds prior to treatment. RBCs are susceptible to excessive biomass growth, particularly when organic loadings are high. If the biomass fails to slough off and an excessively thick (approximately 90–125 mil) biomass layer forms, the shaft and/or disks may be damaged.

All bioremediation systems, including RBCs, are sensitive to temperature variation. Biological activity decreases at temperatures higher than 55°C. Covers should be employed to protect the units from colder climates. Covers are also recommended to inhibit algal growth and to control the release of volatiles (U.S. EPA 1992b).

11.6.3.3 Trickling Filter

A trickling filter (packed media) consists of a bed of coarse materials such as stones, slats, or plastic media, over which wastewater is slowly applied. Trickling filters have been a popular biological treatment process and are common for treating municipal wastewater for BOD removal. Trickling filters do not actually filter, as the name implies. The rocks in a filter measure 25 to 100 mm in diameter and hence the openings are too large to strain out solids.

A common design consists of a packed bed of stones placed from 1 to 3 m deep in a large diameter basin through which the wastewater passes (figure 11.13). The bed of stones provides a vast surface area where microorganisms attach and grow as they feed upon the organic matter. As the wastewater trickles through the bed, microbial growth establishes itself on the surface of

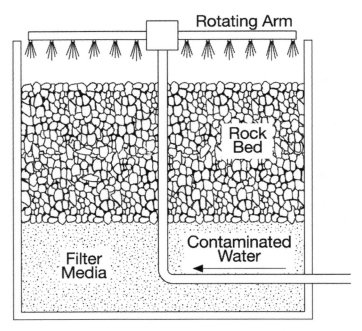

Figure 11.13 **A trickling filter system for treatment of groundwater contaminated with hydrocarbons.**

the support material in a fixed film. The wastewater is typically distributed over the surface of the rocks by a rotating arm. The wastewater passes over the attached microbial layer, providing contact between cells and substrate. Basin diameters may be as much as 60 m.

Excess cellular biomass washes from the rock media regularly and will cause excessive levels of suspended solids in the effluent if not removed. The flow from the filter is passed through a sedimentation basin (secondary clarifier or final clarifier) to allow these solids to settle out. Under high organic loadings, slime growths can be so heavy that they plug the void spaces between the rocks, causing flooding and failure of the system. The volume of void spaces is limited in a rock filter, restricting air circulation and available O_2, and hence the amount of organic waste that can be treated.

The residuals produced in a fixed-film reactor are normally small when compared to those produced in a suspended growth system. Depending on site-specific requirements, it may be necessary for the residuals to undergo additional treatment prior to disposal. Residuals must be disposed appropriately (e.g., land disposal, incineration, solidification).

11.6.4 Land Treatment

Bioremediation via landfarming (bioreclamation, biopiles) is commonly used for treatment of contaminated soil; however, certain industrial sludges and petroleum wastes have also been incorporated into soil for microbial decomposition. Contaminants most commonly treated are petroleum compounds including fuel, lubricating oil, and organic wood preservatives. Other applications include soil contaminated with coal tar wastes, pesticides, and explosives. Until passage of the LDR (40 CFR 2016), petroleum refineries used land treatment to dispose of a wide range of hazardous wastes. Landfarming can be regarded as a combination of biodegradation and soil venting—microbial oxidation reactions occur in combination with volatilization.

A common field installation calls for the affected soil to be excavated and transferred to a prepared location (i.e., a land treatment unit, LTU, or cell) which is designed for optimal control of the process. Treatment involves the installation of layers (lifts) of contaminated soil to the cell. The LTU is usually graded at the base to provide for drainage and lined with clay and/or plastic liners to contain all runoff within the LTU. It may also be provided with sprinklers or irrigation, drainage, and soil water monitoring systems. LTUs are often bordered by soil or clay berms in order to prevent losses from overwatering (figure 11.14).

The level of hydrocarbon contamination suitable for land treatment varies with contaminant type and site conditions. In some cases, soils with higher

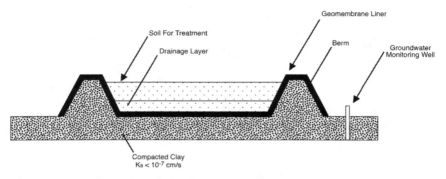

Figure 11.14 Cross section of a land treatment unit.

levels of contaminants than are recommended for land treatment can be mixed with less contaminated soil to achieve desired starting levels. Petroleum contamination as high as 25 percent by weight of soil has been reported as treatable, although 5 percent to 8 percent by weight or less is more readily treated (Philp and Atlas 2005; EPRI 1988; Pope and Matthews 1993).

A major benefit of the land treatment technique is that it allows for very close monitoring of process variables that control the decomposition of hydrocarbons. A typical landfarming field system is designed and implemented as follows:

11.6.4.1 Cell Preparation

The LTU is constructed by first preparing the base. Large debris is removed to protect liners placed upon the base, and grading is necessary in order to control runoff from the LTU. The first layer installed is a liner of compacted clay. A second, optional layer may be a geomembrane liner, for example chemically resistant PVC or HDPE. A drainage system is then installed above the liners for leachate collection. This may consist of a network of perforated PVC piping situated within a highly permeable bedding, for example pea gravel. A sand or soil layer, ranging from 2- to 4-ft thick, is placed above the liner(s) and drainage system to protect them from heavy equipment (figure 11.15). Ditches or berms are installed at the periphery of the cell in order to prevent run-on and to capture runoff. Requirements for the site will vary depending on local regulations.

11.6.4.2 Soil Preparation

The contaminated soil is excavated and screened to remove debris (greater than about 1-in diameter), and placed on the base of the porous sand or soil subbase. Tractors with specialized equipment that can till to depths of 3 ft or more have been used for in situ land treatment. Large augers are now

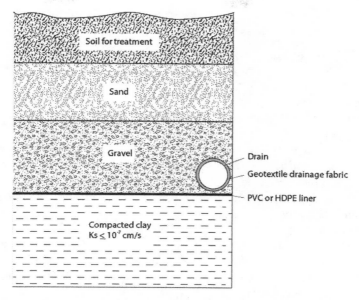

Figure 11.15 Close-up of an LTU liner and leachate collection system.

available that can move soil from 50- to 100-ft depths to the surface for eventual transfer to a LTU.

11.6.4.3 Addition of Nutrients

As with other bioremediation systems described, microorganisms will require a carbon source and nutrients, both for adequate growth and for efficiently carrying out degradation reactions. Simple agricultural fertilizers containing N, P, and K can supply the required macronutrients.

Soil nitrogen can be applied as fertilizer-grade ammonium nitrate or urea, and phosphorus as triple superphosphate. Inorganic nutrients are generally added in mass ratios of approximately 100:10:1 hydrocarbon:N:P. Small amounts of potassium are sometimes added. Ammonium fertilizers are known to decrease soil pH as a consequence of nitrification reactions (equations 11.7 and 11.8). Agricultural fertilizer is often supplied in pellet form, which is suitable for application over large areas. Pelletized application furthermore serves as a slow-release nutrient supply. Water-soluble fertilizers can be applied through irrigation systems, allowing application rates to be closely controlled. This will also allow for immediate availability to microorganisms.

$$2NH_4^+ + 3O_2 \rightarrow 2NO_2^- + 2H_2O + 4H^+ \tag{11.6}$$

$$2NO_2^- + O_2 \rightarrow 2NO_3^- \tag{11.7}$$

Wood chips, sawdust, or straw serve as low-cost sources of carbon. Animal manures are often used to supply both carbon and nutrients. These can be broadcast and tilled into the soil surface. The high organic levels in these amendments increase sorptive properties of soil, thereby decreasing mobility of organic contaminants. Organic amendments also increase soil water-holding capacity, which is beneficial in sandy soils but causes difficulty in clayey soils and in LTUs having poor drainage. An additional benefit of manure application is that it provides an enormous and diverse biomass of heterotrophic microorganisms. Manure should be applied to each lift at the rate of about 3 to 4 percent by weight of soil. The manure should first be sieved and then thoroughly tilled into the soil (Chen et al. 2010; Pope and Matthews 1993).

The manure additive should be analyzed for total N and P content. The operator should also be aware that fresh manure may contain excessive quantities of soluble salts and may also release significant quantities of ammonia, both of which inhibit microbial growth. If salt levels, as indicated by electrical conductivity (EC) readings are high, it may be necessary to leach the soil with nonsaline water to remove excess salts before biodegradation can occur. High levels of sodium may be detrimental to soil structure. Applying calcium supplements such as gypsum, $CaSO_4$, followed by leaching may reduce sodium levels. Leaching of contaminants may also occur at the same time and must be monitored.

Micronutrients (e.g., Cu, Ni, Zn) are sometimes needed to enhance microbial activity in sterile environments (e.g., contaminated subsoils and other geologic strata). Animal manures are a low-cost material that typically can supply a range of micronutrients.

As a result of mineralization reactions, soil pH may need to be adjusted. Common liming agents include CaO and $CaCO_3$, among others. Acidification can be carried out with simple inorganic acids (e.g., sulfuric, phosphoric), elemental sulfur, or aluminum sulfate.

11.6.4.4 Inoculation

Exogenous microbial cultures are becoming increasingly popular for addition to bioremediation units; however, exogenous microorganisms rarely compete well enough with indigenous populations to sustain useful population levels. Additionally, most soils that have been exposed to biodegradable wastes over the long term already contain adequate numbers of indigenous microbial populations that are effective degraders.

11.6.4.5 Tillage and Mixing

Effective land treatment is limited to the top 12 to 24 in of soil; at depths below 12 in the oxygen supply is generally insufficient for treatment by aerobic heterotrophic microbes. Therefore, after placement of soil in the LTU and after addition of nutrients, each lift should be tilled, disked, or plowed. Tillage will enhance oxygen infiltration, mixing of additives, contact of contaminant with microorganisms, and produce a more homogeneous growth media. Tillage generally will require several passes with a disk harrow or similar equipment. The soil is then watered and allowed to incubate. The soil can be tilled at intervals between approximately two weeks and three months, depending on soil and contaminant properties. Tilling should not occur until at least twenty-four hours after irrigation. Tilling more than is necessary for enhanced oxygen incorporation and contaminant mixing may destroy soil structure and compact soil below the treatment zone.

Land treatment may be restricted in clayey soils, especially if they remain relatively moist. This limitation is related to difficulties in oxygen transfer. Clayey soils should be applied in shallower lifts than sandy soils. Permeability, oxygen incorporation, and tilth can be improved by adding organic matter or other bulking agents to the soil. A cover may be placed over the LTU to limit rainfall infiltration.

11.6.4.6 Moisture Control

Soil moisture content should be near the lower end of the recommended range before tilling. Turning of saturated soil may destroy soil structure, reduce oxygen and water incorporation, and limit microbial activity.

Careful maintenance of soil moisture is critical to optimizing bioremediation in LTUs—monitoring soil moisture and scheduling irrigation is essential. If soil is allowed to become excessively dry, cells may lyse and microbial activity will diminish. Too much soil moisture will limit microbial activity by excluding oxygen and favoring anaerobic conditions. Furthermore, since these soils are nonvegetated, excess water may result in erosion of both soil and contaminants. The desired moisture level for the soil to be treated can be determined in the field or laboratory using simple tests. A range of 70 percent to 80 percent field capacity is suggested as an optimum level. A soil is at field capacity when applied water no longer drains by the action of gravity—that is, micropores are filled with water and macropores with air. This condition provides for adequate amounts of both air and water to soil microorganisms.

The addition of a bulking agent such as wood chips, sand, or sawdust increases porosity of the soil and improves drainage.

Surface drainage of the LTU can be critical in the event of excessive irrigation or high rainfall. If soil is saturated for more than a few hours, aerobic microbial activity may be reduced. The contaminated soil layer will take up water from irrigation or rain until the soil nears saturation, at which point excess water will be discharged into the drainage system.

11.6.4.7 Leachate Treatment

Hydrocarbon degradation can be a slow process, relying on a succession of microbial communities. As the microbially catalyzed reactions proceed, leachates produced can be recycled back to the soil in the LTU. Depending on local climate, it may be useful to install a roof or other cover to limit precipitation contacting the cell. This will reduce the quantity of leachate generated for treatment. Liquid storage capacity should be provided for runoff and leachate water for eventual recycling. In many cases leachate and runoff water cannot be discharged without treatment. Monitoring of groundwater and soil may be performed for large operations to ensure that contaminant hydrocarbons are present only within the treatment area.

In some situations, soils that contain petroleum compounds can be applied to the LTU at regular intervals. Reapplication will replenish the carbon supply and maintain biological activity at the desired level.

11.6.4.8 Completion of the Project

Measurement of the success of degradation reactions can be conducted using portable field tools, for example a PID for the nonspecific measurement of gaseous hydrocarbons, to gas chromatographic analysis of soil and soil water for concentrations of specific contaminants and products of decomposition. Several methods for hydrocarbon analysis are listed in chapter 5. Once the desired goals are realized the soil can be returned to the excavated site or applied to other uses.

11.6.4.9 Overview of Landfarming

Landfarming is a relatively inexpensive and low-technology practice, and is not highly labor intensive. This method requires large amounts of land. It also may result in the release of contaminants to air. As with other bioremediation applications, pesticides, halogentated hydrocarbons, metals and inorganic salts will inhibit microbial activity, especially at low pH. Used motor oil may not be appropriate for landfarming due to its relatively high metal and PAH content (see chapter 3).

Many gasoline components possess a high vapor pressure (i.e., are of significant volatility); therefore, land treatment of gasoline-contaminated soil will result in losses by volatilization. A permit for hydrocarbon emissions may therefore be needed. Diesel and kerosene fuels possess some volatiles but weathered forms that are low in volatile components may be appropriate for land treatment. Most of the heavier petroleum products (e.g., fuel oils, lubricating oils, waste sludges, and oils) are also suitable for land treatment. Petroleum products that consist mostly of long-chain hydrocarbons (20+ carbons) and other high molecular weight compounds (asphalts, tars, 5–6 ring PAHs) are not suitable for land treatment because of their biorefractory nature and the difficulty of mixing them with soil.

11.7 BIOREMEDIATION: AN OVERVIEW

Bioremediation has the advantage of being less costly than conventional cleanup methods such as incineration or excavation and landfilling of toxic soil material. Additionally, this technology will degrade and detoxify the contaminant rather than simply transfer it, as in the case of soil vapor extraction or soil flushing. Bioremediation is a relatively simple technology; it can be carried out with minimal disruption to the affected site, few emissions of volatile organics, and minimal health risks to workers on-site. There are disadvantages, however, to its utilization. There is insufficient data regarding products of biological decomposition of several chlorinated and non-chlorinated contaminants. It is also difficult to observe in the field the same successes measured in the laboratory, where conditions are controlled and relatively uniform. Success of the entire process is related to careful control of environmental factors including moisture and oxygen content, nutrient level, and pH and temperature range. Cleanup goals may not be achieved because some contaminants may be only partly biodegradable. Furthermore, as bioremediation proceeds, degradation reactions will slow and the cells may switch to other substrates or stop growing altogether.

11.8 CASE HISTORY 1

The French Limited Superfund Site in Crosby, Texas, is a former industrial waste storage and disposal facility measuring 22.5 acres. Between 1966 and 1971, approximately 70 million gal of industrial wastes from local petrochemical companies were disposed at the site. Wastes included tank bottoms, pickling acids, and off-specification product from petroleum refineries and petrochemical plants. Most of the waste was deposited in an unlined 7.3-acre lagoon. Wastes were also processed in tanks and burned (U.S. EPA 2018a, 1995b).

The lagoon had been established in an abandoned sand pit that filled with water to a 20- to 25-ft depth. Primary contaminants identified in the lagoon included PAHs, halogenated semivolatiles, halogenated volatiles, nonhalogenated volatiles, metals, and nonmetallic elements. The wastes were concentrated in a 4-ft-thick layer of tar-like sludge at the base of the lagoon and in 5 to 6 ft of subsoil (U.S. EPA 1989, 1995b; Hasbach 1993). The thick, viscous, oily layer of sludge consisted of a mixture of petrochemical sludges, kiln dust, and tars (primarily styrene and soils). Specific constituents included PCBs at concentrations up to 616 mg/kg; volatile organics up to 400 mg/kg for an individual contaminant; pentachlorophenol up to 750 mg/kg; semivolatiles up to 5,000 mg/kg for an individual contaminant; and metals up to 5,000 mg/kg. Subsoil material varied from fine-grained silts to coarse sand.

EPA identified approximately ninety companies as PRPs for site cleanup, and in 1983 the PRPs formed a task group that agreed to perform the cleanup. The following remedial alternatives were considered for the French Ltd. Site:

- no action
- encapsulation of contaminants by slurry walls and a multilayered cap
- on-site incineration of sludge and contaminated subsoil
- on-site incineration of sludge and chemical fixation of contaminated subsoil in place
- biological treatment of sludge and contaminated subsoil

EPA proposed incineration as the remedial technology for the sludge and contaminated soil at an estimated cost of $75 to $125 million. The PRP task group subsequently investigated other more cost-effective alternatives. A pilot-scale bioremediation treatability study was conducted in a 0.6-acre section of the lagoon. As a result of the study, the Record of Decision replaced incineration with in situ biodegradation for remediation of the site (U.S. EPA 1995b, 1988; DEVO 1992). Biological treatment was selected because it was considered capable of meeting cleanup goals within a reasonable period and at a lower cost than incineration. However, if bioremediation was found to not be successful in the lagoon, the ROD called for the use of incineration.

The slurry phase bioremediation system used at French Ltd. was designed to stimulate indigenous microorganisms to oxidize hydrocarbon wastes via aeration, pH control, and nutrient addition. The tar-like sludge was sheared and converted to a mixed liquor using centrifugal pumps. A layer of subsoil was also scraped off and introduced into the liquor using subsoil mixers. Four mixers were installed for shear mixing of sludges, and four more provided for shear mixing of lagoon bottom subsoils. The mixed liquor contained about

5 to 10 percent solids. The sludge and subsoils were treated separately; this prevented the sludge from coating soil particles and maximized surface area available for treatment, thereby enhancing treatment effectiveness.

The lagoon was divided into two treatment cells of approximately equal volume. Cells were created by installing a sheet piling wall across the lagoon, allowing for sequential remediation. The benefits of this strategy included limiting air emissions, reducing the amount of capital equipment to purchase, and allowing for process improvements during the operation.

The treatment cells were designed to hold a total mixed liquor volume of 34 million gal (17 million gal in each cell), and to maintain a minimum dissolved oxygen concentration of 2.0 mg/L. Based on treatability studies, an oxygen uptake rate of 0.30 mg/L/min was considered appropriate for the aeration supply. Oxygen requirements for each cell were calculated to be about 2,500 lb/hr (U.S. EPA 1995b, 1988).

The main components of the slurry bioremediation process included a MixFlo aeration system, a liquid oxygen supply unit, a chemical feed system, and dredging and mixing equipment (figure 11.16). A pure oxygen aeration unit was selected over an air-based setup in order to limit organic emissions. The MixFlo technology has higher transfer efficiencies than air-based aeration systems (90% as opposed to 30%) and uses high-purity oxygen. The combination of higher transfer efficiency and high-purity oxygen reduces the amount of organic off-gases from the treatment process (Hasbach 1993). The MixFlo system dissolves oxygen in a two-stage process. First, water is pumped from the treatment area and pressurized. Pure oxygen is then injected into the water. The resulting two-phase mixture passes through a pipeline contractor where 60 percent of the injected oxygen dissolves. In the second stage, the oxygen/water mixture is reinjected into the treatment area.

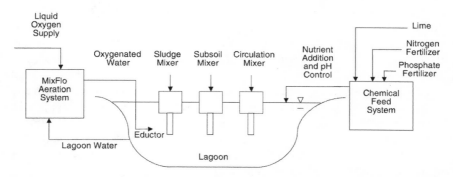

Figure 11.16 Flow diagram of the treatment process for the French Ltd. lagoon.
Source: U.S. EPA 1995b.

At the French Ltd. site, oxygen was injected via eight pipeline contractors into the mixed liquor. Furthermore, raft-mounted self-powered circulation mixers were utilized.

Batch systems for chemical addition were used to control pH and nutrient chemistry of the mixed liquor during treatment. Hydrated lime was diluted on-site to 15 percent concentration. To offset nutrient losses, N was added as hydrated urea (46% N by weight) and P as liquid ammonium phosphate. The system was designed to add batches of up to 1,500 gal of chemicals to the lagoon at several locations.

Treatment performance was monitored by analyzing subsoil, sludge, and mixed liquor samples. Five indicator compounds were reduced in concentration over the course of treatment. Benzene, for example, was reduced from 608.0 mg/kg to 4.4 mg/kg in one cell and from 393.3 mg/kg to 5.2 mg/kg in the second cell.

Cleanup criteria were achieved within ten months of treatment for the first cell and eleven months for the second cell. For individual constituents, cleanup goals were met earliest for vinyl chloride (i.e., four months in the first cell, one month in the second) and total PCBs (four months in the first cell, one month in the second). Benzo[a]pyrene required the longest treatment time to meet cleanup goals. Concentrations of vinyl chloride were reduced to BDLs in the first cell and 6.6 mg/kg in the second; benzene was reduced to 4.4 mg/kg in the first cell and 5.2 mg/kg in the second; and benzo[a]pyrene to 6.0 mg/kg in the first cell and 6.8 mg/kg in the second. PCBs were biodegraded in the slurry phase system to concentrations below action levels established for the site (table 11.7). Apparently the arsenic and metals imparted no deleterious effects on activity of hydrocarbon-degrading organisms.

An air monitoring program was established to monitor potential releases of VOCs from the lagoon. Automated instrumentation was placed at the periphery of the treatment cell. The data revealed no excesses over the established criteria for releases of VOCs; total VOC concentrations ranged from 0.3 to 1.6 ppm, well below the action level of 11 ppm specified in the ROD. These data, in combination with operating data, indicate that organic compounds including vinyl chloride, benzene, and benzo[a]pyrene were removed from the lagoon via biodegradation.

After soil and sludge cleanup objectives were achieved, reverse osmosis was used to treat the surface water in the lagoon. Forty million gallons of water were processed through the system and ultimately discharged to the San Jacinto River. The emptied lagoon was backfilled with clean soil. The site was then planted with grass and native vegetation and contoured to drain away from the lagoon (U.S. EPA 1995b; Collins and Miller 1994).

The total cost of remediating the soil and sludge in the lagoon at French Ltd. was $49 million including costs for pilot studies, technology

Table 11.7 Concentration of contaminant compounds over time, first treatment cell

Day	Vinyl Chloride	Benzene	Benzo[a] Pyrene	Total PCBs	Arsenic
			mg/kg		
Cleanup Criteria	43.0	14.0	9.0	23.0	7.0
96	314.8	608.0	BDL	77.0	9.2
110	29.4	162.3	BDL	6.4	20.3
124	32.5	119.8	BDL	N/A	121.2
145	0.5	10.8	BDL	11.4	125.4
166	0.8	25.6	BDL	15.8	N/A
187	1.2	41.9	14.4	13.4	17.9
201	BDL	18.8	27.4	10.3	84.7
215	BDL	16.3	BDL	13.2	61.8
229	BDL	10.9	28.0	3.1	10.2
239	1.2	7.3	33.7	8.7	1.2
253	4.8	12.1	26.3	14.7	1.4
272	4.3	7.5	3.1	7.4	BDL
275	BDL	7.3	8.7	6.8	BDL
289	BDL	6.0	7.4	4.6	BDL
293	BDL	5.8	4.8	6.0	1.1
295	BDL	4.4	6.0	7.9	NR

BDL = below detection limit; NR = not reported.
Source: U.S. EPA 1995b.

development, project management, EPA oversight, and backfilling the lagoon (table 11.8). About 55 percent of project costs ($26,900,000) were for activities directly associated with treatment, such as solids/liquids/vapor/gas preparation and handling, pads/foundations/spill control, mobilization/setup, startup/testing/permits, training, and operation (three years).

About 34 percent of project costs were for before-treatment activities and 11 percent for after-treatment activities. The $26,900,000 in costs for

Table 11.8 Breakdown of project costs for the French Ltd. cleanup activity

Project	Cost ($)
Development and pilot-scale work	12,200,000
Floodwall	2,300,000
Operation, maintenance, analytical	22,900,000
Dewatering	1,000,000
Fixation	400,000
Technical support	2,900,000
Administrative	3,100,000
Demobilization	1,900,000
EPA oversight	2,300,000
TOTAL	49,000,000

Source: U.S. EPA 1995b.

activities directly associated with treatment converts to $90/ton of sludge and soil treated (300,000 tons treated) (U.S. EPA 1995b, 1988).

11.9 CASE HISTORY 2

The Burlington Northern Superfund site, located in Baxter and Brainerd, Minnesota, operated a railroad tie treating plant between 1907 and 1985. The preserving process involved pressure-treating wood using a heated creosote/coal tar or creosote/fuel oil mixture. Wastewater generated from the process was disposed into two shallow, unlined surface impoundments. The first impoundment measured approximately 60,000 ft^2 and was filled with sludge and buried under clean fill in the 1930s. A second impoundment was used from the 1930s until 1982. Soil and groundwater beneath these impoundments became contaminated as a result of waste leaching. Soil at three additional areas at Burlington Northern was also determined to be contaminated (U.S. EPA 2018b, 1995a; Remediation Technologies, Inc. 1995a, 1995b).

The site was found to be contaminated with PAHs and other nonhalogenated semivolatile organic compounds including naphthalene, acenaphthene, phenanthrene, anthracene, fluoranthene, pyrene, benzo[a]anthracene, chrysene, and others. Total PAH concentrations for so-called visibly contaminated soils in the surface impoundments were as high as 70,633 mg/kg, with individual PAHs measuring up to 21,319 mg/kg (acenaphthene), 7,902 mg/kg (phenanthrene), and 10,053 mg/kg (fluoranthene) (table 11.9).

As part of the mandated remediation activity, Burlington Northern was required to excavate and treat soils and sludges that were visibly contaminated and that contained free oils that could migrate to groundwater.

Three alternatives for treatment were considered:

- land treatment of contaminated soil and sludges;
- incineration of soil and sludges;
- land treatment of soil and incineration of sludges.

On-site land treatment was selected because it was considered to be protective of public health and the environment, and was the lowest-cost alternative. Additionally, tests consisting of six pilot-scale test plots and six bench-scale reactors demonstrated the feasibility of using land treatment for the soil and sludges (U.S. EPA 1995a; ERT 1985).

A land treatment unit (LTU) was constructed at the site. The unit was constructed over one former surface impoundment after the visibly contaminated soil and sludges were removed. The area available for treatment measured

Table 11.9 Average PAH concentrations for visibly contaminated soils in the surface impoundments, Burlington Northern Superfund site

Compound	Impoundment 1	Impoundment 2
	mg/kg	
Total 2-ring PAHs	15,565	31,464
Total 3-ring PAHs	5,895	11,839
Total 4- and 5-ring PAHs	12,522	27,330
Total PAHs	33,982	70,633
Benzene extractables	66,100	112,500
Total phenols	16	65

Source: Remediation Technologies, Inc. 1995b.

approximately 255 x 450 ft (about 115,000 ft^2). The following layers were installed at the base of the LTU:

- 100 mm thick high-density polyethylene membrane liner (for the bottom and the side slopes)
- 18 in of silty sand
- 6 in of gravel
- 24 in of clean, silty sand

The bottom of the LTU sloped 0.5 percent. The LTU was enclosed by containment berms to control the flow of run-on and runoff. A gravel layer served as a leachate collection system during treatment. Leachate collection drains 2-ft (0.6 m) wide were placed in the gravel layer. All drains were lined with gravel, and perforated pipe wrapped with filter fabric was installed to collect leachate. This plumbing carried leachate to a sump.

Soil and sludge excavated from the surface impoundments and other contaminated areas at the site were stockpiled adjacent to the LTU. Dump trucks transported the contaminated materials from the stockpile to the cells. Initially, approximately 1,100 to 1,500 yd^3 of soil and sludge were spread over the LTU to a depth of 6 to 8 in each year.

Land treatment was conducted annually from May through October, and the system was operated for nine seasons. The treatment area was tilled weekly to a depth of 12 in with a tractor-mounted rototiller. A major purpose of mixing was to promote microbial activity in the current year's lift. A 24-in. ripper was used on limited occasions to break up the compacted soil layer beneath the tillage zone. The LTU was irrigated to maintain a soil moisture content of approximately 10 percent by weight. Soil pH was maintained between 6.2 and 7.0 using liming materials, and the C:N:P ratio was maintained near 100:2:1 with application of cattle manure (Remediation Technologies, Inc. 1995b; U.S. EPA 1993).

Leachate from the LTU, collected in the sump, was discharged to an onsite storage tank. Some of the leachate was applied to the LTU as irrigation water while the remainder was discharged to a local sewer system. Upon completion of the land treatment operation a cover was placed over the LTU.

To assess LTU treatment performance, each lift of contaminated soil and sludge was sampled immediately after application and then monthly through the end of the treatment season. Samples were analyzed for methylene chloride-extractable (MCE) hydrocarbons and PAHs. At the end of the last treatment season, samples were collected at four depths in the LTU to assess residual concentrations of MCE hydrocarbons and PAHs.

The cleanup goal for total PAHs in soil and sludge for all nine treatment seasons was achieved. Total PAHs in soil before treatment ranged from 626 to 17,871 mg/kg, and from 564 to 4,326 mg/kg after treatment. The average value for total soil PAHs after treatment was 1,854 mg/kg, well below the cleanup goal (table 11.10).

The cleanup goal for MCE hydrocarbons (21,000 mg/kg) was not met in any treatment season. Concentrations of soil MCE hydrocarbons before treatment ranged from 26,000 to 89,000 mg/kg, and from 22,000 to 48,000 mg/kg after treatment. The average value after treatment was 33,000 mg/kg. Because the cleanup goal for MCE hydrocarbons was not met at the end of the treatment period, Burlington Northern was required to carry out a contingency procedure of placing a cover over the treatment area to prevent infiltration of liquids through the treatment zone.

Residual concentrations of MCE hydrocarbons and PAHs did not vary substantially with depth in the LTU. Through the top 56 in of the unit, concentrations of MCE hydrocarbons varied less than 10 percent with depth, and concentrations of total PAHs varied about 26 percent with depth.

Table 11.10 Total PAH concentrations in the LTU at start and end of each treatment season

Season	Before Treatment	After Treatment
	mg/kg	
1	16,160	1,895
2	17,871	4,326
3	12,931	3,145
4	10,887	2,518
5	6,008	1,275
6	6,097	948
7	626	564
8	13,471	1,220
9	3,544	795
Average	9,733	1,854

Source: Remediation Technologies, Inc. 1995b.

Table 11.11 Residual concentrations of MCE hydrocarbons and PAHs in the LTU at completion of land treatment activity

Compounds	0–8 in	8–32 in	32–56 in	56–66 in
		mg/kg		
MCE hydrocarbons	26,900	24,800	25,300	450
Total 2-ring PAHs	65.13	45.95	49.5	0.02
Total 3-ring PAHs	225	128	199	0.15
Total 4-ring PAHs	382	276	256	0.17
Total 5-ring PAHs	123.5	154.8	226.7	0.225
Total PAHs	795.63	608.35	731.2	0.565

Source: Remediation Technologies, Inc. 1995b.

Additionally, contaminants in the LTU soils did not migrate to uncontaminated layers below the unit.

PAH treatment efficiency, measured as percent reduction in average concentration before and after treatment, decreased with increasing number of ring structures in the PAH molecule. Two-ring PAHs were reduced by an average of 96 percent, 3-ring PAHs by 92 percent, and 4- and 5-ring PAHs by 60 percent (table 11.11). Two-ring PAHs were reduced to concentrations below analytical detection limits for two of the nine treatment seasons.

With the exception of naphthalene and 2-methylphenol, the range of concentrations measured in the leachate was below groundwater action levels. Naphthalene measured as high as 590 µg/L (action level 30 µg/L) and 2-methylphenol measured as high as 87 µg/L (action level 30 µg/L) (U.S. EPA 1995a).

Operation and maintenance activities are ongoing at the site. In 2012, reevaluation of the extent of soil and groundwater contamination remaining at the site took place (U.S. EPA 2018b).

QUESTIONS

1. List the electron acceptor(s) required in the metabolism of an organic compound during aerobic respiration and during anaerobic respiration.
2. Choose the correct answer. All other factors being equal, in situ bioremediation would be most effective for a site contaminated with: (a) Cd^{2+}; (b) DDT; (c) Cr^{6+}; (d) PAHs; (e) gasoline; (f) > C_{100} alkanes.
3. Choose the correct answer. A biorefractory molecule: (a) is not amenable to bioremediation; (b) may be a heavily chlorinated hydrocarbon; (c) is recalcitrant; (d) includes PCBs and PAHs; (e) all of the above.
4. During in situ bioremediation, well water may be used for introducing nutrients. For which specific compounds/elements may this water require treatment?

5. In order to increase the DO content of infiltrating water during ISB, H_2O_2 at levels of __ can be added.
6. List and discuss the common limitations to bioremediation for site remediation.
7. Complete the chemical reaction showing the biologically mediated oxidative decomposition of hexadecane to CO_2 and H_2O.
$CH_3(CH_2)_{14}CH_3 \rightarrow$ _____ $\rightarrow$ _____ $\rightarrow$ _____ $\rightarrow CO_2 + H_2O$
   a. RCOH   b. RCH_4 c. RCOR   d. RCHO
   e. RCH_3 f. $RC = CH_2$ g. RCOOH
8. Branched alkanes biodegrade more quickly in soils compared to straight-chain alkanes. True or false? Explain your answer.
9. For what specific purpose(s) do soil bacteria carry out oxidative degradation of organic materials?
10. During slurry bioremediation, high loading rates of organic-contaminated soil plus the effects of mixing and microbial processes can cause rapid changes in slurry pH and N levels. Explain the mechanisms for these phenomena.
11. During in situ bioremediation, it is important to remove free product; such molecules may become toxic to microbes if allowed to accumulate. Explain.
12. Oxygen availability is a key limiting factor during in situ bioremediation. True or false?
13. Of the organic molecules listed below, determine which: have low water solubility; are most toxic to microorganisms; will experience slow beta oxidation; are removed more rapidly by volatilization rather than by biodegradation.
    - n-alkanes, C_5–C_9
    - n-alkanes above C_{22}
    - cycloalkanes $< C_{10}$
    - highly condensed aromatics
    - branched alkanes
    - PCBs

REFERENCES

Abdel-Moghny, T., R. S. A. Mohamed, E. El-Sayed, S. M. Aly, and M. Gamal Snousy. 2012. Effect of soil texture on remediation of hydrocarbons-contaminated soil at el-Minia District, Upper Egypt. *ISRN Chemical Engineering* 2012: 13. doi:10.5402/2012/406598.

Adrion, A. C., D. R. Singleton, J. Nakamura, D. Shea, and M. D. Aitken. 2016. Improving polycyclic aromatic hydrocarbon biodegradation in contaminated soil through low-level surfactant addition after conventional bioremediation. *Environmental Engineering Science* 33(9): 659–70.

Aggarwal, P. K., J. L. Means, and R. E. Hinchee. 1991. Formulation of nutrient solutions for in-situ bioremediation. In: *In-Situ Bioremediation, Applications and Investigations for Hydrocarbon and Contaminated Site Remediation*, edited by R. E. Hinchee, and R. F. Olfenbuttel. Boston: Butterworth-Heinemann.

Alexander, M. 1977. *Introduction to Soil Microbiology*, 2nd ed. New York: Wiley.

Anderson, J. P. E. 1984. Herbicide degradation in soil: Influence of microbial biomass. *Soil Biology and Biochemistry* 16(5): 483–89.

Andersson, B. E., S. Olsson, T. Hentry, L. Welinder, and P. A. Olsson. 2000. Growth of inoculated white-rot fungi and their interactions with the bacterial community in soil contaminated with polycyclic aromatic hydrocarbons, as measured by phospholipids fatty acids. *Bioresource Technology* 73(1): 29–36.

Arora, N. K. 2015. *Plant Microbes Symbiosis: Applied Facets*. Springer.

Aulenta, F., V. Tandoi, M. Potalivo, M. Majone, and M. P. Papini. 2006. Anaerobic bioremediation of groundwater containing a mixture of 1,1,2,2-tetrachloroethane and chloroethenes. *Biodegradation* 17(3): 193–206.

Baxter, J., and S. P. Cummings. 2006. The impact of bioaugmentation on metal cyanide degradation and soil bacteria community structure. *Biodegradation* 17(3): 207–17.

Bianchi-Mosquera, G. C., R. M. N. Allen-King, and D. M. Mackay. 1994. Enhanced degradation of dissolved benzene and toluene using a solid oxygen-releasing compound. *Groundwater Monitoring and Remediation* 9(1): 120–28.

Bossert, I., and R. Bartha. 1984. The fate of petroleum in soil ecosystems. In: *Petroleum Microbiology*, edited by R. M. Atlas. New York: Macmillan.

Brox, G. 1993. Bioslurry treatment. In: *Proceedings Applied Bioremediation*. Fairfield, NJ.

Brubaker, G. R. 1995. The boom in situ bioremediation. *Civil Engineering-ASCE* 65(10): 38–41.

Chen, H. Y., C. M. Kao, J. K. Liu, K. Takagi, and R. Y. Surampalli. 2010. Clean up of petroleum-hydrocarbon contaminated soils using enhanced bioremediation system: Laboratory feasibility study. *Journal of Environmental Engineering* 136(6): 597–606.

Chikere, C. B., G. Chijioke Okpokwasili, and B. O. Chikere. 2011. Monitoring of microbial hydrocarbon remediation in the soil. *3 Biotech* 1(3): 117–38.

Code of Federal Regulations. 2016. *Land Disposal Restrictions*, Vol. 40. Washington, DC: U.S. Government Printing Office.

Cole, G. M. 1994. *Assessment and Remediation of Petroleum Contaminated Sites*. Boca Raton, FL: Lewis Publishers.

Collina, E., A. Franzetti, F. Gugliersi, M. Lasagni, D. Pitea, G. Bestetti, and P. Di Gennaro. 2005. Napthalene biodegradiation kinetics in an aerobic slurry-phase bioreactor. *Environment International* 31(2): 167–71.

Collins, M., and K. Miller. 1994. Reverse osmosis reverses conventional wisdom with Superfund cleanup success. *Environmental Solutions* 7(September): 9.

Cookson, J. T. 1995. *Bioremediation Engineering. Design and Application*. New York: McGraw-Hill.

Cortez, S., P. Teixeira, R. Oliveira, and M. Mota. 2013. Bioreactors: Rotating biological contactors. In: *Encyclopedia of Industrial Biotechnology*. doi: 10.1002/9780470054581.eib650.

Das, N., and P. Chandran. 2011. Microbial degradation of petroleum hydrocarbon contaminants: An overview. *Biotechnology Research International SAGE-Hindawi Access to Research*. doi: 10.4061/2011/941810.

Das, A. C., P. Sukul, D. Mukherjee, A. Chakravarty, and G. Sen. 2005. A comparative study on the dissipation and microbial metabolism of organophosphate and carbamate insecticides in orchaqualf and fluvaquent soils of West Bengal. *Chemosphere* 58(5): 579–84.

Dettmers, D. L., K. L. Harris, L. N. Petersen, G. D. Mecham, J. S. Rothermel, T. W. Macbeth, K. S. Sorenson Jr., and L. O. Nelson. 2006. Remediation of a TCE plume using a three-component strategy. *Practice Periodical of Hazardous, Toxic and Radioactive Waste Management* 10(2): 116–25.

DEVO Enterprises, Inc. 1992. *French Limited: A Successful Approach to Bioremediation*. Washington, DC.

Ecologix. 2018. *EcoDisk Rotating Biological Contactor (RBC)*. https://www.ecologix systems.com/ecodisk-rbc.php.

Electric Power Research Institute. 1988. *Remedial Technologies for Leaking Underground Storage Tanks*. Chelsea, MI: Lewis.

ERT (Environmental Research & Technology). 1985. *Treatment Demonstration Report. Creosote Contaminated Soils*. Prepared for Burlington Northern Railroad. Prepared by Environmental Research & Technology, Inc. Document D245.

Envirex, Inc. n.d. *Rex Biological Contactors: For Proven, Cost-Effective Options in Secondary Treatment*. Bulletin 315-13A-51/90-3M. https://www.evoqua.com/en/ brands/Envirex.

Eweis, J. B., S. J. Ergas, D. P. Y. Chang, and E. D. Schroeder. 1998. *Bioremediation Principles*. Boston: McGraw-Hill.

Fleming, E. J., I. Cetinić, C. S. Chan, D. W. King, and D. Emerson. 2014. Ecological succession among iron-oxidizing bacteria. *The ISME Journal* 8(4). doi:10.1038/ ismej.2013.197.

Flocco, C. G., N. C. M. Gomes, W. M. Cormack, and K. Smalla. 2009. Occurrence, diversity of naphthalene dioxygenase genes in soil microbial communities from the Maritime Antarctica. *Environmental Microbiology* 11(3): 700–14.

Goi, A., N. Kulik, and M. Trapido. 2006. Combined chemical and biological treatment of oil contaminated soil. *Chemosphere* 63(10): 1754–63.

Green-Ruiz, C. 2006. Mercury(II) removal from aqueous solutions by nonviable Bacillus sp. from a tropical estuary. *Bioresource Technology* 97(15): 1907–11.

Hasbach, A. 1993. Biotreatment of PCB sludges cuts cleanup costs. *Pollution Engineering* May 15.

Hettiaratchi, J. P. A., P. L. Amaatya, E. A. Jordan, and R. C. Joshi. 2001. Slurry phase experiments as screen protocol for bioremediation of complex hydrocarbon waste.

Practice Periodical of Hazardous, Toxic, and Radioactive Waste Management 5(2): 88–97.

Jeong, H. Y., S.-C. Jun, J.-Y. Cheon, and M. Park. 2017. A review on clogging mechanisms and managements in aquifer storage and recovery (ASR) applications. *Geosciences Journal*. doi:10.1007/s12303-017-0073-x.

Kim, H. S., and W. J. Weber Jr. 2005. Polycyclic aromatic hydrocarbon behavior in bioactive soil slurry reactors amended with a nonionic surfactant. *Environmental Toxicology and Chemistry* 24(2): 268–76.

Koenigsberg, S. 1997. Enhancing bioremediation: New magnesium peroxygen compounds accelerates natural attenuation. *Environmental Protection* 8: 2. https://www.osti.gov/biblio/482211.

Krishna, K. R., and L. Philip. 2005. Bioremediation of Cr(VI) in contaminated soils. *Journal of Hazardous Materials* 121(1–3): 109–17.

Langwaldt, J. H., and J. A. Puhakka. 2000. On-site biological remediation of contaminated groundwater: A review. *Environmental Pollution* 107(2): 187–97.

Lee, T. H., M. Ike, and M. Fujita. 2002. A reactor system combining reductive dechlorination with cometabolic oxidation for complete degradation of tetrachloroentylene. *Journal of Environmental Sciences* 14(4): 445–50.

Master, E. R., W. W. Mohn, V. W.-M. Lai, B. Kuipers, and W. R. Cullen. 2002. Sequential anaerobic-aerobic treatment of soil contaminated with weathered aroclor 1260. *Environmental Science and Technology* 36(1): 100–3.

McGrath, R., and I. Singleton. 2000. Pentachlorophenol transformation in soil: A toxicological assessment. *Soil Biology and Biochemistry* 32(8–9): 1311–14.

McLean, E. O. 1982. Soil pH and lime requirement. In: *Methods of Soil Analysis*, 2nd ed., edited by A. L. Page, R. H. Miller, and D. R. Keeney. Madison, WI: American Society of Agronomy.

Meade, T., and E. M. D'Angelo. 2005. [14C]Pentacholorphenol mineralization in the rice rhizosphere with established oxidized and reduced soil layers. *Chemosphere* 61(1): 48–55.

NAVFAC (Naval Facilities Engineering Command). 2013. *Best Practices for Injection and Distribution of Amendments*. Technical Report TR-NAVFAC-EXWC-EV-1303. Port Hueneme, CA.

Neilson, A. H., and A.-S. Allard. 2008. *Environmental Degradation and Transformation of Organic Chemicals*. Boca Raton, FL: CRC Press.

Nnamchi, C. I., J. A. N. Obeta, and L. I. Ezeogu. 2006. Isolation and characterization of some polycyclic aromatic hydrocarbon degrading bacteria from Nsukka soils in Nigeria. *International Journal of Environmental Science and Technology* 3(2): 181–90.

Norris, R. D., K. Dowd, and C. Maudin. 1993. The use of multiple oxygen sources and nutrient delivery systems to effect in situ bioremediation of saturated and unsaturated soils. In: *Symposium on Bioremediation of Hazardous Wastes: Research, Development, and Field Evaluations*. EPA/600/R-93/054. Cincinnati: U.S. EPA.

Norris, R. D., and J. E. Matthews. 1994. *Handbook of Bioremediation*. Boca Raton, FL: CRC Press.
Opatken, E. J., and H. K. Bond. 1988. Stringfellow leachate treatment with RBC. *Environmental Progress* 7: 23–31.
O'Shaughnessy. 1982. *Treatment of Oil Shale Retort Wastewater using Rotating Biological Contactors*. Water Pollution Control Federation, 55th Annual Conference, St. Louis.
Peng, R., A. Xiong, Y. Sue, X. Fu, F. Gao, W. Zhao, Y. Tian, and Q. Yao. 2008. Microbial biodegradation of polyaromatic hydrocarbons. *FEMS Microbiological Review* 32(6): 927–55.
Philp, J. C., and R. M. Atlas. 2005. Bioremediation of contaminated soils and aquifers. In: *Bioremediation: Applied Microbial Solutions for Real-World Environmental Cleanup*, edited by R. M. Atlas, and J. Philp. Washington, DC: ASM Press.
Piotrowski, M. R. 1989. Bioremediation: Testing the Waters. *Civil Engineering* 59(8) (August): 51.
Pope, D. F., and J. E. Matthews. 1993. *Bioremediation Using the Land Treatment Concept*. EPA/600/R-93/164. Washington, DC: U.S. EPA, Office of Research and Development.
Remediation Technologies, Inc. 1995a. *Remedial Action Report for the BNSF Former Tie Treating Plant, Brainerd, Minnesota*. Prepared for Burlington Northern Santa Fe Railroad, November.
———. 1995b. *Treatment Completion Report for the BNRR Former Tie Treating Plant Brainerd, Minnesota*. Prepared for Burlington Northern Railroad, Overland Park, Kansas, Fort Collins, CO. May.
Richards, L. A. 1965. Physical condition of water in soil. In: *Methods of Soil Analysis*, edited by C. A. Black. Madison, WI: American Society of Agronomy.
Robles-Gonzalez, I. V., F. Fava, and H. M. Poggi-Varaldo. 2008. A review on slurry bioreactors for bioremediation of soils and sediments. *Microbial Cell Factories* 7(5). doi:10.1186/1475-2859-7-5.
Rojas-Avelizapa, N. G., T. Roldán-Carrillo, H. Zegarra-Martínez, A. M. Muñoz Colunga, and L. C. Fernández-Linaresa. 2007. A field trial for an ex-situ bioremediation of a drilling mud-polluted site. *Chemosphere* 66(9): 1595–600.
Sims, J. L., R. C. Sims, R. R. Dupont, J. E. Matthews, and H. H. Russell. 1993. *In Situ Bioremediation of Contaminated Unsaturated Subsurface Soils*. EPA/540/S-93/501. Washington, DC: U.S. EPA, Office of Solid Waste and Emergency Response.
Singh, A., N. Parmar, and R. C. Kuhad. 2011. *Bioaugmentation, Biostimulation, and Biocontrol*. New York: Springer.
Song, X., and E. A. Seagren. 2011. In situ bioremediation in heterogeneous porous media: Dispersion-limited scenario. *Environmental Science and Technology* 42(16): 6131–40.
Sui, H., X. Li, G. Huang, and B. Jiang. 2006. A study on cometabolic bioventing for the in situ remediation of trichloroethylene. *Environmental Geochemistry and Health* 28(1–2): 147–52.
Sulfita, J. M., and G. W. Sewell. 1991. *Anaerobic Biotransformation of Contaminants in the Subsurface*. Environmental Research Brief. EPA/600/M-90/024. Ada, OK:

U.S. Environmental Protection Agency, Robert S. Kerr Environmental Research Laboratory.

Thomas, R. A. P., D. E. Hughes, and P. Daily. 2006. The use of slurry phase bioreactor technology for the remediation of coal tars. *Land Contamination and Reclamation* 14(2): 235–40.

Tuhela, L., S. A. Smith, and O. H. Tuovinen. 1993. Microbiological analysis of iron-related biofouling in water wells and a flow-cell apparatus for field and laboratory investigations. *Groundwater* 31(6): 982–88.

Umrania, V. V. 2006. Bioremediation of toxic heavy metals using acidothermophilic autotrophs. *Bioresource Technology* 97(10): 1237–42.

U.S. EPA. 1983. *EPA Guide for Identifying Cleanup Alternatives at Hazardous Waste Sites and Spills: Biological Treatment.* EPA/600/3-83/063. Washington, DC: Office of Emergency and Remedial Response.

———. 1984. *Review of In-Place Treatment Techniques for Contaminated Surface Soils. Vol. 2, Background Information for In Situ Treatment.* EPA/540/2-84/003b. Cincinnati.

———. 1985. *Handbook: Remedial Actions at Waste Disposal Sites.* Rev. ed. EPA/625/6-85/006. Cincinnati.

———. 1987. *Data Requirements for Selecting Remedial Action Technology.* EPA/600/2-87/001. Washington, DC: Office of Emergency and Remedial Response.

———. 1988. *Superfund Record of Decision.* Texas: French Limited, March.

———. 1989. *Superfund LDR Guide #6A: Obtaining a Soil and Debris Treatability Variance for Remedial Actions.* OSWER Directive 9347.3-06FS.

———. 1990a. *Enhanced Bioremediated Utilizing Hydrogen Peroxide as a Supplemental Source of Oxygen.* EPA/600/2-90/006. Ada, OK: U.S. Environmental Protection Agency, Robert S. Kerr Environmental Research Laboratory.

———. 1990b. *Handbook on In Situ Treatment of Hazardous Waste-Contaminated Soils.* EPA/540/2-90/002. Cincinnati, OH.

———. 1990c. *Slurry Biodegradation.* EPA/540/2-90/016. Washington, DC: Office of Emergency and Remedial Response.

———. 1992a. *A Citizen's Guide to Using Indigenous and Exogenous Microorganisms in Bioremediation.* EPA/542/F-92/009. Washington, DC: Office of Solid Waste Emergency Response.

———. 1992b. *Rotating Biological Contactors.* EPA/540/S-92/007. Washington, DC: Office of Emergency and Remedial Response.

———. 1993. *Five-Year Review Report.* Burlington Northern Brainerd/Baxter Minnesota. U.S. EPA, Region V, Chicago, IL, January 27, 1993.

———. 1994. *In Situ Biodegradation Treatment.* EPA/540/S-94/502. Washington, DC: Office of Emergency and Remedial Response.

———. 1995a. *Land Treatment at the Burlington Northern Superfund Site, Brainerd/Baxter, Minnesota. Cost and Performance Report.* Washington, DC: Office of Solid Waste and Emergency Response, Technology Innovation Office.

———. 1995b. *Slurry-Phase Bioremediation at the French Limited Superfund Site Crosby, Texas. Cost and Performance Report.* Washington, DC: Office of Solid Waste and Emergency Response, Technology Innovation Office.

———. 2005. *A Review of Biofouling Controls For Enhanced In Situ Bioremediation of Groundwater.* https://www.google.com/url?sa=t&rct=j&q=&esrc=s&source=

web&cd=1&ved=0ahUKEwjNifGVxvHZAhWF7YMKHT4VDncQFgguMAA&url=https%3A%2F%2Fclu-in.org%2Fdownload%2Fcontaminantfocus%2Fdnapl%2FTreatment_Technologies%2FER-0429-WhtPaper.pdf&usg=AOvVaw04CELXl9QFyu4v_FqcVpvu.

———. 2018a. *Superfund Site: French, Ltd., Crosby, TX.* https://cumulis.epa.gov/supercpad/cursites/csitinfo.cfm?id=0602498.

———. 2018b. *Superfund Site: Burlington Northern (Brainerd/Baxter Plant) Brainerd/Baxter, MN. Cleanup Activities.* https://cumulis.epa.gov/supercpad/SiteProfiles/index.cfm?fuseaction=second.cleanup&id=0503688.

Vance, D. B. 2008. *Groundwater Injection and Problem Prevention.* http://2the4.net/gwinject.htm.

Vira, A., and S. Fogel. 1991. Bioremediation: The treatment for tough chlorinated hydrocarbons. *Biotreatment News* 1: 8.

Walker Process Corporation. *EnviroDisc Rotating Biological Contactor.* Bulletin 11-S-88. www.walker-process.com.

White, D. M., R. L. Irvine, and C. R. Woolard. 1998. The use of solid peroxides to stimulate growth of aerobic microbes in tundra. *Journal of Hazardous Materials* 57(1–3): 71–8.

Young, T. S. M., M. C. Morley, and D. D. Snow. 2006. Anaerobic biodegradation of RDX and TCE: Single- and dual-contaminant batch tests. *Practice Periodical of Hazardous, Toxic, and Radioactive Waste Management* 10(2): 94–101.

Chapter 12

Phytoremediation

What is a weed? A plant whose virtues have not yet been discovered.
—Ralph Waldo Emerson, *Fortune of the Republic*, 1878

12.1 INTRODUCTION

Plant-based remediation systems can serve as a cost-effective treatment approach or contaminated soils, including those whose properties impede the success of conventional technologies (e.g., low permeability, saturation, dense structure, mixtures of contaminants). Phytoremediation is a low-cost, low-technology process defined as the engineered use of green plants to extract, accumulate, and/or detoxify environmental contaminants. Phytoremediation employs common plants including trees, vegetable crops, grasses, and even annual weeds to treat heavy metals, inorganic ions, radioactives, and organic compounds. When the appropriate plants are cultivated in contaminated soil, the root system functions as a dispersed uptake and/or treatment system.

Certain plants have been identified which can take up and concentrate metals and other inorganics from soil into leaves, stalks, seeds, and roots. Organic compounds can be degraded or immobilized in the root zone or incorporated into shoot tissues and metabolized. Relevant mechanisms involve the biological, chemical, and physical processes associated with uptake, storage, and metabolism of substrates by the plant and/or the microorganisms that occur in the root zone (figure 12.1) (Hinchman et al.

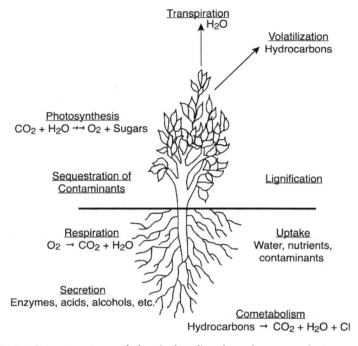

Figure 12.1 Oxygen, water, and chemical cycling through a green plant.

1997). An example of a simple phytoremediation system in use for years is the *constructed wetland*, in which aquatic plants such as cattails or water hyacinths are cultivated to remove contaminants (metals, nitrate, etc.) from municipal or industrial wastewater.

Phytoremediation is useful for soil contaminated to shallow depth. The technology can work well in low-permeability soils, where most other strategies have a low success rate. It can also be used in combination with conventional cleanup practices (e.g., pump-and-treat). Phytoremediation is an alternative to harsher remediation technologies such as solidification/stabilization, incineration, and soil washing, which destroy the biological portion of the soil, alter its chemical and physical properties, and create a relatively inert material. Phytoremediation can benefit the soil, leaving an improved, self-sustaining soil ecosystem at a fraction of the cost of current technologies. As a result of using green plants over the long term, there are obvious improvements to the aesthetics of the affected area.

Certain plants possess a remarkable capability to survive and even flourish in soils designated hazardous. In many cases, common agronomic practices can be utilized to foster an environment where plants serve as a feasible soil treatment mechanism.

12.2 PHYTOEXTRACTION OF METALS

Most metals that pose environmental hazards typically occur naturally in soils and plants in trace quantities, that is, a few mg/kg (see chapter 2). Trace metals considered essential for plant growth include cobalt (Co), chromium (Cr), copper (Cu), iron (Fe), manganese (Mn), molybdenum (Mo), nickel (Ni), vanadium (V), and zinc (Zn). Many metals are toxic above 'micro' concentrations; therefore, plants have developed mechanisms to regulate cellular concentrations of metals, primarily by regulating uptake from soil and translocation within the plant. For most metals, uptake occurs primarily through the roots, where the majority of mechanisms to prevent metal toxicity are located. For example, roots possess carrier systems that can take up ions selectively from soil.

The specific mechanisms that enable certain plants to incorporate and tolerate high quantities of toxic elements are not yet clear. In one classification system proposed by Baker (1981), categories of plant uptake mechanisms include accumulators, indicators, and excluders. *Accumulators* survive in metal-rich soil by concentrating metals in aboveground tissue. Accumulation often occurs in a single organlle or in metabolized forms so they are less harmful to the plant. Accumulators concentrate high levels of toxic elements independently of the soil concentration. *Indicator plants* control the translocation of toxic elements from roots to shoots, so that the concentration in the soil is reflected as a proportional concentration in the aboveground biomass. *Geobotanical indicators* are plants whose composition reflects the composition of the parent soil/rock formation so well that they can be used to identify mineral deposits. *Excluders* restrict the amount of toxic element that is transferred to aboveground biomass (Phusantisampan et al. 2016; Baker 1981).

More than one definition of a *hyperaccumulator* is in use; however, the basic meaning is consistent. Hyperaccumulators are plants that take up toxic elements and accumulate them in aboveground biomass at levels many times the typical concentrations, with little or no adverse effect to the plant (table 12.1).

Hyperaccumulators occur in over thirty-four different families. The Brassicaceae family is relatively rich in these unusual plants, in particular the genera *Alyssum* and *Thlaspi* (Verbruggen et al. 2009). Some hyperaccumulators of Ni and Zn may contain as much as 5 percent each on a dry weight basis (Blaylock et al. 1997; Brown et al. 1995). An outstanding example of hyperaccumulation is found in the latex of the New Caledonian tree *Sebertia acuminata*, which accumulates more than 20 percent Ni. Other examples include 10 percent Zn accumulation by pennycress (*Thlaspi calaminare*), 10 percent Ni accumulation by alyssum (*Alyssum bertolonii*), up to 3 percent Cr in *Pimela suteri* and broom tea tree (*Leptospermum scoparium*), and up to 3 percent uranium by *Uncinia leptostachya* and *Coprosma arborea*. The Indian mustard plant, *Brassica juncea*, can accumulate 3.5 percent dry weight of

Table 12.1 Selected species of metal hyperaccumulators

Metal	Plant Species	Percentage of Metal in Dry Weight of Leaves (%)	Native Location
Cd	*Thlaspi caerulescens*	< 1	Europe
Co	*Haumaniastrum robertii*	1	Zaire
Cr	*Brassica juncea*	< 1	India
Cu	*Aeolanthus biformifolius*	1	Zaire
Ni	*Phyllanthus serpentinus*	3.8	New Caledonia
	Alyssum bertoloni and Fifty other species of alyssum	> 3	Southern Europe and Turkey
	Sebertia acuminate	25 (in latex)	New Caledonia
	Stackhousia tryonii	4.1	Australia
Pb	*Brassica juncea*	< 3.5	India
	Ambrosia artemisiifolia	< 1	North America, Europe
Zn	*Arabidopsis halleri*	< 1	Eurasia and Northern Africa
	Thlaspi species	< 3	Europe
	Viola species	1	Europe

leaves with Pb (Kabata-Pendias 2001). Poplar trees (*Populus* spp.), hemp dogbane (*Apocynum* sp.), and common ragweed (*Ambrosia artemisiifolia*) also accumulate significant quantities of Pb (Khanna and Khanna 2011; Sutherson 1997). The Zairean hyperaccumulator *Haumaniastrum katangense* has been cropped on soil contaminated with radioactive Co (Baker and Brooks 1989). Several researchers (Arapis 2006; Fesenko et al. 2003; Furmann et al. 2002; Dushenkov et al. 1999; Salt and Kay 1999) have isolated higher plants that remove ^{137}Cs from soil. Accumulation of up to 1 percent Hg has been reported for white birch (*Betula papyrifera*) (Negri and Hinchman 1996).

More than 400 plant species have been reported so far that hyperaccumulate metals (McIntyre 2003; Baker et al. 2000). A substantial number of species have the capacity to accumulate two or more elements (Yang et al. 2002, 2004; He et al. 2002). Baker and Brooks (1989) identified 145 hyperaccumulators of Ni, 26 of Co, 24 of Cu, and 8 of Mn.

Almost all metal hyperaccumulating species in use today were discovered on metal-rich mineral outcroppings. Such plants are endemic to these soils, indicating that hyperaccumulation is a physiologic adaptation to metal stress (Verbruggen et al. 2009). The majority of hyperaccumulating species discovered thus far are restricted to a few limited geographical locations.

A simple and common application of phytoremediation is *phytoextraction*, which involves the use of hyperaccumulating plants to transfer metals from the soil and concentrate them in roots and aboveground shoots. In certain cases, contaminants are concentrated thousands of times higher in the plant than in soil. Following harvest of the extracting crop, the metal-rich plant biomass can be digested (e.g., composted) or ashed to reduce its volume, and the resulting material can be processed as an 'ore' to recover the contaminant (e.g., valuable heavy metals, radionuclides). If recycling the metal is not economically feasible, the small amount of ash (compared to the original plant biomass or the large volume of contaminated soil) can be disposed appropriately.

Phytoextraction is based on the ability of plants and their associated rhizospheres to solubilize, absorb, and concentrate dilute contaminants. Plant root exudates comprise a critical component of the rhizosphere. These secretions contain chelating agents (e.g., citric and acetic acid) that render both nutrients and contaminants more mobile in soil.

Numerous plants, both woody and nonwoody, have been identified with the capacity to solubilize, translocate and accumulate metals into aboveground biomass (figure 12.2). Some grasses accumulate high levels of metals in their shoots without exhibiting toxic effects (Fowler et al. 2004; Pang et al. 2003; Garcia et al. 2004; Pichtel and Salt 1998). However, their low biomass production results in relatively low yield of metals. The breeding of

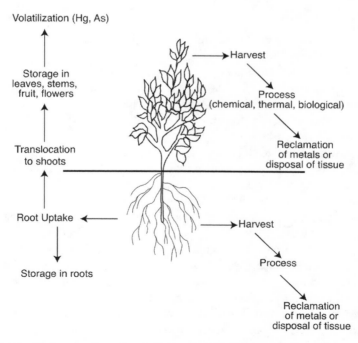

Figure 12.2 **Phytoextraction of metallic contaminants from soil.**

hyperaccumulating plants that produce large amounts of biomass would make metal extraction more effective.

Many crop plants accumulate metals in aboveground shoots and roots. This is particularly true for certain leafy vegetables (e.g., cabbage, lettuce, spinach) and root crops (potato, carrot), respectively (Channon et al. 2005; Tandi et al. 2004; Mantovi et al. 2003), although other crops such as corn (*Zea mays* L.) and soybean (*Glycine max*) have been shown to be effective (Muchuweti et al. 2006; Liu et al. 2005; Huang et al. 1997; Blaylock et al. 1997). Using crop plants to extract metals from soil is practical by virtue of high biomass production and a relatively fast rate of growth. Another benefit is that cultivation of crop plants is simple and straightforward using common agricultural equipment. While extraction by crops has advantages from a remediation standpoint, there exists a potential threat to the food chain. Regulations established under RCRA for hazardous waste (40 CFR 1996) include limits on the amounts of Cd, Pb, and other metals that can be applied to soils grown for crops.

The use of trees can result in extraction of significant quantities of metals because of their high biomass production. Trees are also appropriate because of their genetic variability, established cultivation practices, high degree of public acceptability, and their contribution to site stability by preventing migration of heavy metals by leaching, wind dispersion, or erosion (Pulford and Watson 2003). Growing woody plants for phytoextraction also provides economic returns via obtaining woody biomass which can eventually be used for producing energy (Laureysens et al. 2004). In addition to removing soil metals, trees serve to establish green belts in brownfields and/or urban areas. The use of trees in phytoremediation may create concerns about falling leaves, however. When metal-laden leaves fall and disperse, recirculation of metals back to the site and migration off-site by wind can occur.

Phytoextraction by the fast-growing, high biomass-producing Salicaceae species, poplar (*Populus* sp.) and willow (*Salix* sp.) has been considered a promising approach for treatment of contaminated soil.

It is strongly preferred that metal accumulation occur in shoot rather than root tissue in order to simplify tissue harvest (figure 12.3). In most plants the roots sequester the majority of the contaminant that is removed from soil. Techniques and equipment for harvesting roots, including young tree roots, are being evaluated for use with hyperaccumulating plants. Sequestration of heavy metals in roots as opposed to translocation to aboveground plant parts reduces the possibility of dispersion of contaminants via food chains by wildlife, domestic animals, or birds. However, root harvest is energy-intensive and slow compared to shoot harvest. Furthermore, there is an increased exposure hazard to workers involved in root harvest.

Figure 12.3   Poplar trees planted for the extraction of zinc, copper, and nickel from a former industrial waste site.

12.2.1 Genetic Engineering for Phytoextraction

Most metal hyperaccumulators tend to be small and grow very slowly. Such properties are clearly not beneficial for metal phytoextraction at the field scale; additionally, small plants limit the applicability of conventional agronomic practices such as mechanical harvest (Tong et al. 2004).

Conventional breeding approaches have been suggested to modify plants for metal extraction; however, results have not been consistent. Biotechnology has the potential to overcome this limitation by allowing direct gene transfer (Kramer and Chardonnens 2001). Recent research has shown that manipulation of plant characteristics such as metal tolerance is possible. One strategy is to have the plant detoxify heavy metals entering the cytoplasm by inactivation via chelation, or conversion to a less toxic form, and/or compartmentalization into a specific cellular organelle (Yang et al. 2005).

Rugh et al. (1998) modified yellow poplar trees with two bacterial genes, merA and merB, to detoxify methyl-Hg from contaminated soil. Pilon-Smits et al. (1999) overexpressed the ATP-sulfurylase (APS) gene in Indian mustard. The transgenic plants accumulated three times more Se than wild plants.

For effective use of biotechnology to design transgenic plants capable of efficient phytoremediation, a comprehensive knowledge of the genetic basis for hyperaccumulation is essential, especially those processes in hyperaccumulators that have enhanced metal tolerance, uptake, translocation, and accumulation abilities and thus are the most promising source of potential phytoremediation genes (Yang et al. 2005).

12.2.2 Technical Feasibility

Success of phytoextraction of soil metals is variable and depends on:

1. **Concentration of contaminant(s).** Phytoremediation is best used at sites with low-to-moderate metal content. Excess quantities of soil metals will impair normal plant physiologic processes. What is considered 'excessive' is a function of plant physiology and soil conditions. Plants may experience slowed vegetative growth, impaired maturation, or death on exposure to excessive concentrations. Specific examples of damage by metals include root blockage by the formation of metal precipitates, interference with uptake of metals required for plant growth (e.g., Cd^{2+} can substitute for Zn^{2+} in some situations), or enzyme dysfunction from metal saturation. Heavily contaminated soils may not permit normal plant growth without addition of amendments.

2. **Depth of contamination.** Plant roots must be in intimate contact with the metal-enriched soil in order to be effective. Rooting depth for many nonwoody species can be up to 24 in, depending on the plant and soil type. The roots of certain prairie grass species can reach 16 ft (MARC 2018). Certain deep-rooted trees (e.g., poplars) can extend roots up to 10 ft.

3. **Chemical form(s) of the metal.** The capability of a plant to accumulate metals is dependent in large part on the chemistry of the soil system in which the plants are established. Soil metals occur in numerous forms, with varying degrees of availability to plants (see chapter 2). Heavy metals that have aged in place for decades tend to crystallize into nonreactive or partly reactive forms. In certain situations, elemental metal may occur at a site. For example, sites have been assessed in which Pb ingots and metallic Pb from automobile batteries are clearly visible in soil (Pichtel et al. 2000). Metal in this form is obviously unavailable to plants except over extremely long periods.

Of paramount importance to uptake is the actual chemical *species*, or fraction, of that metal. Soil fractionation procedures in combination with plant uptake bioassays may serve to assess metal bioavailability (Aikpokpodion et al. 2012; Pichtel and Salt 1988; Tessier et al. 1979; Sims and Kline 1991; Petruzzelli 1989; Sposito et al. 1982). Attempts have been made to partition metals into chemically distinct forms via sequential extraction using selective reagents. For example, soils have been extracted sequentially with H_2O, KNO_3, NaOH, Na_2EDTA, and hot HNO_3. These reagents are expected to remove the soluble, exchangeable, organic-bound, carbonate-bound, and residual soil metals, respectively. The order of plant-available metals is:

soluble = exchangeable > organic-bound = carbonate-bound >> residual

Such bench-scale analysis of soil metals may provide useful indices of phytoremediation success. Fractionaon of soil metals was also discused in section 7.5.1 "Contaminant Forms and Concentrations."

4. **Presence of potentially interfering contaminants.** Types and amount of soil organic matter affect the availability of metals. Some may strongly chelate and immobilize metals, rendering them only slowly extractable by the plant root.

Candidate sites for phytoremediation usually have multiple contaminants. In the United States, more than 80 percent of the metal-contaminated Superfund, Department of Defense, and Department of Energy sites are also contaminated by organic pollutants (Ensley 2000). The presence of NAPL in soil will slow plant growth by interfering with availability and uptake of soil water. NAPL and other hydrocarbons may also be directly toxic to the plant and/or act as a surfactant/dispersant.

5. **Soil chemical and physical properties.** The general soil chemical milieu can strongly influence metal uptake. The amount of biomass that can be produced is one of the limiting factors affecting phytoremediation; therefore, optimizing fertility should be considered for increased productivity of the selected plant species. Amendment of soils to change pH, nutrient composition, or microbial activity can be selected in treatability studies. Likewise, soil moisture status must be considered. Soil texture and structure will influence infiltration rate, available moisture, and potential for leaching and runoff.

6. **Plant factors.** The effectiveness of a phytoremediation program is determined by a number of plant attributes. Plant species native to the target area should be given high priority as they are adapted to the local climate, insects, and diseases. Any plant used as a phytoremediator must be able to tolerate high concentrations of the metal of concern in addition to any other pollutants present at the affected site (Peer et al. 2005).

Other important plant attributes include tolerance to soil conditions (pH, salinity, structure, permeability); rate of uptake; transpiration rate; biomass production; root type, fibrosity, rooting depth, and harvestability; duration of growth (i.e., annual, biennial, perennial); and dormancy (Shimp et al. 1993).

7. **Degree of site preparation**. Many contaminated sites have served as waste dumps and may contain unconsolidated and large wastes. Bulky debris should be removed prior to tillage and planting. Tillage, surface drainage, irrigation, and controls for run-on and runoff may be needed at the site.

As mentioned above, many hyperaccumulator species are not suitable for phytoremediation in the field due to low biomass production and slow growth. It has therefore been suggested to use high biomass species such as maize (*Zea mays* L.), pea (*Pisum sativum* L.), oat (*Avena sativa* L.), canola (*Brassica napus* L.), and barley (*Hordeum vulgare* L.), while optimizing plant and soil management practices to enhance metal uptake (Shen et al. 2002; Ajwa et al. 1999; Ebbs and Kochian 1998). Greenhouse-scale studies have shown that certain crop plants are capable of promising rates of phytoextraction. Maize, alfalfa, sorghum, and sunflower (*Helianthus annuus*) were found to be effective due to their rapid growth rate and substantial biomass production (Niu et al. 2007; U.S. EPA 2000). The highest bioaccumulation of lead among crop plants is reported for leafy vegetables (especially lettuce) grown near nonferrous metal smelters where plants are exposed to lead sources in both soil and air. In these locations, lettuce contained up to 0.15 percent Pb (dry weight) (Smith et al. 1995). Lead has also been found to accumulate in certain crops such as oats grown on biosolids-amended soils (Azevedo Silveira et al. 2003; Pichtel and Anderson 1997).

12.2.3 Agronomic Practices for Enhancing Phytoextraction

As with production of any crop plant, metal-accumulating plants respond favorably to establishment of optimum soil chemical and physical characteristics and dedicated plant husbandry. For example, application of appropriate fertilizer materials correlates with increased biomass production. Some precautions are worth noting, however. For example, the addition of phosphorus fertilizer during phytoextraction can inhibit Pb uptake due to precipitation as pyromorphite and chloro-pyromorphite (Chaney et al. 2000). Foliar application of P is one means to circumvent this problem.

Planting practices also affect success of metal uptake from soil. As stated above, the degree of metal extraction depends in part on the total quantity of biomass produced. Plant density (number of plants/m^2) influences biomass production as it affects both yield per plant and yield per hectare. In general, higher planting density tends to minimize yield per plant and maximize yield per hectare. Density may also affect the pattern of plant growth and development. For example, at higher stand density, plants compete more aggressively for light. As a result, more nutrients and energy may be allocated to vegetative growth as opposed to developmental processes (e.g., flowering and reproduction). Additionally, distance between plants may affect the structure of the root system with subsequent effects on metal uptake (Lasat 2000).

Growth of weeds and proliferation of disease may decrease yields; therefore, crops cultivated for soil remediation should be rotated. If phytoremediation is anticipated to last for only a short period (e.g., two to three years) monoculture may be acceptable. However, for longer-term applications (as is the case for most metal phytoextraction projects), successful metal cleanup probably will not be achieved with only one species. Plant rotation is even more important when multiple crops per year are to be grown (Lasat 2000).

12.2.4 Treatment of Biomass

The total quantity of contaminant that a plant removes over a growing season depends on contaminant concentration in the harvested biomass multiplied by biomass production. Metal-enriched plant residues can be treated as hazardous waste or recycled as metal ore (Sas-Nowosielska et al. 2004). In the latter case, reclamation as 'bio-ore' is accomplished by smelting or acid extraction of the dried or ashed tissue. Some researchers (Keller et al. 2005) have investigated whether thermal treatment could be a feasible option for evaporatively separating metals from plant residues. Gasification (i.e., pyrolysis) was found to be a better method than incineration to increase volatilization and, subsequently, recovery, of Cd and Zn from plants.

12.2.5 Benefits of Phytoextraction

Some of the major benefits of phytoextraction include:

1. Potential for the production of green belts. This is especially valuable in urban brownfield areas. There are examples of woodlots being cultivated on urban metal-rich sites. Green belts provide the additional benefit of serving as wildlife habitat.
2. A low-cost practice. Operators are needed for soil tillage, fertilizing, and seeding. No specialized field equipment is required; standard plows, planters, and harvesters work effectively. Plant tissue is harvested and processed as needed.
3. Versatility. Plants may be successful at sites that are not suited to heavy equipment or vehicles (e.g., saturated soils).

12.2.6 Disadvantages of Phytoextraction

Some significant disadvantages of phytoextraction include:

1. Longer times are required for remediation compared with other technologies. Phytoextraction is not suitable for sites posing an immediate public health or environmental threat.
2. The potential exists for release of biomass via wind dispersal of leaves or transport through the food chain.
3. Careful screening and pilot-scale tests are recommended in order to ensure compatibility of plants with the affected soil material.

4. There is no single plant that will successfully remove all metals from soil. The chemical composition of soil at hazardous sites is often highly complex (i.e., many metals may occur); therefore, several species or genera may be needed in order for all target metals to be treated.

12.3 ORGANIC CONTAMINATION

Phytoremediation of organic (e.g., hydrocarbon) contaminated soils involves: (1) uptake of the contaminant by the plant followed by storage or metabolism, and/or (2) degradation in the plant rhizosphere (figure 12.4). The two mechanisms may occur simultaneously.

For *phytodegradation* of organics to be successful, contaminants must be available for uptake and metabolism by the plant or by its associated microbial populations. Bioavailability is a function of the solubility of the compound in water versus in hydrocarbons, soil type (e.g., organic matter content, types and amounts of clay minerals) and age of the contaminant.

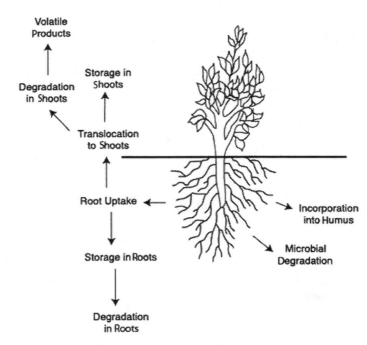

Figure 12.4 Phytoremediation of organic contaminants from soil.

The study of soil-applied pesticides in recent decades provides useful data regarding the feasibility of phytoremediation of organics (Lunney et al. 2004). A common parameter used to predict plant uptake from soil is the octanol-water partitioning coefficient (K_{ow}). This coefficient was discussed in section 7.4, "Flushing Organics From Soil."

$$K_{ow} = [\text{contaminant in octanol}] / [\text{contaminant in water}] \quad (12.1)$$

Contaminants with a log K_{ow} < 1 are considered highly water-soluble, can accumulate in plants, and are typically mobile in both xylem and phloem. Those with log K_{ow} values of approximately 1 to 4 are taken up by roots and are considered xylem-mobile. Compounds in this range, many of which are environmental pollutants, should be amenable to phytoexraction or phytodegradation. Compounds with log K_{ow} values > 4 are strongly sorbed to roots and are only minimally translocated to the shoot. Remediation technologies that either harvest roots or rely on degradation reactions at root surfaces may be applicable for treatment of these compounds.

Once transported within the plant, compounds are either sequestered, bound to plant structural constituents, metabolized, or carried to the leaves and volatilized through the stomata. The pesticide industry has studied extensively the metabolism of various organics within plants; the industry exploits the differences between the metabolic capacities of crop and weed plants. Desirable (i.e., crop) plants metabolize the selective herbicide into a nontoxic compound; however, weeds do not and consequently die. Such metabolic capacities of plants may be adapted for remediation of contaminated soils; in other words, plants may absorb pollutants from soil and metabolize them into nontoxic materials or incorporate them into stable cellular constituents (e.g., cellulose or lignin). In efforts to broaden the uses of currently registered herbicides, researchers are extending this degradative capacity by incorporating microbial or mammalian genes into the plant genome (Inui 2005; Schaffner et al. 2002; Cunningham and Berti 1993).

12.3.1 The Rhizosphere and Associated Microbial Consortia

Not all phytodegradation processes occur within the plant. For example, plant enzymes released into soil may produce catalytic effects and thus be useful. The metabolic capacity, and consequently remediation success of plant-associated microbial systems is also being studied, including those within the rhizosphere, among plant-leaf microflora, and endophytic (within plant) organisms. In the rhizosphere, accelerated rates of degradation for many pesticides as well as trichloroethylene and petroleum hydrocarbons have been observed (Menon et al. 2004; Xia et al. 2003; Costa et al. 2000; Schnoor et al. 1995).

While some plants are known to metabolize certain hydrocarbon compounds within their biomass, microorganisms act on a wider range of substrates, carry out more complex and difficult degradative steps, and generally transform the contaminant molecule to a simpler final product than do plants alone. Some microbial populations known to carry out significant reactions occur in the plant rhizosphere. The rhizosphere is a metabolically active zone directly adjacent to the plant root; it possesses a high density of heterotrophic microbial populations and is enriched in plant exudates including simple sugars, carbohydrates, alcohols, acids, and enzymes. As a result of the rich microbial diversity and nutrient content, this is also a zone where organic contaminants that are normally poor microbial substrates can be microbially degraded via *co-metabolism*. Co-metabolism is the process by which a compound that cannot support the growth of microorganisms is degraded when another, more readily metabolized substrate is present. The root zone, where up to 25 percent of all the crop biomass can be sloughed off, is an ideal environment for this to occur.

Plant-assisted biodegradation is a function of: (1) the composition of the microbial consortia present in the rhizosphere; (2) root exudates that may act as supplemental substrates; (3) types and amounts of N present in soil water, via decaying organic matter, or fixation of atmospheric N_2; (4) oxygen transfer to the soil; and (5) kinetics of microbial degradation (Shimp et al. 1993).

Research in rhizosphere-induced phytoremediation of organics is addressing the appropriate choice of plant host, rooting patterns, and associated microflora (Liu et al. 2015; Xia et al. 2003; Dominguez-Rosado et al. 2004). Numerous studies have focused on the composition of root exudates, exudation of specific compounds to induce microbial reactions, inoculation of the rhizosphere with microbes that are efficient degraders, and alternative methods of sequestering the metabolically more active microbes into plant tissue. The potential of ectomycorrhizal associations to facilitate cleanup of soil contaminated with persistent organic pollutants has also been studied (Meharg and Cairney 2000).

Liu et al. (2015) found that coneflower (*Echinacea purpurea*), tall fescue (*Festuca arundinacea* Schred), Fire Phoenix (a modified *F. arundinacea*), and alfalfa (*Medicago sativa* L.) possess the potential for remediation of PAH-contaminated soils. Other researchers have determined that a number of grass species will promote microbial degradation of PAHs (Parrish et al. 2005; Rezek et al. 2008; Smith et al. 2006).

12.4 PHYTOSTABILIZATION

Contaminants that are tightly sorbed to soil particles and resist uptake by microbes or plants are not suitable for phytoextraction or phytodegradation; however, *phytostabilization* of the site may be suitable. Phytostabilization uses plants to limit the mobility and bioavailability of metals in soil. Ideally, phytostabilizing plants tolerate high levels of soil metals and should be able to immobilize them by sorption, precipitation, complexation, or oxidation-reduction reactions. Additional stabilization can occur by raising soil pH. Phytostabilizing plants also should exhibit low accumulation of metals in shoots to eliminate the possibility that residues in shoots would pose an environmental or health hazard. In addition to immobilizing soil metals, phytostabilizing plants can also stabilize the soil surface to minimize erosion.

Since many sites contaminated with metals lack established vegetation, metal-tolerant plants are used for revegetation to limit soil erosion and metal leaching. This approach, however, is technically defined as containment rather than remediation. Some scientists and engineers consider phytostabilization to be an interim measure to be utilized until a suitable remediation technology is determined to be feasible or cost- effective. Others, however, are developing phytostabilization as a standard protocol of metal remediation technology, especially for sites at which metal removal is not economically feasible (Phusantisampan et al. 2016).

Following field studies by a group in the United Kingdom, three grasses were made commercially available for phytostabilization: *Agrostis tenuis*, cv Parys for Cu wastes, *Agrostis tenuis*, cv Coginan for acid Pb and Zn wastes, and *Festuca rubra*, cv Merlin for calcareous Pb and Zn wastes (Smith and Bradshaw 1979). Pichtel and Salt (1998) found *Agrostis capillaris* L. var. "Heriot," *Festuca ovina* L., *F. rubra* L. var "Boreal," *Lolium perenne* L., and *Phleum pratense* L. var. "Scots" to be suitable in phytostabilizing soil from several UK sites including an abandoned Pb mine, an abandoned metalworks facility, and a Cr-contaminated dyeworks facility.

12.5 FUTURE DEVELOPMENT

Increased rates of metal uptake, increased translocation to aboveground biomass, and higher yields of harvested plant biomass are under study via selective breeding and genetic engineering. Recent testing has revealed that genetically altered species of poplar, rice, peanut, and others can take up mercuric ions from soil and convert them to metallic mercury, which is transpired through the leaves (Ruiz and Daniell 2009; Heaton et al. 2003). Testing of plants may lead to identification of species that have metal accumulation qualities that exceed those now documented.

A U.S. DOE report entitled "Summary Report of a Workshop on Phytoremediation Research Needs" noted three key areas of research and development for plant-based treatment of soil contaminated with metals:

1. Mechanisms of uptake, transport, and accumulation of metals: Research is needed to develop better understanding of the use of physiological, biochemical, and genetic processes in plants that relate to tolerance of contaminant-enriched sites, and translocation and decomposition and of contaminants.
2. Genetic evaluation of hyperaccumulators: Research must continue in screening for plants growing in soils that contain high levels of metals, and for evaluating them for specific traits useful in phytoremediation.
3. Field evaluation and validation: Field testing of promising plant species is needed to accelerate implementation of phytoremediation technologies. Standardization of field-test protocols and subsequent application of test results to real-world situations also are needed.

12.6 SUMMARY

Several distinct advantages (e.g., liability, economic, aesthetic) may accrue while remediating a contaminated site while at the same time minimizing disturbance to the surface (table 12.2). Research in applying and enhancing phytoremediation continues to expand (table 12.3), as many current engineering technologies for treating contaminated surface soils are costly, energy-intensive, and disruptive to a site. Phytoremediation, when fully developed, could result in marked cost savings and in restoration of sites by a noninvasive, solar-driven, in situ method that actually benefits a soil and, in many cases, can be aesthetically pleasing.

12.7 CASE HISTORY

The Aberdeen Proving Grounds (Edgewood, MD) was established as a U.S. Army weapons testing facility in 1918. Weapons testing and munitions disposal have resulted in extensive soil and groundwater contamination at the proving grounds.

The installation is divided into two sections, the Edgewood and the Aberdeen areas. Within the former, the J. Fields Toxic Pits Site had been used for years as an open burning facility for munitions and chemical agents. Large

Table 12.2 Advantages and disadvantages of phytoremediation technology

Type of Phytoremediation	Advantages	Disadvantages
Phytoextraction by trees	High biomass production	Potential for off-site migration and leaf transportation of metals to surface. Metals are concentrated in plant biomass and must eventually be disposed.
Phytoextraction by grasses	High accumulation	Low biomass production; slow process. Metals are concentrated in plant biomass and must eventually be disposed.
Phytoextraction by crops	High biomass and increased growth rate	Potential threat to food chain through ingestion by herbivores. Metals are concentrated in plant biomass and must eventually be disposed.
Phytostabilization	No disposal of contaminated biomass	Remaining liability issues, including maintenance required for indefinite period of time.
Phytodegradation in rhizosphere	No disposal of contaminated soil	Limited to hydrocarbons only.
Phytovolatilization	Limited application (As, Hg)	Possibility of releasing toxins to the atmosphere. Permit may be required.

volumes of chlorinated solvents were disposed there as well. As a result, a plume of chlorinated solvents has formed in the aquifer below the pits. Concentrations of total VOCs in the groundwater range from less than 20,000 µg/L to over 220,000 µg/L (table 12.4) (Chappell 1997).

Due to the contamination present, the Edgewood area was placed on the Superfund NPL. Several technologies were considered for cleaning soil and groundwater at the site. Soil washing, vapor extraction, and capping were considered for the soil, and pump-and-treat and air sparging for groundwater. All were eliminated from consideration due to potential difficulties with perched water tables and the potential for encountering unexploded bombs buried on-site. Pumping and treating the water would be difficult because of the high concentrations of contaminants combined with strict discharge regulations. Soil excavation was not considered feasible due to cost. The site was eventually considered for a pilot-scale phytoremediation system (Chappell 1997; Tobia and Compton 1997).

Table 12.3 Phytotechnology use at superfund sites contaminated with chlorinated solvents, metals, explosives, and pesticides

Site Name	Site Location	ROD Date	Contaminants	Phytotechnology Status
Aberdeen Pesticide Dumps	NC	06/04/1999	Dieldrin, hexachlorobenzene, hexachlorohexane	Ongoing (1999–)
Aberdeen Proving Ground	MD	09/27/2001	1,1,2,2-tetrachloroethane; 1,1,2-trichloroethane; 1,1-DCE; 1,2-dichloroethane, 1,2-DCE; PCE; TCE; VC	Ongoing (1996–)
Argonne National Laboratory West 1	ID	09/29/1998	Cesium-137, silver, mercury, chromium	Complete (1999–2002)
Atlas Tack Corporation	MA	03/10/2000	Benzene, chromium, copper, cyanide, mercury, nickel, zinc	Completed (2007)
Boarhead Farm	PA	11/18/1998	Benzene, cadmium, nickel TCE	Designed/not installed (2003)
Bofors-Nobel Inc.	MI	ROD amendment: 07/16/1999	3,3-dichlorobenzidine; acetone; arsenic; VC; PCE; aniline; benzene; toluene; zinc	Pilot completed (1999–2002); ongoing
Combustion, Inc.	LA	05/28/2004	DCA, PCB, benzene, lead mercury, nickel, silver, toluene, toluene diisocyanate, toluene diamine	Ongoing (2002–)
Del Monte Corporation	HI	09/25/2003	Ethylene dibromide; 1,2-dibromo-3-chloropropane; 1,2-dichloropropane; 1,2,3-trichloropropane (pesticides)	Ongoing (1998–)
East Palo Alto	CA	RCRA site	Arsenic, sodium	Ongoing
Fort Dix	NJ	Field demonstration Lead	Completed (1997–2002)	
Fort Wainwright	AK	06/27/1997	Aldrin, DDD, DDT, dieldrin	Completed (1997–2001)
Middlefield-Ellis	CA	1989; 2010	TCE; metals (various)	Ongoing
Naval Undersea Warfare Station	WA	09/28/1998	TCA, halogenated volatiles	Ongoing
Palmerton	PA	12/04/1987	Zinc and other metals	Ongoing
Tibbetts Road	NH	09/28/1998	TCE	Ongoing

Source: U.S. Environmental Protection Agency 2018a; 2018b; 2018c; 2018d; 2005.

Table 12.4 Primary contaminants at Aberdeen Proving Grounds' J. Fields phytoremediation site

Contaminant	Concentration (μg/L)
1,1,2,2-tetrachloroethane	170,000
Trichloroethene (TCE)	61,000
cis-1,2-dichloroethene (c-DCE)	13,000
Tetrachloroethene (PCE)	9,000
Trans-1,2-dichloroethene (t-DCE)	3,900
1,1,2-trichloroethane (TCA)	930

Sources: Chappell 1997; Tobia and Compton 1997.

The U.S. DOD and EPA jointly funded pilot-scale applications of phytoremediation. At the Fields site, hybrid poplars were planted over a shallow plume of chlorinated solvents in order to hydraulically contain the contaminants and treat groundwater.

The phytoremediation strategy employed at the Fields site began with an assessment for phytotoxicity of on-site pollutants and to determine any nutrient deficiencies that would affect tree growth. A total of 183 hybrid poplars (*P. trichocarpa* × *deltoides* [HP-510]) were planted over the areas of highest pollutant concentration around the leading edge of the plume (refer to figure 12.4). In order to promote root growth into the saturated zone, each tree was planted with a plastic pipe surrounding its upper roots. A drainage system was installed to remove rainwater and promote root tropism to groundwater.

Extensive monitoring took place to determine the fates of the pollutants, tree transpiration rates, and best methods for monitoring phytoremediation (table 12.5). The sampling design at the site involved collecting soil, transpiration gases, and tissue from tree roots, shoots, stems, and leaves. Results were used to determine concentrations of contaminants and their metabolites along the plant translocation pathway (Chappell 1997).

Table 12.5 Monitoring methods at Aberdeen Proving Grounds' J. Fields site

Type of Analysis	Parameters Tested
Plant growth measurements and visual observations	Diameter, height, health, pruning, replacement
Groundwater and vadose zone sampling and analysis	Fourteen wells and four lysimeters to sample for VOCs, metals, and nutrients
Soil sampling and analysis	Biodegradation activity, VOCs, metals
Tissue sampling and analysis	Degradation products, VOCs
Plant sap flow measurements	Correlate sap flow data to meteorological data
Transpirational gas sampling and analysis	Various methods

Sources: Chappell 1997; Tobia and Compton 1997.

Eight monitoring wells were in place at the time of tree planting and five more were added. Two pairs of lysimeters were also installed. Tree sap flow rates were monitored in order to determine pumping rates of the trees. An onsite weather monitor was used during sampling to correlate tree evapotranspiration rates with weather fluctuations. Groundwater monitoring data indicates that the trees are pumping large quantities of groundwater—there is a 2-ft depression in the water table beneath the trees in comparison to earlier data (Chappell 1997). Tree tissue samples indicate the presence of trichloroacetic acid (TCAA), a breakdown product of TCE. Other researchers (Newman et al. 1997) have also detected TCAA in hybrid poplar tissue in a greenhouse study. Chlorinated solvents (TCE and 1,1,2,2-tetrachloroethane) are being transpired. The J. Fields site experienced about 10 percent tree loss during the first year. Some loss was due to the transplant process, and deer damaged many. The cost for installation of 183 trees was $15,000. Costs of monitoring varied due to experimenting with numerous monitoring techniques at the site.

QUESTIONS

1. Explain why the plant rhizosphere, as compared to nonvegetated soil, may result in enhanced hydrocarbon decomposition.
2. What are the appropriate classes of plants (e.g., trees, weeds) to use for phytoremediation? Is one class of plant better suited for phytoextraction than another? For rhizosphere-enhanced degradation?
3. Discuss the method(s) to treat metal-enriched plant tissue once a hyperaccumulator plant has been harvested.
4. Most plants are unsuitable for phytoremediation if the contaminant occurs beyond 2 to 4 ft or more below the surface. True or false? Explain your answer.
5. The chemical form of a metal in the soil (e.g., soluble, bound to organic matter, bound to carbonates) significantly affects uptake by plants. Explain and provide an example.
6. All other factors being equal, soil Pb is much easier for plants to take up compared to Cd. True or false? Explain.
7. How could the presence of oily wastes interfere with plant uptake of Cd, Zn, and Cr?
8. Discuss one major disadvantage of phytoremediation technology for the cleanup of soil heavily contaminated with plutonium-239.
9. Phytoremediation is typically not well suited for low-permeability soils. True or false? Discuss.
10. In order to promote plant uptake of soil Pb, how can the soil be amended or otherwise managed? Be specific.

REFERENCES

Aikpokpodion, P. E., L. Lajide, and A. F. Aiyesanmi. 2012. Metal fractionation in soils collected from selected cocoa plantations in Ogun State, Nigeria. *World Applied Sciences Journal* 20(5): 628–36.

Ajwa, H. A., G. S. Banuelos, and H. F. Mayland. 1999. Selenium uptake by plants from soils amended with inorganic materials. *Journal of Environmental Quality* 27: 1218–27.

Anderson, T. A., and B. T. Walton. 1992. *Comparative Plant Uptake and Microbial Degradation of Trichlorethylene in the Rhizospheres of Five Plant Species—Implications for Bioremediation of Contaminated Surface Soils*. Oak Ridge, TX: Oak Ridge National Laboratory. Environmental Science Division, Pub. 3809. ORNL/TM-12017.

Arapis, G. D. 2006. Root and foliar uptake of 134Cs by three tobacco plant varieties. *Revue d'écologie (La Terre et la vie)* 60(4): 333–40.

Azevedo Silveira, M. L., L. Reynaldo Ferracciú AlleoniI, and L. R. Guimarães Guilherme. 2003. Biosolids and heavy metals in soils. *Scientia Agricola* 60(4). doi:10.1590/S0103-90162003000400029.

Baker, A. J. M. 1981. Accumulators and excluders—Strategies in the response of plants to heavy metals. *Journal of Plant Nutrition* 3(1–4): 643–54.

Baker, A. J. M., and R. R. Brooks. 1989. Terrestrial higher plants which hyperaccumulate metallic elements—A review of their distribution, ecology and phytochemistry. *Biorecovery* 1: 81–126.

Baker, A. J. M., S. P. McGrath, R. D. Reeves, and J. A. C. Smith. 2000. Metal hyperaccumulator plants: A review of the ecology and physiology of a biological resource for phytoremediation of metal-polluted soils. In: *Phytoremediation of Contaminated Soil and Water,* edited by N. Terry and G. Banuelos. Boca Raton, FL: Lewis.

Banks, K. M., and A. P. Schwab. 1993. Dissipation of polycyclic aromatic hydrocarbons in the rhizosphere. In: *Symposium on Bioremediation of Hazardous Wastes: Research, Development and Field Evaluations*. EPA/600/R-93/054. Washington, DC: U.S. Environmental Protection Agency.

Banuelos, G. S., H. A. Ajwa, B. Mackey, L. Wu, C. Cook, S. Akohoue, and S. Zambruzuski. 1997. Evaluation of different plant species used for phytoremediation of high soil selenium. *Journal of Environmental Quality* 26(3): 639–46.

Blaylock, M. J., D. E. Salt, S. Dushenkov, O. Zakharova, C. Gussman, Y. Kapulnik, B. D. Ensley, and I. Raskin. 1997. Enhanced accumulation of Pb in Indian Mustard by soil applied chelating agents. *Environmental Science and Technology* 31(3): 860–65.

Brown, S. L., R. L. Chaney, J. S. Angle, and A. J. M. Baker. 1995. Zinc and cadmium uptake by hyperaccumulator *Thlaspi caerulescens* grown in nutrient solution. *Soil Science Society of America Journal* 59(1): 125–31.

40 CFR. 1996. *Part 264. Standards for Owners and Operators of Hazardous Waste Treatment, Storage and Disposal Facilities*. Washington, DC: U.S. Government Printing Office.

Chamon, A. S., M. Rahman, W. E. H. Blum, M. H. Gerzabek, M. N. Mondol, and S. M. Ullah. 2005. Influence of soil amendments on heavy metal accumulation in crops on polluted soils of Bangladesh. *Communications in Soil Science and Plant Analysis* 36(7–8): 907–24.

Chaney, R. L., Y. M. Li, S. L. Brown, F. A. Homer, M. Malik, J. S. Angle, A. J. M. Baker, R. D. Reeves, and M. Chin. 2000. Improving metal hyperaccumulator wild plants to develop commercial phytoextraction systems: Approaches and progress. In: *Phytoremediation of Contaminated Soil and Water*, edited by N. Terry and G. Banuelos, 129–58. Boca Raton, FL: CRC Press.

Chappell, J. 1997. *Phytoremediation of TCE Using Populus*. Status report prepared for the U.S. EPA Technology Innovation Office. Washington, DC. clu-in.com/phytoTCE.htm.

Costa, R. M., N. D. Camper, and M. B. Riley. 2000. Atrazine degradation in a containerized rhizosphere system. *Journal of Environmental Science and Health. Part B: Pesticides, Food Contaminants, and Agricultural Wastes* 35(6): 677–87.

Cunningham, S. D., and W. R. Berti. 1993. Remediation of contaminated soils with green plants: An overview. *In Vitro Cellular and Developmental Biology - Plant* 29(4): 207–12.

Dominguez-Rosado, E., J. Pichtel, and M. Coughlin. 2004. Phytoremediation of soil contaminated with used motor oil: I. Laboratory and growth chamber studies. *Environmental Engineering Science* 21(2): 157–68.

Dushenkov, S., B. Sorochinsky, A. Mikheev, A. Prokhnevsky, and M. Ruchko. 1999. Phytoremediation of radiocesium-contaminated soil in the vicinity of Chernobyl, Ukraine. *Environmental Science and Technology* 33(3): 469–75.

Ebbs, S. D., and L. V. Kochian. 1998. Phytoextraction of zinc by oat (*Avena sativa*), barley (*Hordeum vulgare*) and Indian mustard (*Brassica juncea*). *Environmental Science and Technology* 32(6): 802–6.

Entry, J. A., N. C. Vance, M. A. Hamilton, D. Zabowsky, L. S. Watrud, and D. C. Adriano. 1996. Phytoremediation of soil contaminated with low concentrations of radionuclides. *Water, Air, and Soil Pollution* 88: 167–76.

Ensley, B. 2000. Rationale for use of phytoremediation. In: *Phytoremediation of Toxic Metals: Using Plants to Clean up the Environment*, edited by I. Raskin and B. Ensley. New York: Wiley Interscience.

Fesenko, S. V., R. Avila, D. Klein, E. Lukaus, N. V. Sukhova, N. I. Sanzharova, and S. I. Spirdonov. 2003. Analysis of factors determining accumulation of ^{137}Cs by woody plants. *Russian Journal of Ecology* 34(5): 309–13.

Fowler, D., D. Branford, R. Donovan, P. Rowland, U. Skiba, E. Nemitz, and F. Choubedar. 2004. Measuring aerosol and heavy metal deposition on urban woodland and grass using inventories of 210Pb and metal concentrations in soil. *Water, Air, and Soil Pollution: Focus* 4(2–3): 483–99.

Fuhrmann, M., M. M. Lasat, S. D. Ebbs, L. V. Kochian, and J. Cornish. 2002. Uptake of cesium-137 and strontium-90 from contaminated soil by three plant species; Application to phytoremediation. *Journal of Environmental Quality* 31(3): 904–9.

Garcia, G., A. Faz, and M. Cunha. 2004. Performance of Piptatherum miliaceum (Smilo grass) in edaphic Pb and Zn phytoremediation over a short growth period. *International Biodeterioration and Biodegradation* 54(2–3): 245–50.

He, B., X. E. Yang, Y. Z. Wei, Z. Q. Ye, and W. Z. Ni. 2002. A new lead resistant and accumulating ecotype—*Sedum alfredii* H. *Acta Botanica Sinica* 44(11): 1365–70.

Heaton, A. C. O., R. B. Meagher, C. L. Rugh, T. Kim, and N. J. Wang. 2003. Toward detoxifying mercury-polluted aquatic sediments with rice genetically engineered for mercury resistance. *Environmental Toxicology and Chemistry* 22(12): 2940–47.

Hinchman, R. R., M. C. Negri, and E. G. Gatliff. 1997. *Phytoremediation: Using Green Plants to Clean Up Contaminated Soil, Groundwater, and Wastewater.* 12th Annual Conference on Contaminated Soils. University of Massachusetts, Amherst, MA, October 18–23.

Huang, J. W., J. Chen, W. R. Berti, and S. C. Cunningham. 1997. Phytoremediation of lead-contaminated soils: Role of synthetic chelates in lead phytoextraction. *Environmental Science and Technology* 31(3): 800–5.

Inui, H., and H. Ohkawa. 2005. Herbicide resistance in transgenic plants with mammalian p450 monooxygenease genes. *Pest Management Science* 61(3): 286–91.

Kabata-Pendias, A. 2001. *Trace Elements in Soils and Plants,* 3rd ed. Boca Raton, FL: CRC Press.

———. 2010. *Trace Elements in Soil and Plants,* 4th ed. Boca Raton, FL: CRC Press.

Kabata-Pendias, A., and B. Szteke. 2015. *Trace Elements in Abiotic and Biotic Environments.* Boca Raton, FL: CRC Press.

Keller, C., C. Ludwig, F. Davoli, and J. Wochele. 2005. Thermal treatment of metal-enriched biomass produced from heavy metal phytoextraction. *Environmental Science and Technology* 39(9): 3359–67.

Khanna, S., and P. Khanna. 2011. Assessment of heavy metal contamination in different vegetables grown in and around urban areas. *Research Journal of Environmental Toxicology* 5(3): 162–79.

Kramer, U., and A. N. Chardonnens. 2001. The use of transgenic plants in the bioremediation of soils contaminated with trace elements. *Applied Microbiology and Biotechnology* 55(6): 661–72.

Lasat, M. M. 2000. *The Use of Plants for the Removal of Toxic Metals from Contaminated Soil.* Prepared for The U.S. Environmental Protection Agency. clu-in.org/download/remed/lasat.pdf.

Lasat, M. M., M. Fuhrmann, S. D. Ebbs, J. E. Cornish, and L. V. Kochian. 1998. Phytoremediation of a radiocesium-contaminated soil: Evaluation of cesium-137 bioaccumulation in the shoots of three plant species. *Journal of Environmental Quality* 27(1): 165–69.

Laureysens, I., R. Blust, L. Temmerman, C. Lemmens, and R. Ceulemans. 2004. Clonal variation in heavy metal accumulation and biomass production in a poplar coppice culture: I. Seasonal variation in leaf, wood and bark concentrations. *Environmental Pollution* 131(3): 485–94.

Liu, R., Y. Dai, and L. Sun. 2015. Effect of rhizosphere enzymes on phytoremediation in PAH-contaminated soil using five plant species. *PLoS.* doi:10.1371/journal.pone.0120369.

Liu, H., A. Probst, and B. Liao. 2005. Metal contamination of soils and crops affected by the Chenzhou lead/zinc mine spill (Hunan, China). *Science of the Total Environment* 339(1–3): 153–66.

Lunney, A. I., B. A. Zeeb, and K. J. Reimer. 2004. Uptake of weathered DDT in vascular plants: Potential for phytoremediation. *Environmental Science and Technology* 38(22): 6147–54.

Mantovi, P., G. Bonazzi, E. Maestri, and N. Mamiroli. 2003. Accumulation of copper and zinc from liquid manure in agricultural soils and crop plants. *Plant and Soil* 250(2): 249–57.

MARC (Mid-America Regional Council). 2018. *Native and Non-native Root Comparison Chart.* https://cfpub.epa.gov/npstbx/files/KSMO_KnowYourRoots.pdf (Accessed February 19, 2018).

McIntyre, T. 2003. Phytoremediation of heavy metals from soils. *Advances in Biochemical Engineering: Biotechnology* 78: 97–123.

Meharg, A. A., and J. W. G. Cairney. 2000. Ectomycorrhizas—Extending the capabilities of rhizosphere remediation? *Soil Biology and Biochemistry* 32(11–12): 1475–84.

Menon, P., M. Gopal, and R. Prasad. 2004. Dissipation of chlorpyrifos in two soil environments of semi-arid India. *Journal of Environmental Science and Health, Part B: Pesticides, Food Contaminants, and Agricultural Wastes* 39(4): 517–31.

Muchuweti, M., R. Zvauya, M. D. Scrimshaw, J. N. Lester, J. W. Birkett, and E. Chinyanga. 2006. Heavy metal content of vegetables irrigated with mixtures of wastewater and sewage sludge in Zimbabwe: Implications for human health. *Agriculture, Ecosystems and Environment* 112(1): 41–8.

Negri, M. C., and R. R. Hinchman. 1996. Plants that remove contaminants from the environment. *Laboratory Medicine* 27(1): 36–40.

Newman, L., S. Strand, J. Duffy, G. Ekuan, M. Raszaj, B. Shurtleff, J. Wilmoth, P. Heilman, M. Gordon, and M. P. Gordon. 1997. Uptake and biotransformation of trichloroethylene by hybrid poplars. *Environmental Science and Technology* 31(4): 1062–67.

Niu, Z.-X., L.-A. Sun, T.-H. Sun, Y.-S. Li, and H. Wang. 2007. Evaluation of phytoextracting cadmium and lead by sunflower, ricinus, alfalfa and mustard in hydroponic culture. *Journal of Environmental Sciences* 19(8): 961–67.

Pang, J., M. H. Wong, G. S. Y. Chan, J. Zhang, and J. Liang. 2003. Physiological aspects of vetiver grass for rehabilitation in abandoned metalliferous mine wastes. *Chemosphere* 52(9): 2003.

Parrish, Z. D., M. K. Banks, and A. P. Schwab. 2005. Assessment of contaminant lability during phytoremediation of polycyclic aromatic hydrocarbon impacted soil. *Environmental Pollution* 137(2): 187–97.

Peer, W. A., I. R. Baxter, E. L. Richards, J. L. Freeman, and A. S. Murphy. 2005. *Phytoremediation and Hyperaccumulator Plants.* https://link.springer.com/content/pdf/10.1007%2F4735_100.pdf.

Petruzzelli, G. 1989. Recycling wastes in agriculture: Heavy metal bioavailability. *Agriculture, Ecosystems, and Environment* 27(1–4): 493–503.

Phusantisampan, T., W. Meeinkuirt, P. Saengwilai, J. Pichtel, and R. Chaiyarat. 2016. Phytostabilization potential of two ecotypes of *Vetiveria zizanioides* in cadmium-contaminated soils: Greenhouse and field experiments. *Environmental Science and Pollution Research International* 23(19): 20027–38.

Pichtel, J., and M. Anderson. 1997. Trace metal bioavailability in municipal solid waste and sewage sludge composts. *Bioresource Technology* 60(3): 223–29.

Pichtel, J., and C. A. Salt. 1998. Vegetative growth and trace metal accumulation on metalliferous wastes. *Journal of Environmental Quality* 27(3): 618–24.

Pichtel, J., K. Kuroiwa, and H. T. Sawyerr. 2000. Distribution of Pb, Cd and Ba in soils and plants of two contaminated sites. *Environmental Pollution* 110: 171–78.

Pilon-Smits, E. A. H., S. Hwang, C. M. Lytle, Y. Zhu, J. C. Tai, R. C. Bravo, Y. Chen, T. Leustek, and N. Terry. 1999. Overexpression of ATP sulfurylase in Indian mustard leads to increased selenate uptake, reduction, and tolerance. *Plant Physiology* 119(1): 123–32.

Pulford, I., and S. Watson. 2003. Phytoremediation of heavy metal-contaminated land by trees—A review. *Environment International* 29(4): 529–40.

Rezek, J., C. Wiesche, M. Mackova, F. Zadrazil, and T. Macek. 2008. The effect of ryegrass (*Lolium perenne*) on decrease of PAH content in long term contaminated soil. *Chemosphere* 70(9): 1603–8.

Rugh, C., H. D. Wilde, N. M. Stack, D. M. Thompson, A. O. Summers, and R. B. Meagher. 1996. Mercuric ion reduction and resistance in transgenic *Arabidopsis thaliana* plants expressing a modified bacterial MerA gene. *Proceedings of the National Academy of Sciences of the United States of America* 93(8): 3182–87.

Rugh, C. L., J. F. Seueoff, R. B. Meagher, and S. A. Merkle. 1998. Development of transgenic yellow poplar for mercury phytoremediation. *Nature Biotechnology* 16(10): 925–28.

Ruiz, O. N., and H. Daniell. 2009. Genetic engineering to enhance mercury phytoremediation. *Current Opinion in Biotechnology* 20(2): 213–19.

Salt, C. A., and J. W. Kay. 1999. The seasonal pattern of radiocaesium partitioning within swards of *Agrostis capillaris* at two defoliation intensities. *Journal of Environmental Radioactivity* 45(3): 219–34.

Salt, D. E., M. Blaylock, N. P. B. A. Kumar, V. Dushenkov, B. D. Ensley, I. Chet, and I. Raskin. 1995. Phytoremediation: A novel strategy for the removal of toxic metals from the environment using plants. *Biotechnology* 13(5): 468–74.

Sas-Nowosielska, A., J. M. Kuperberg, K. Krynski, R. Kucharski, E. Malstrokkowski, and M. Pogrzeba. 2004. Phytoextraction crop disposal—An unsolved problem. *Environmental Pollution* 128(3): 373–79.

Schaffner, A., B. Messner, C. Langebartels, and H. Sandermann. 2002. Genes and enzymes for in-planta phytoremediation of air, water and soil. *Acta Biotechnologica* 22(1–2): 141–52.

Schnoor, J. L., L. A. Licht, S. C. McCutcheon, N. L. Wolfe, and L. H. Carreira. 1995. Phytoremediation of organic and nutrient contaminants. *Environmental Science and Technology* 29(7): 318–23.

Shen, Z.-G., X.-D. Li, C.-C. Wang, H.-M. Chen, and H. Chua. 2002. Lead phytoextraction from contaminated soil with high-biomass plant species. *Journal of Environmental Quality* 31(6): 1893–900.

Shimp, J. F., J. C. Tracy, L. C. Davis, E. Lee, W. Huang, L. E. Erickson, and J. L. Schnoor. 1993. Beneficial effects of plants in the remediation of soil and groundwater contaminated with organic materials. *Critical Reviews in Environmental Science and Technology* 23(1): 41–77.

Sims, J. T., and J. S. Kline. 1991. Chemical fractionation and plant uptake of heavy metals in soils amended with co-composted sewage sludge. *Journal of Environmental Quality* 20(2): 387–95.

Smith, R. A. H., and A. D. Bradshaw. 1979. The use of metal tolerant plant populations for the reclamation of metalliferous wastes. *The Journal of Applied Ecology* 16(2): 595–612.

Smith, M. J., T. H. Flowers, H. J. Duncan, and J. Alder. 2006. Effects of polycyclic aromatic hydrocarbons on germination and subsequent growth of grasses and legumes in freshly contaminated soil with aged PAHs residues. *Environmental Pollution* 141(3): 519–25.

Smith, L. A., J. L. Means, A. Chen, B. Alleman, C. C. Chapman, J. S. Tixier, S. E. Brauning, A. R. Gavaskar, and M. D. Royer. 1995. *Remedial Options for Metals—Contaminated Soils.* Boca Raton, FL: CRC Press.

Solhi, M., H. Shareatmadari, and M. A. Hajabbasi. 2005. Lead and zinc extraction potential of two common crop plants, *Helianthus annuus* and *Brassica napus. Water, Air, and Soil Pollution* 167(1–4): 59–71.

Sposito, G., L. J. Lund, and A. C. Chang. 1982. Trace metal chemistry in arid-zone field soils amended with sewage sludge: I. Fractionation of Ni, Cu, Zn, Cd and Pd in solid phases. *Soil Science Society of America Journal* 46: 260–64.

Sutherson, S. S. 1997. *Remediation Engineering: Design Concepts.* Boca Raton, FL: CRC Press.

Tandi, N. K., J. Nyamangara, and C. Bangira. 2004. Environmental and potential health effects of growing leafy vegetables on soil irrigated using sewage sludge and effluent: A case of Zn and Cu. *Journal of Environmental Science and Health, Part B: Pesticides, Food, Contaminants and Agricultural Wastes* 39(3): 461–71.

Tessier, A., P. G. C. Campbell, and M. Bisson. 1979. Sequential extraction procedure for the speciation of particulate trace metals. *Analytical Chemistry* 51(7): 844–51.

Tobia, R., and H. Compton. 1997. *Phytoremediation of TCE in Groundwater Using Populus.* Status report prepared for the U.S. EPA Technology Innovation Office. Washington, DC.

Tong, Y. P., R. Kneer, and Y. G. Zhu. 2004. Vacuolar compartmentalization: A second-generation approach to engineering plants for phytoremediation. *Trends in Plant Science* 9(1): 7–9.

U.S. Environmental Protection Agency. 1983. *Process Design Manual for Land Application of Municipal Sludge.* EPA/625/1-83-016. Cincinnati.

———. 1997. *Recent Developments for In Situ Treatment of Metal Contaminated Soils.* EPA-542-R-97-004. Office of Solid Waste and Emergency Response.

———. 2000. *An Overview of the Phytoremediation of Lead and Mercury.* Washington, DC: Office of Solid Waste and Emergency Response Technology Innovation Office. clu-in.org.

———. 2005. *Use of Field-Scale Phytotechnology for Chlorinated Solvents, Metals, Explosives and Propellants, and Pesticides.* EPA 542-R-05-002. Washington, DC: Office of Solid Waste and Emergency Response. www.clu-in.org.

———. 2018a. *Sangamo Weston, Inc./twelve-mile Creek/Lake Hartwell PCB contamination Pickens, SC.* https://cumulis.epa.gov/supercpad/cursites/csitinfo.cfm?id=0403252.

———. 2018b. *Bofors Nobel, Inc. Muskegon, MI.* https://cumulis.epa.gov/supercpad/cursites/csitinfo.cfm?id=0502372.

———. 2018c. *Case Summary: Third-Party Agreement Supports Cleanup and Redevelopment at the Middlefield-Ellis-Whisman (MEW) Study Area.* https://www.epa.gov/enforcement/case-summary-third-party-agreement-supports-cleanup-and-redevelopment-middlefield-ellis.

———. 2018d. *Palmerton Zinc Pile, Palmerton, PA.* https://cumulis.epa.gov/supercpad/SiteProfiles/index.cfm?fuseaction=second.cleanup&id=0300624#Status.

Verbruggen, N., C. Hermans, and H. Schat. 2009. Molecular mechanisms of metal hyperaccumulation in plants. *New Phytologist* 181(4): 759–76.

Xia, H., L. Wu, and Q. Tao. 2003. A review on phytoremediation of organic contaminants. *Chinese Journal of Applied Ecology* 14(3): 457–60.

Yang, X., Y. Feng, Z. He, and P. J. Stoffella. 2005. Molecular mechanisms of heavy metal hyperaccumulation and phytoremediation. *Journal of Trace Elements in Medicine and Biology* 18(4): 339–53.

Yang, X. E., X. X. Long, W. Z. Ni, and C. X. Fu. 2002. *Sedum alfredii* H: A new Zn hyperaccumulating plant first found in China. *China Science Bulletin* 47(19): 1634–37.

Yang, X. E., X. X. Long, H. B. Ye, Z. L. He, P. J. Stoffella, and D. V. Calvert. 2004. Cadmium tolerance and hyperaccumulation in a new Zn-hyperaccumulating plant species (*Sedum alfredii* Hance). *Plant and Soil* 259(1–2): 181–9.

Chapter 13

Innovative Technologies

Every great advance in science has issued from a new audacity of imagination.

—John Dewey

Ideas are like rabbits.

—John Steinbeck

13.1 INTRODUCTION

Increasingly sophisticated technologies to address remediation of contaminated soil, substrata, and groundwater continue to come available worldwide. There is a broad range of applicability, feasibility, and cost concerns for each technology. Applications address many contaminant types and levels of sophistication. This chapter will provide only a brief view of some promising methods. Chapter 14 is devoted entirely to another emerging technology, *nanoremediation*.

13.2 ELECTROKINETIC REMEDIATION

Electrokinetic remediation, also referred to as *electrokinetic soil processing*, *electromigration*, *electrochemical decontamination*, or *electroreclamation*, involves the application of low-density direct current between electrodes placed in soil to mobilize contaminants occurring as charged species. This is, therefore, a possible separation and removal technique for radionuclides,

Table 13.1 Performance of electrokinetic remediation at five field sites in Europe

Site Description	Volume (ft^3)	Contaminants	Initial Concentration (mg/kg)	Final Concentration (mg/kg)
Former paint factory	8,100 peat/clay soil	Cu	1,220	< 200
		Pb	> 3,780	< 280
Operational galvanizing plant	1,350 clay soil	Zn	> 1,400	600
Former timber plant	6,750 heavy clay soil	As	> 250	< 30
Temporary landfill	194,400 sand	Cd	> 180	< 40
Military air base	68,000 clay	Cd	660	47
		Cr	7,300	755
		Cu	770	98
		Ni	860	80
		Pb	730	108
		Zn	2,600	289

Source: United States Environmental Protection Agency 1997.

metals, and some organic contaminants from saturated or unsaturated soils, slurries, and sediments (table 13.1).

Electrodes can be installed horizontally or vertically (depending on the location and shape of the contaminant plume) in deep, directionally drilled tunnels or in trenches around sites contaminated by leaking USTs; by spillage from industrial processes; and by leachate from agricultural fields, landfills, and mine tailings (Lageman 1993).

The principle of electrokinetic remediation relies on application of a low-intensity direct current through soil between two or more electrodes. Water occurring in soil pores possesses electrical conductivity due to presence of salts. Current is applied in the range of mA/cm^2 of cross-sectional area between electrodes or an electric potential difference on the order of a few volts/cm across the electrodes. The current mobilizes charged species, particles, and ions in the soil by electromigration (transport of charged chemical species under an electric gradient); electro-osmosis (transport of pore fluid under an electric gradient); electrophoresis (movement of charged particles under an electric gradient); and electrolysis (chemical reactions influenced by the electric field) (Cameselle et al. 2013; Rodsand and Acar 1995).

As depicted in figure 13.1, groundwater and/or a processing fluid (supplied externally through the boreholes that contain the electrodes) serves as the conductive medium. The additives in the processing fluid, the products of electrolysis reactions at the electrodes, and the dissolved contaminant species are transported by conduction under electric fields. This transport, when coupled with a removal phase of sorption, precipitation/dissolution, and

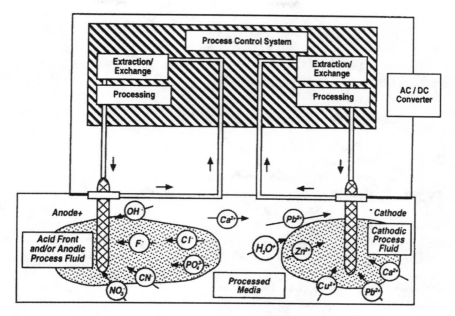

Figure 13.1 Schematic of single electrode configuration and geometry used in electrokinetic remediation. *Source*: U.S. Environmental Protection Agency 1995a.

volatilization/complexation, provides the mechanism for the electrokinetic remediation process.

Electrolysis reactions dominate at each electrode. Solution pH varies at the electrodes as a result of electrolysis of water molecules. Oxidation occurs at the anode, where an acid front is generated if water is the primary pore fluid present. Reduction occurs at the cathode and produces a base front. The solution becomes acidic at the anode because hydrogen ions are produced and oxygen gas is released, and the solution becomes basic at the cathode, where hydroxyl ions are generated and hydrogen gas is released. At the anode, the pH can decrease to < 2.0, and increase at the cathode to > 12 depending on the current applied (U.S. EPA 1997a; Jacobs et al. 1994; Acar and Alshawabkeh 1993).

$$\text{Anode} \quad 2H_2O - 4e^- \rightarrow O_{2(g)} + 4H^+ \quad (13.1)$$

$$\text{Cathode} \quad 2H_2O + 2e^- \rightarrow H_{2(g)} + 2OH^- \quad (13.2)$$

The acid front eventually migrates from anode to cathode. Movement of the acid front by advective forces results in desorption of contaminants from the soil. This migration also results in temporary soil acidification. In some cases metallic electrodes may dissolve as a result of electrolysis and introduce corrosion products. However, if inert electrodes such as carbon, graphite, or platinum are used, no residue will be introduced into the soil. Water or some other suitable salt solution may be added to the treatment zone to enhance mobility of the contaminant and increase the effectiveness of the technology. For example, addition of a buffer solution can serve to alter and subsequently maintain pore fluid pH at a desired level.

Contaminants arriving at the electrode(s) are removed by any of several methods, either at the electrodes or by pumping the processing fluid from the soil. For example, ions may be electroplated or precipitated at the electrode. Alternatively, the ion-enriched soil water may be removed by wells installed upgradient of the electrode. The water may be treated by evaporation/condensation, passage through ion exchange columns, or via electrochemical techniques or simple chemical precipitation (e.g., a calcium hydroxide solution will precipitate metals from the process fluid). Several environmental variables affect the overall migration of ions, particles, and/or fluids between electrodes. These include soil mineralogy, pore fluid composition and conductivity, electrochemical properties of the species in the pore fluid, and the porosity and tortuosity of the porous medium (Cameselle et al. 2013; Acar and Alshawabkeh 1993).

Before electrokinetic remediation is undertaken at a site, field and laboratory screening tests must be conducted to determine whether the particular site is amenable to the treatment technique. Issues to address include:

- Field conductivity. Spatial variability of geologic layers should be delineated because buried insulating material can cause variations in electrical conductivity of the soil. Such variations will affect the overall ability of the system to mobilize contaminants. In addition, the presence of deposits that exhibit very high electrical conductivity (e.g., solid metallic wastes), which may render the technique ineffective, should be identified.
- Chemical analysis of water. Pore water should be analyzed for dissolved anions and cations as well as for concentrations of contaminants. In addition, electrical conductivity and pH must be measured.
- Chemical analysis of soil. The buffering capacity and geochemistry of the soil should be determined. Soil pH is highly relevant as it affects the valence, and hence the solubility and sorption of contaminant ions.
- Bench-scale tests. Myriad physical and chemical reactions of soil are interrelated; therefore, it is valuable to conduct bench-scale tests to predict the performance of electrokinetic remediation at the field scale. Transport,

removal rates, and residual level of contamination can be assessed for different removal scenarios.

13.2.1 Field application of Electrokinetic Remediation

A waste disposal site at the Naval Air Weapons Station in China Lake, California, was subjected to electrokinetic treatment. Between 1947 and 1978 two unlined waste lagoons received wastewater discharge from electroplating and metal finishing activities. A plating shop discharged approximately 95 million gal of plating rinse solution into the lagoons. Additionally, up to 60,000 gal of waste photographic fixer solution and small quantities of organic solvents and rocket fuel were disposed in the lagoons. Surface sampling determined levels of chromium and cadmium as high as 25,100 mg/kg and 1,810 mg/kg, respectively (USAEC 2000).

The area of study measured approximately one-half acre. Three cathodes were centered between six anodes. The objective was to concentrate the contaminants in the center of the treated area around the cathodes. The anode-cathode spacing was 15 ft (4.5 m) and each was inserted 10 ft (3 m) deep. A constant voltage of 60V was applied for twenty days and then was reduced to 45V for six months.

After six months of treatment 78 percent of the soil volume was treated to below natural levels of Cr (initial Cr concentration was between 180 and 1100 mg/kg). The majority of Cd was removed in approximately 70 percent of the soil between the electrodes (initial concentrations ranged between 5 and 20 mg/kg).

The chloride content of the surrounding soil and groundwater resulted in continuous production of chlorine gas, which inhibited the formation of the pH front between the electrodes; hence, movement of metals was restricted. Additionally, by-products were generated due to the chemical amendments used. The study was transitioned to a pilot-scale project and the electrokinetic system was restarted with a reduced number of electrodes, thus inducing a higher current density and making contaminant removal possible in the brackish conditions (Gent et al. 2004).

Recent experiments show that electrokinetic remediation can be used in combination with other remediation techniques such as pump-and-treat, biodegradation, and vacuum extraction. When combined with electrical heating, the electrokinetic technique can be used to remove polar and nonpolar organic compounds from soil and groundwater (Cameselle et al. 2013; Lageman 1993).

A variation of the electrokinetic process involves the installation of an electrokinetic fence. The electrokinetic phenomena that occur when soil is electrically charged can be used to 'fence off' hazardous sites. Electrokinetic

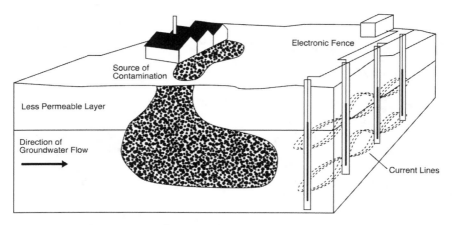

Figure 13.2 Schematic view of an electrokinetic fence enclosing a contaminant plume. *Source*: U.S. Environmental Protection Agency 1997a.

fencing combines containment and remediation. With an electrokinetic fence, it is possible to capture electrically charged (polar) contaminants while treated water passes through the fence. The fence can also increase soil temperatures in the zone within the fence to accelerate biodegradation processes. Electrokinetic fences can be installed both horizontally and vertically and at any depth (figure 13.2).

13.3 CHEMICAL OXIDATION

This technology involves the stimulation of oxidation/reduction (redox) reactions which ultimately result in chemical conversion of hazardous contaminants to nonhazardous or less toxic compounds that are more stable and less mobile. In a chemical context, redox reactions involve the transfer of electrons from one element to another; specifically, one reactant is oxidized (it loses electrons) and one is reduced (gains electrons).

Oxidation processes tend to be more commonly used in site remediation. Several oxidants are used for decomposing contaminants including potassium or sodium permanganate, hydrogen peroxide, ozone, sodium persulfate, and Fenton's reagent. Each has its inherent advantages and limitations. Although applicable to soil contamination, these oxidizers have been applied primarily toward remediating groundwater.

Several important issues to address in oxidant selection for site cleanup include (U.S. EPA 2006):

1. Is the oxidant capable of degrading the contaminant of concern? Is a catalyst or other additive required to increase effectiveness?

2. What is the soil oxidant demand (SOD)? This is a measure of how the naturally occurring materials in soil will affect the performance of the oxidant. For nonselective oxidants, high SOD will increase the cost of cleanup, as more oxidant will be required.
3. What is the pH of the soil/ groundwater system? Some oxidants require an acidic environment to function optimally. If the soil is basic, an acid may need to be applied.
4. How will the decomposition rate of the oxidant affect application strategy? Some unreacted oxidants may remain in the subsurface for weeks to months, while others naturally decompose within hours of injection.

The delivery system depends upon depth of the contaminant, the physical state of the oxidant (gas, liquid, solid), and its decomposition rate. Backhoes, trenchers, and augers have been used to incorporate liquid and solid oxidants into contaminated soil and sludge. Liquids can be delivered either by gravity through wells and trenches or by injection. In the vadose zone, pressurized injection of liquids or gases, either through the screen of a well or through the probe of a direct push rig, forces oxidant into the formation. The rig offers a cost-effective means of delivering oxidant, and if needed, the hole can be converted to a well for later injections or monitoring.

Potassium permanganate and other solid phase chemical oxidants have been added by hydraulic or pneumatic fracturing.

Several oxidants have been capable of achieving high treatment efficiencies (e.g., > 90%) for unsaturated aliphatic (e.g., trichloroethylene [TCE]) and aromatic compounds (e.g., benzene), with very fast reaction rates (90% destruction within minutes). The simplified stoichiometric reaction for peroxide degradation of TCE is given in equation 13.3.

$$3H_2O_2 + C_2HCl_3 \rightarrow 2CO_2 + 2H_2O + 3HCl \qquad (13.3)$$

Chemical oxidation usually requires multiple applications. Field applications affirm that matching the oxidant and in situ delivery system to the contaminants of concern and site conditions is the key to successful implementation (U.S. EPA 2000). Table 13.2 lists the reactivities of common oxidants with contaminants encountered at field sites.

13.4 STEAM ENHANCED RECOVERY

The Steam Enhanced Recovery Process (SERP) is an in situ technology designed to remove volatile and semivolatile organic contamination using steam to provide heat and pressure to the affected media. The process is applicable to treatment of contaminated soil and groundwater. The process works

Table 13.2 Reactivity of oxidants with common soil and groundwater contaminants

Oxidant	High	Moderate	Low
Ozone	PCE, TCE, DCE, VC, MTBE, CB, PAHs, Phenols, Explosives, PCBs, Pesticides	BTEX, CH_2Cl_2	CT, $CHCl_3$
Hydrogen peroxide	PCE, TCE, DCE, VC, CB, BTEX, MTBE, Phenols	DCA, CH_2Cl_2, PAHs, Explosives	TCA, CT, $CHCl_3$, PCBs, Pesticides
Calcium peroxide	PCE, TCE, DCE, VC, CB	DCA, CH2Cl$_2$	CT, $CHCl_3$
Fenton's Reagent	PCE, TCE, DCE, VC, CB, BTEX, MTBE, Phenols	DCA, CH_2Cl_2, PAHs, Explosives	TCA, CT, $CHCl_3$, PCBs, Pesticides
Potassium/sodium permanganate	PCE, TCE, DCE, VC, TEX, PAHs, Phenols, Explosives	Pesticides	Benzene, DCA, CH_2Cl_2, TCA, CT, CB, $CHCl_3$, PCBs
Sodium persulfate	PCE, TCE, DCE, VC, CB, BTEX, Phenols	DCA, CH_2Cl_2, $CHCl_3$, PAHs, Explosives, Pesticides	TCA, CT, PCBs
Sodium persulfate (Hot)	All CVOCs, BTEX, MTBE, PAHs, Phenols, Explosives, PCBs, Pesticides		

BTEX = benzene, toluene, ethylbenzene, xylene; CB = chlorobenzenes; CT = carbon tetrachloride; CVOCs = chlorinated volatile organic compound; DCA = dichloroethane; DCE = dichloroethene; MTBE = methyl-tert-butyl ether; PAHs = polycyclic aromatic hydrocarbons; PCBs = polychlorinated biphenyls; PCE = perchloroethylene; TCA = trichloroethane; TCE = Trichloroethene; VC = vinyl chloride.
Source: US EPA 2006; IRTC 2005; Brown 2003.

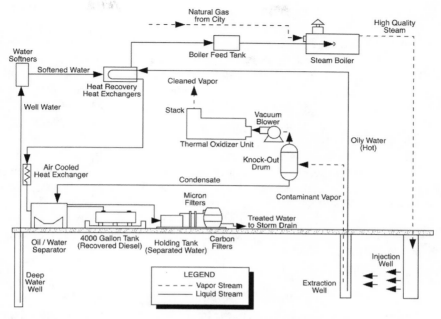

Figure 13.3 Steam Enhanced Recovery Plant treatment train. *Source*: U.S. Environmental Protection Agency 1995b.

by injecting steam through injection wells constructed to a depth at or below the contaminant plume. Extraction wells are operated under vacuum to create a pressure gradient in the soil to draw out liquids, vapor, and contaminants. Liquid and vapor streams removed by the extraction wells are directed to an aboveground liquid and vapor treatment system (figure 13.3) (U.S. EPA 2006).

Site geology is important in determining whether SERP will be applicable. Site requirements for effective operation include:

1. The contamination must consist of volatile and/or semivolatile compounds such as those found during spilled fuel events.
2. The soil must have moderate to high permeability.
3. There must be a confining layer below the depth of contamination. This can take the form of a continuous low-permeability layer such as a bedrock aquiclude, or a water table (for LNAPL compounds).
4. A low-permeability surface layer may be needed to prevent steam breakthrough for shallow treatment applications.

The removal of volatile and semivolatile contamination from soil by SERP occurs via several mechanisms. High-temperature steam (approximately 250°F) heats soil to the steam temperature in a pattern radiating from the injection wells toward extraction wells, following pressure gradients. As the soil heats, contaminants that have boiling points lower than that of water will vaporize. The vapor is then pushed ahead of the steam front. This results in a band of liquid contaminant that is formed just ahead of the advancing steam. When the steam front reaches an extraction well, the vapor, liquid, and contaminants are removed.

The Rainbow Disposal site in Huntington Beach, California, was contaminated by a spill of diesel fuel, which is composed primarily of longer chain hydrocarbon compounds (C_8 or heavier; see chapter 3). Diesel compounds, although less dense than water, are heavier than those in most other petroleum-based fuels (e.g., gasoline or jet fuel) and are less volatile and more viscous. These properties make diesel a more difficult contaminant to remove from soil than most other petroleum-based fuels.

SERP was applied at the Rainbow Disposal site which covers an area of 2.3 acres. The arrangement of process wells was designed to treat the entire area concurrently. Thirty-five steam injection wells and thirty-eight vapor/liquid extraction wells were constructed in the treatment area. Wells were placed in a repeating pattern of four injection wells surrounding each extraction well. The distance between adjacent injection well/extraction well clusters was approximately 45 ft; between adjacent wells of the same type, spacing was approximately 60 ft. Well spacing for a site was determined based on soil permeability in the treatment area, size of the area, and depth and concentration of contaminants. About 700 gal of diesel fuel were collected in liquid form, and approximately 15,400 gal were oxidized in the vapor treatment system (U.S. EPA 1995b). Significant quantities of fuel could not be recovered from this site, however.

In a similar study, a Dynamic Underground Stripping process was developed to recover gasoline contamination from a subsurface plume. The process uses steam injection, vacuum extraction, and electrical heating to effect contaminant removal from soil and groundwater. This system was more successful than SERP, as gasoline contains more volatile contaminants, the electric heating enhanced volatilization, and the latter system used more effective monitoring of the steam zone for operational control (U.S. EPA 1995b).

Innovative Technologies 355

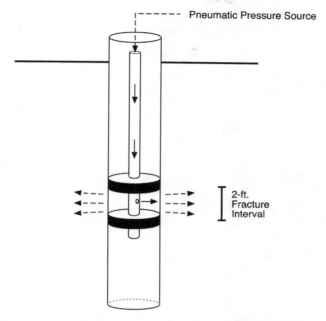

Figure 13.4 A pressure injector used to fracture subsurface strata. *Source*: U.S. Environmental Protection Agency 1993c.

13.5 PNEUMATIC FRACTURING AND HOT GAS INJECTION

As discussed in chapter 9, soil vapor extraction (SVE) is an accepted method for removal of volatile hydrocarbons from the vadose zone. A primary limitation to SVE technology, however, is that the affected zone formation must be sufficiently permeable for air to flow and mobilize volatile contaminants into the airstream. A method has been devised to facilitate cleanup of soil and rock formations having poor air permeability, for example, shales and clay. The method involves injecting short bursts (< 1 min) of compressed air into the formation, causing it to fracture at weak points. These fractures, which occur mostly in the horizontal direction in clay and shale formations, enlarge and extend existing fissures and/or create new fissures. Where these fractures connect an extraction well with an air injection well or other source of air, they promote increased flow through the formation and increase its permeability. Greater airflow allows increased quantities of trapped and adsorbed organics to be removed by volatilization. An additional benefit of this technique is that the creation and extension of fractures provides access to areas of the formation that were not previously accessible to treatment (U.S. EPA 1993c).

Fracturing is conducted over narrow depth intervals using a proprietary lance equipped with rubber packers that are expanded by pressurization with air (figure 13.4). The effect of the pressure pulse is concentrated and the design minimizes the propagation of vertical fractures by providing resistance above and below. Once fracturing has been successful at several intervals, the permeability of the formation is significantly increased. Wells are installed as with conventional SVE technology; for example, vacuum extraction from a central well with surrounding wells that are either air-injected or open to the atmosphere, or air injection into a central well with vacuum extraction from surrounding wells. Following pneumatic fracturing, the radius of influence for vapor extraction is extended, and in situ removal of VOCs is enhanced. Hot gas injection into bedrock accelerates VOC removal by vapor extraction, particularly when used simultaneously with pneumatic fracturing. Hot gas can be generated by catalytic oxidation of extracted VOCs and can attain temperatures of approximately 1,000°F (538°C).

In order to remove VOCs from extracted air before it is exhausted to the atmosphere, GAC columns may be installed as a final step. Alternate means for gas removal such as catalytic oxidation may be cost-effective at higher concentrations (> 50 ppm_v).

In a system devised by Accutech and the Hazardous Substance Management Research Center at the New Jersey Institute of Technology, a vacuum extraction system was developed consisting of compressors, a manifold, a water knock-out vessel, and compressor/vacuum blowers. Two GAC adsorption drums (55 gal) were installed in series to remove VOCs from extracted air before it was exhausted to the atmosphere (U.S. EPA, 1993).

Additional practical concerns with pneumatic fracturing technology include the necessity of employing high-temperature grouts when installing well casings that will be exposed to extreme heat during the hot gas injection phase. Furthermore, if hot exhaust gases from catalytic oxidation of VOCs are directly injected into the formation, it must be demonstrated that the gases are not contaminated (e.g., with HCl from destruction of chlorinated VOCs) and will not adversely affect groundwater or the formation (U.S. EPA 1993c).

13.6 RADIO-FREQUENCY HEATING

Radio-frequency heating (RFH) uses a high-frequency alternating electric field for in situ heating of soil and removal of hydrocarbon vapors. The technology depends on the presence of material with unevenly distributed electrical charges. The application of an electric field promotes vibration of polar molecules which creates mechanical heat. A range of radio frequencies (e.g., 6.78 MHz, 13.56 MHz, 27.12 MHz, and 40.68 MHz plus seven higher frequencies) regulated and assigned by the Federal Communications Commission can be used in industrial, scientific, or medical applications (FCC 2006).

An RFH system usually consists of:

- a three-phase power supply;
- a radio-frequency source with an oscillator that generates a low-power current at the desired frequency, several serial amplifiers that increase the strength of the oscillator current, and a final amplifier that delivers the current at the prescribed output level;
- an applicator system consisting of electrodes or antennae;
- a monitoring control system;
- a grounded metal shield over the treatment area;
- a vapor collection and treatment system (Haliburton 1995).

Rows of applicator electrodes are placed in the ground to the depth of the contaminant plume. The electrodes can be installed with conventional drilling equipment or via direct push. In some designs, the electrodes are also used to recover soil gas and heated vapors; in others, wells are placed specifically for soil vapor extraction and act as electromagnetic sinks to prevent heating

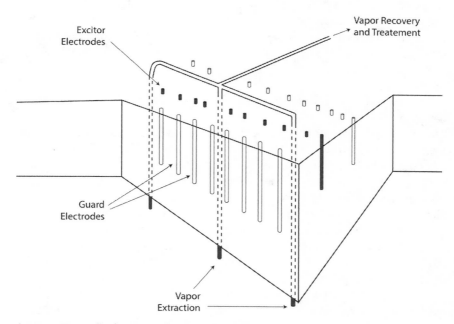

Figure 13.5 Radio-frequency heating with electrodes.

beyond the treatment zone (figure 13.5). Heating is both radiative and conductive, with soil near the applicator electrodes heating fastest (the radio-frequency wave gets weaker the farther from the electrode due to energy absorption).

In saturated conditions, RFH boils the water in the immediate vicinity of the applicator electrode and does not heat the treatment zone to an effective temperature. If the water table is shallow, dewatering may be necessary (Edelstein et al. 1996; Davis 1997).

The antenna method requires the placement of vapor recovery wells around the treatment area. Drilled or pushed applicator boreholes are lined with a fiberglass casing or other nonconductive nonpolar material that will withstand the temperatures expected. The antennae are lowered into the applicator holes to an appropriate depth, and the heating is initiated. The antenna can be lowered or raised as needed.

Both antenna and electrode systems monitor the heat distribution in the subsurface (usually with thermocouples) to ensure target temperatures are obtained throughout the treatment zone. Depending upon the contaminants of concern, RFH can obtain temperatures over 250°C and possibly as high as 400°C (Davis 1997; U.S. EPA 1997b). These temperatures allow the system to treat both VOCs and many SVOCs.

The vapor extraction system consists of conventional extraction wells. Also, for safety and prevention of potential interference with local radio transmissions, a grounded metal shield is usually employed over the treatment area. Metal structures will absorb RF energy; therefore, RFH is not applicable for sites that contain excessive metallic debris or other conductive objects.

13.7 IN SITU VITRIFICATION

The in situ vitrification (ISV) process is designed to treat soil, sludges, sediments, and mine tailings contaminated with a variety of organic and inorganic contaminants. The technology uses joule heating to melt the soil and waste matrix, destroying organic compounds in the process and encapsulating inorganic constituents in a monolithic and leach-resistant form. In joule heating, electric current flows through the material, thereby transferring heat energy.

The typical ISV arrangement involves a square array of four graphite electrodes spaced up to 18 to 20 ft apart. This allows formation of a maximum melt width of about 35 to 40 ft and a maximum melt depth of approximately 20 ft. Electrode spacing is partly dependent on soil characteristics; the electrodes are lowered gradually as the melt progresses. Figure 13.6 shows a typical ISV equipment layout and figure 13.7 a side view of a system designed for a field situation.

Innovative Technologies 359

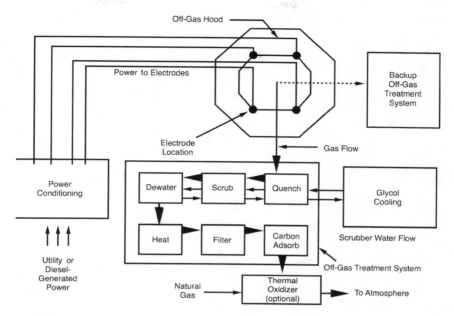

Figure 13.6 Components of in situ vitrification technology. *Source*: U.S. Environmental Protection Agency 1995c.

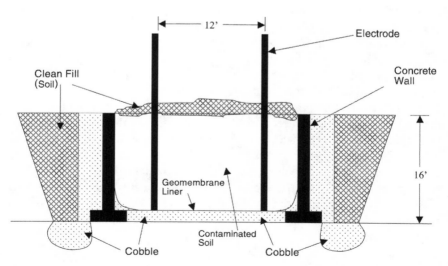

Figure 13.7 Side view of ISV reaction zone. *Source*: U.S. Environmental Protection Agency 1995c.

A conductive mixture of flaked graphite and glass frit is placed just below the soil surface between the electrodes to act as a starter path. A layer of insulation is applied to cover the soil. The starter path facilitates the flow of current between electrodes until the soil reaches a temperature and viscosity sufficient to conduct the current and cause melting. At this point the soil warms to approximately 2,900°F to 3,600°F (1,600°C to 2,000°C), well above the melting (fusion) temperature of soils (2,000°F to 2,500°F or 1,100°C to 1,400°C). The graphite and glass starter path is eventually consumed by oxidation.

Upon melting, most soils become electrically conductive; thus, the molten mass becomes the primary conductor and heat transfer medium. As a result of joule heating, soil viscosity decreases. The melt begins to grow within the soil matrix, extending laterally as well as downward. Power is maintained at levels sufficient to overcome heat losses from the surface and to surrounding soil. Convection currents within the melt aid in heating of the melt. Heat is transferred to adjacent soil by conduction from the melt. The melt grows outward to a width approximately 50 percent wider than the electrode spacing. The molten zone becomes roughly a cube with slightly rounded corners on the bottom and sides; this shape reflects the higher power density around the electrodes.

During processing, ISV eliminates the soil voids, resulting in volume reduction. Additional volume reduction occurs since soil components such as humus plus many organic contaminants are removed as gases and vapors. Overall volume reduction (typically 20% to 50%) results in subsidence above the melt (Thompson et al. 2000; Weston 1988).

Possible fates of contaminants resulting from ISV processing include chemical and thermal destruction, removal to the off-gas treatment system, chemical and physical incorporation within the residual product, lateral migration ahead of the advancing melt, and escape to the biosphere. Organic compounds exposed to the advancing thermal gradient are either drawn into the melt or migrate into the dry zone where they are vaporized and ultimately pyrolyzed. Only a small fraction of vapor passes through the melt itself. Organic pyrolysis products are typically gaseous; because of the high viscosity of the molten material, these gases move slowly through the melt, usually on a path adjacent to the electrodes, toward the upper melt surface. While some gases dissolve in the molten mass, the remainder move to the surface; those that are combustible react in the presence of air. Pyrolysis and combustion products are collected in an off-gas collection hood and subsequently treated in the off-gas treatment system (figure 13.8). Because of the high temperature of the melt, no residual organic contaminants are expected to remain in their original form within the vitrified product.

The behavior of inorganic materials upon exposure to the advancing thermal gradient is similar to that of organics. Inorganic compounds thermally

Figure 13.8 Off-gas treatment system. *Source*: Reproduced with kind permission of AMEC Corp., Washington, DC.

decompose or otherwise react directly with the melt. Typically, any metals present are incorporated into the vitrified residual.

Immobilization occurs when contaminants are incorporated into the glass network or encapsulated (surrounded) by the glass. If significant quantities of nonvolatile metals are present, they may sink to the bottom of the melt and concentrate.

An off-gas collection hood covers the processing area. Air flow through the hood is controlled to maintain a vacuum, which prevents escape of fugitive emissions from the hood and the ground surface/air interface. Air provides oxygen for combustion of pyrolysis products and organic vapors. An induced draft blower draws off-gases from the hood into the gas treatment system. The off-gas is treated by quenching, pH-controlled scrubbing, mist elimination, particulate filtration, and activated carbon adsorption.

Once power to the electrodes is turned off, the melt begins to cool. In most cases no attempts are made to force cooling of the melt; slow cooling is expected to produce a vitreous (amorphous) and microcrystalline solid (figure 13.9). Removal of the hood is normally accomplished within twenty-four hours after power to the electrodes is discontinued. The used graphite electrodes are severed near the melt surface and are left within the treated monolith. After the off-gas hood is removed and the electrodes are severed, the subsidence volume is filled to the desired depth with clean backfill.

Figure 13.9 Vitreous material occurring after ISV treatment. *Source*: U.S. Department of Energy.

There have been few commercial applications of ISV. The ISV process has been operated for test and demonstration purposes at the pilot scale and at full scale at the following sites: (1) Parsons Chemical site (MI); (2) DOE Hanford (WA) Nuclear Reservation; (3) DOE Oak Ridge (TN) National Laboratory; and (4) DOE Idaho National Engineering Laboratory (U.S. EPA 2018). More than 170 tests at various scales have been performed on a broad range of waste types in soils and sludge.

QUESTIONS

1. Describe the process and mechanism of electrokinetic remediation. Can this technology be effectively combined with phytoextraction? With phytostabilization? Explain.
2. List and discuss possible interferences with utilization of electrokinetic remediation for soil treatment. What tests should be conducted to determine whether a site is amenable to electrokinetics?
3. What potential hazards and/or drawbacks exist when using H_2O_2 for treatment of a hydrocarbon plume?
4. In situ vitrification is carried out at a site contaminated with PCBs, mercury and $PbO_{(s)}$. What are the ultimate fates of chlorine, mercury and lead in this ISV process?

REFERENCES

Acar, Y. B., and A. N. Alshawabkeh. 1993. Principles of electrokinetic remediation. *Environmental Science and Technology* 27(13):2638–47.

Brown, R. 2003. *In Situ Chemical Oxidation: Performance, Practice, and Pitfalls. AFCEE Technology Transfer Workshop*, February 25, 2003, San Antonio, TX. http://www.cluin.org/download/ techfocus/chemox/4_brown.pdf.

Cameselle, C., S. Gouveia, D. E. Akretche, and B. Belhadj. 2013. Advances in electrokinetic remediation for the removal of organic contaminants in soils. Chapter 9. In: *Organic Pollutants – Monitoring, Risk and Treatment*, edited by M. N. Rashed. InTech.

Davis, E. 1997. *Ground Water Issue: How Heat Can Enhance In-Situ Soil and Aquifer Remediation: Important Chemical Properties and Guidance on Choosing the Appropriate Technique*. EPA 540/S-97/502. U.S. EPA. Office of Research and Development, 18 p. http://www.cluin.org/down load/remed/heatenh.pdf.

Edelstein, W., et al. 1996. *Radiofrequency Ground Heating System for Soil Remediation, Patent Number 5,484,985*. General Electric Company. http://patft1.u spto.gov/netacgi/nph-Parser?Sect1=PTO1&Sect2=HITOFF&d=PALL&p=1&u= %2Fnetahtml%2FPTO%2Fsrchnum.htm&r=1&f=G&l=50&s1=5,484,985.PN .&OS=PN/5,484,985&RS=PN/5,484.

FCC. 2006. *Code of Federal Regulations: Title 47–Telecommunication Chapter I– Federal Communications Commission Part 18–Industrial, Scientific, and Medical Equipment*. http://www.access. gpo.gov/nara/cfr/waisidx_05/47cfr18_05.html.

Gent, D. B., R. M. Bricka, A. N. Alshawabkeh, S. L. Larson, G. Fabian, and S. Granade. 2004. Bench and field-scale evaluation of chromium and cadmium extraction by electrokinetics. *Journal of Hazardous Materials* 110(1–3):53–62.

Haliburton NUS Environmental Corporation. 1995. *Installation Restoration Program Technical Evaluation Report for the Demonstration of Radio Frequency Soil Decontamination at Site S-1*. Air Force Center for Environmental Excellence. Brooks Air Force Base, TX.

ITRC. 2005. *Technical and Regulatory Guidance for In Situ Chemical Oxidation of Contaminated Soil and Groundwater*, 2nd ed., 172 p. http://www.itrc web.org/ Documents/ISCO-2.pdf.

Jacobs, R. A., M. Z. Sengun, R. E. Hicks, and R. F. Probstein. 1994. Model and experiments on soil remediation by electric fields. *Journal of Environmental Science and Health. Part A: Environmental Science and Engineering* 29(9):1933–39.

Lageman, R. 1993. Electroreclamation: Applications in the Netherlands. *Environmental Science and Technology* 27(13):2648–50.

Rodsand, T., and Y. B. Acar. 1995. Electrokinetic extraction of lead from spiked Norwegian marine clay. *Geoenvironment 2000* 2:1518–34.

Thompson, L., P. Ombrellaro, N. Megalos, D. Osborne, and D. Timmons. 2000. *Final Results from the In Situ Vitrification Treatment at Maralinga*. WM'00 Conference, February 27–March 2, Tucson, AZ.

USAEC (U.S. Army Environmental Command). 2000. *In Situ Electrokinetic Remediation of Metal Contaminated Soils Technology Status Report*. US Army Environmental Center. Report Number SFiM-/1U.:-CI:CR-99022.

U.S. EPA (U.S. Environmental Protection Agency). 1993. *Accutech Pneumatic Fracturing Extraction and Hot Gas Injection, Phase I*. EPA/540/AR-93/509. Washington, DC: Office of Research and Development.

———. 1995a. *Emerging Technology Bulletin: Electrokinetic Soil Processing. Electrokinetics, Inc*. EPA/540/F-95/504. Washington, DC: Office of Solid Waste and Emergency Response.

———. 1995b. *In Situ Steam Enhanced Recovery Process Hughes Environmental Systems, Inc.* EPA/540/R-94/510. Washington, DC: Office of Research and Development.

———. 1995c. *Geosafe Corporation In Situ Vitrification. Innovative Technology Evaluation Report.* EPA/540/R-94/520. Washington, DC: Office of Research and Development.

———. 1997a. *Recent Development for In Situ Treatment of Metal Contaminated Soils.* EPA/542/R-97/004. Washington, DC: Office of Solid Waste and Emergency Response.

———. 1997b. *Analysis of Selected Enhancements for Soil Vapor Extraction.* EPA 542/R-97/007. Office of Solid Waste and Emergency Response, 246 pp. http://www.cluin.org/download/remed/ sveenhmt.pdf.

———. 2000. *In Situ Chemical Oxidation for Remediation of Contaminated Soil and Ground Water by Robert L. Siegrist, Author.* EPA 542-N-00-006. Issue No. 37. Washington DC: Office of Solid Waste and Emergency Response.

———. 2006. *In Situ Treatment Technologies for Contaminated Soil: Engineering Forum Issue Paper.* https://www.epa.gov/remedytech/situ-treatment-technologies-contaminated-soil-engineering-forum-issue-paper.

———. 2018. *Geosafe Corporation (In Situ Vitrification).* https://clu-in.org/products/site/complete/democomp/geosafe.htm.

Weston, Roy F., Inc. 1988. *Remedial Technologies for Leaking Underground Storage Tanks.* Chelsea, MI: Lewis.

Chapter 14

Nanotechnology in Site Remediation

O amazement of things—even the least particle!
  —Walt Whitman, "Song at Sunset," *Leaves of Grass* (1897)

The particles of a dew-drop and the masses of a planet are moulded and controlled by the same force.
  —Sir William Withey Gull

14.1 INTRODUCTION

Nanoscale materials (aka nanomaterials) are defined as particles having at least one dimension measuring between 1 and 100 nm (NNI 2006). Nanomaterials are commonly grouped into three categories, based on origin: natural, incidental, and engineered. Natural nanomaterials include clays, organic matter, and iron oxides occurring in soil and water (Klaine et al. 2008). Incidental nanoscale materials enter the environment through atmospheric emissions, industrial activities such as welding, and via agricultural operations (Klaine et al. 2008; U.S. EPA 2008). Engineered or manufactured nanoscale materials are designed to possess specific properties which are ultimately applied to commercial, industrial, or environmental uses (U.S. DHHS 2006; U.S. EPA 2007).

Nanoremediation technology involves the application of reactive nanomaterials for immobilization, destruction, and/or detoxification of contaminants in soil, sediment, groundwater, and surface water. Such materials possess properties that enable chemical reactions, including catalysis, to transform pollutants to nonhazardous forms.

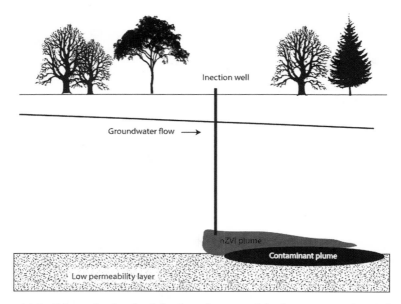

Figure 14.1 Schematic showing injection of nanoparticles into a contaminant plume.

Nanoremediation can be carried out ex situ or in situ at the affected site. In the former case, soil or groundwater is removed from the subsurface, reacted with nanoparticles (NPs) in a dedicated vessel, and returned to the ground. For in situ nanoremediation, a specially tailored recipe of nanomaterials is injected directly into the affected portion of the subsurface for treatment (figure 14.1).

14.2 NANOPARTICLE TYPES FOR REMEDIATION

The majority of laboratory-scale research and field application of nanoscale materials for remediation has focused on nanoscale zerovalent iron (nZVI) and associated products. However, other nanoscale materials are in the research and development stages for use in site remediation (U.S. EPA 2016). Some major types of NPs being synthesized and tested are as follows (Rana and Kalaichelvan 2013):

- Metal NPs (e.g., elemental Ag, Au, and Fe);
- Oxides (or binary compounds), for example, TiO_2, ZnO and Fe oxides;
- Complex compounds (alloys, composites, nanofluids, etc., consisting of two or more elements), for example, cobalt and zinc iron oxide;
- Fullerenes (Buckminster fullerenes, carbon nanotubes, and nanocones);

- Quantum dots (or q-dots);
- Organic polymers (dendrimers, polystyrene, etc.).

Several of these materials are discussed below.

14.2.1 Zerovalent iron (nZVI)

In recent decades, project managers have taken advantage of the properties of elemental iron (Fe^o) to degrade chlorinated solvent plumes in groundwater (Mueller and Nowack 2010). An example of an in situ treatment technology for chlorinated solvent plumes is the installation of a trench filled with macroscale zerovalent iron (ZVI) to form a permeable reactive barier (PRB) (ITRC 2005) (see chapter 10). Nanoscale ZVI provides the same benefits to soil and groundwater remediation as does macroscale ZVI; however, the substantially greater surface area per volume of material provides more reactive sites, allowing for more rapid degradation of contaminants as compared with macroscale particles (U.S. EPA 2008b).

14.2.2 Bimetallic nanoscale particles (BNPs)

BNPs consist of particles of elemental Fe or other metals joined to a metal catalyst such as platinum, gold, nickel, or palladium. BNPs have been used successfully for remediation of contaminated soil and groundwater. The combination of metals increases the kinetics of oxidation-reduction reactions, thereby catalyzing the necessary contaminant transformations.

Palladium-gold NPs can catalyze the reduction of trichloroethene (TCE) to ethane without producing hazardous intermediates such as vinyl chloride (Nutt et al. 2005). Palladium-Fe BNPs are commercially available and are currently the most common of the BNPs (U.S. EPA 2016). In laboratory tests, BNPs containing 99.9 percent Fe and < 0.1 percent Pd achieved contaminant degradation two orders of magnitude greater than microscale Fe particles alone (Zhang and Elliot 2006).

In field remediation situations, BNPs are commonly incorporated into a slurry for injection and can be injected by gravity or by pressure feed (Gill 2006).

14.2.3 Emulsified zerovalent iron (EZVI)

Emulsified zerovalent iron consists of nanoscale ZVI surrounded by an oily emulsion membrane. The exterior membrane is formulated from food-grade surfactant and biodegradable oil. The interior consists of water and ZVI particles (ESTCP 2006) (figure 14.2).

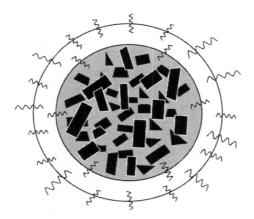

Figure 14.2 Generic structure of an EZVI particle.

The exterior membrane is hydrophobic, with properties similar to those of DNAPL compounds such as chlorinated hydrocarbons. As a result, EZVI particles are readily miscible with DNAPL contaminants. When a TCE molecule, for example, contacts the emulsion membrane, the TCE partitions into the oil and then diffuses into the interior of the emulsion droplet, where it comes into contact with the ZVI and is chemically degraded. A concentration gradient is established by migration of the TCE molecules into the interior aqueous phase of the particle and by concurrent release of products out of the droplet, further promoting degradation reactions (O'Hara et al. 2006).

In addition to the abiotic degradation of chlorinated hydrocarbons associated with ZVI, the outer layer of vegetable oil and surfactant serves to sequester the chlorinated compounds for subsequent biodegradation (ESTCP 2006). Another potential benefit of EZVI over nZVI for site remediation is that the hydrophobic membrane surrounding the nZVI protects it from other groundwater constituents, for example inorganic compounds that might oxidize it.

EZVI is commercially available and has been used to remediate chlorinated solvents.

14.2.4 Nanotubes

Nanotubes are engineered molecules typically formulated from carbon (figure 14.3). Nanotubes are easily polymerized and have properties of being electrically insulating and highly electronegative. The majority of the studies published thus far have focused on carbon nanotubes, fullerenes, and photocatalytic particles (Nature 2018; Watlington 2005).

Carbon nanotubes have been recognized for their ability to adsorb chlorinated dibenzodioxins much more strongly than conventional activated carbon

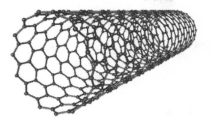

Figure 14.3 Carbon nanotube structure. *Source*: Michael Ströck.

(Long 2001; Watlington 2005). Nanotubes have also been made from TiO_2 and hold potential for photocatalytic degradation of chlorinated hydrocarbons (Chen et al. 2005). Laboratory-scale research has shown that TiO_2 nanotubes are particularly effective at high temperatures. Xu et al. (2005) found TiO_2 nanotubes capable of reducing methyl orange concentrations by more than 50 percent in three hours. Remediation using TiO_2 is considered an ex situ strategy, as illumination is required. Treatment must therefore occur in a reactor designed for this purpose (Tratnyek and Johnson 2006).

Chen et al. (2005) reported the production of nanotubes containing platinum on the inside and TiO_2 on the outside. The outside of the tube acts as an oxidizing surface, while the inside acts as a reductive surface. Chen et al. (2005) tested the ability of the nanotubes to decompose toluene and found the nanotube catalyst to have a much higher destruction rate than non-nanotube material (Watlington 2005).

14.2.5 Dendrimers

Dendrimers, also known as arborols and cascade molecules, are repetitively branched molecules having reactive termini (figure 14.4). These are three-dimensional, highly branched, globular macromolecules consisting of a core, branches, and end groups, all covalently bound. It is typically symmetric around the core, and often possesses a spherical three-dimensional morphology. Common dendrimer shapes include spheres, cones, and disc-like structures (Mamadou et al. 2005; Mishra and Clark 2013).

Both architecture and composition of dendrimers can be carefully controlled. Dendrimer molecules terminate in a range of functional groups that can be modified to enhance specific chemical reactions. Interior chemistry can likewise be formulated to carry out desired chemical functions. Molecular design parameters including size, shape, surface/interior chemistry, flexibility, and topology can be closely ordered (Cagin 2005; Watlington 2005). Functionalization can allow particles to be soluble in certain media or bind selected molecules (Watlington 2005).

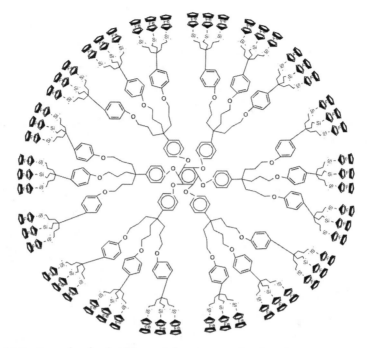

Figure 14.4 Example of a dendrimer: A ferrocene molecule.

Dendrimers have been applied for complexation of heavy metals from water and wastewater; the molecules have outstanding adsorption properties (Savage and Diallo 2005). Recent studies have addressed removal of Cu, Cr, and Pb. Poly(amidoamine) (PAMAM) dendrimers removed Cu^{2+} from aqueous solution (Diallo et al. 1999) and extracted Pb from contaminated soil (Xu and Zhao 2006). Modifications of the PAMAM terminal groups impacted the type of metal ions which were sorbed. By preparing composites of TiO_2 and PAMAM, Cu^{2+}, Ni^{2+}, and Cr^{3+} were removed from simulated wastewater (Barakat et al. 2013).

Several investigations have utilized dendrimers for treatment of hydrocarbons. Triano et al. (2015) showed that benzyl ether dendrimers removed 94 percent of pyrene from an aqueous solution after thirty minutes. Dendrimers containing modified benzyl ether or cyclohexane functional groups removed up to 75 percent of pyrene, while particles with benzyl amine resulted in a 77 percent decrease in pyrene. Ceramic filters impregnated with PAMAM and poly(ethyleneimine) (PEI) dendrimers extracted trihalomethanes, pesticides, and PAHs from water (Arkas et al. 2006). PAMAM and PEI dendrimers were also shown to solubilize phenanthrene (Geitner et al. 2012).

14.2.6 Ferritin

Ferritin is a protein produced during mammalian metabolism that assists cells in storing iron in tissues. Research with ferritin has shown that it can reduce the toxicity of contaminants such as Cr and technetium in surface water and groundwater (Temple University 2004). Like TiO_2, ferritin has photocatalytic properties. In one study, application of visible light allowed ferritin to reduce Cr(VI) to the less toxic Cr(III), which precipitated out of solution (U.S. EPA 2008a). It has been reported that ferritin can also be used for remediation of chlorinated hydrocarbons (Moretz 2004). Ferritin is being researched for in situ use (U.S. EPA 2016; Temple University 2004).

14.2.7 Metalloporphyrinogens

Metalloporphyrinogens are complexes of metals with naturally-occurring organic porphyrin molecules. Examples of biological metalloporphyrinogens are hemoglobin and vitamin B12. Metalloporphyrins have been found capable of reducing chlorinated hydrocarbons such as TCE, PCE, and carbon tetrachloride under anoxic conditions to remediate contaminated soil and groundwater (Sivasubramanian 2016; Fulekar and Pathak 2017). Laboratory studies using metalloporphyrinogens are being performed for potential in situ remediation of groundwater (Sivasubramanian 2016; Dror et al. 2005).

14.2.8 Swellable Organically Modified Silica (SOMS)

Swellable organically modified silica (commercial name Osorb™) is a nanomaterial that swells and captures small organic compounds such as acetone, ethanol, and those occurring in gasoline, natural gas, pharmaceuticals, and solvents. When organic compounds come into contact with SOMS, the mass will swell, thus capturing the organics. The SOMS material is hydrophobic and does not absorb water (Edmiston 2009).

Palladium-Osorb is a metal-glass material used in ex situ systems for remediation of chlorinated VOCs (Edmiston 2010). Palladium NPs are incorporated into the Osorb to remove halogenated VOCs from groundwater. A hydrogen gas source supplies a proton that completes the reductive dehalogenation of a contaminant, for example, TCE, inside the palladium-Osorb and reduces the TCE to ethane and salts (ABS Materials 2010a). Pilot-scale testing of an ex situ system at an Ohio site has shown continuous reduction of TCE from 5800 mg/l to non-detect. The manufacturer reports that palladium-Osorb is a catalytic material that does not decompose or need replacement. An aboveground unit has been designed to treat 10 gal per minute of produced water (ABS Materials 2010a), and the unit has been modified to treat oil sheens and mousses that result from oil spills (Boccieri 2010).

Researchers are also using nanotechnology to develop membranes for water treatment, desalination, and water reclamation. These membranes incorporate a variety of nanoscale materials including those composed of ZVI, alumina, and gold (Theron et al. 2008). Carbon nanotubes can be aligned to form membranes with nanoscale pores to filter organic contaminants from groundwater (Mauter et al. 2008; Meridian Institute 2006).

14.3 FIELD APPLICATION OF NANOSCALE MATERIALS

Nanotechnology offers flexibility for both in situ and ex situ remediation. NPs are utilized in ex situ slurry reactors for treatment of contaminated groundwater, soil, and sediment. Alternatively, particles can be anchored onto a solid matrix such as carbon, zeolite, or ceramic membrane for enhanced treatment of water, wastewater, or gaseous process streams (Shipley et al. 2011).

14.3.1 In situ

For in situ treatment it is necessary to create either a reactive zone in which relatively immobile NPs are emplaced (figure 14.5) or a reactive NP plume that migrates to the contaminated zone. In cases of contaminated surface soil, NPs can be incorporated using conventional agricultural equipment (Mueller and Nowack 2010).

Subsurface injection of nanoscale Fe particles has been shown to effectively degrade chlorinated organics such as trichloroethylene to environmentally harmless compounds. The technology also holds promise for immobilizing heavy metals and radionuclides (Mansoori et al. 2008).

Application of nanoscale materials for in situ treatment is site-specific. The method of injection and distribution of injection points are a function of type and distribution of contaminants, local geologic conditions, and form of the nanomaterials injected.

Injection of nanoscale iron is typically performed using direct injection through gravity feed or under pressure (EPA 2005). Some options for injection of nanoscale iron include direct push technology or via wells (e.g., temporary or permanent). The direct push method involves driving push rods progressively deeper into the subsurface. The method allows nanomaterials to be injected without the need to install permanent wells (Butler 2000; U.S. EPA 2016).

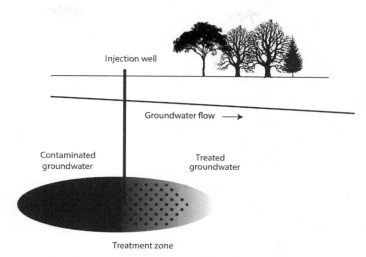

Figure 14.5 Schematic showing groundwater remediation via creation of a treatment zone by injection of nanomaterials. the reactive zone may be formed by a series of injections of nZVI or other nanoparticles.

Other methods to apply nanoscale material for in situ treatment include pressure pulse technology, pneumatic fracturing, and hydraulic fracturing. Pressure pulse technology uses large-amplitude pulses of pressure to inject the nZVI slurry into porous media at the water table; the pressure then excites the media and increases fluid level and flow (OCETA 2003). Pneumatic fracturing injection is a high-pressure injection technique using compressed gas that allows liquids and vapors to be transported quickly through the newly created channels (see chapter 13). Pneumatic fracturing uses compressed gas to create a fracture network of preferential flow paths in rock around the injection point to allow liquids and vapors to be transported quickly through the fractured rock; this technology improves access to contaminants and allows liquids to flow freely (Pneumatic Fracturing Inc. 2008; Zhang 2003).

Two scenarios of groundwater treatment using nZVI are shown in this chapter. Treatment of DNAPL contamination by injection of nanoscale material is shown in figure 14.1; in figure 14.5, a reactive treatment zone is formed by a series of injections of nZVI. These injections create overlapping zones of particles that become lodged within the native aquifer material (Tratnyek and Johnson 2006).

Research is ongoing to improve methods of injection to allow nanoscale materials to maintain their reactivity and increase access to contaminants by achieving wider distribution in the subsurface. For example, creating nZVI on-site could reduce the amount of oxidation the iron undergoes between production and use, thereby reducing potential losses in reactivity (U.S. EPA 2016).

14.4 CASE HISTORY

Soil near a former dry cleaning facility at the Parris Island Marine Corps facility, South Carolina, was found to be contaminated with chlorinated hydrocarbons. Four aboveground storage tanks were situated alongside the facility. The tanks were installed in 1988 following the removal of an underground storage system where hydrocarbon cleaning solvents had been stored.

In 1994 one of the tanks was overfilled with PCE and an unknown quantity flowed into the concrete catch basin. The PCE overflow was not collected, and heavy rainfall washed it onto the surrounding soil (ESTCP 2010).

PCE and its degradation products (TCE, *cis*-1,2-dichloroethene [cDCE], 1,2-dichloroethene [1,2-DCE] and vinyl chloride [VC]) were detected in surface and subsurface soil and groundwater in three potential source areas during the site investigation. Polycyclic aromatic hydrocarbons were detected in locations nearby the site but at relatively low concentrations.

The affected area occurred within a shallow, unconfined surficial aquifer. The aquifer consists of permeable, fine to medium, Pleistocene-age sand, with lenses of finer-grained silty clay and clayey sand, to a depth of 17 ft (5.2 m). The shallow aquifer is underlain by a 1- to 3-ft thick layer of peat. The water table generally ranges from 3- to 5-ft BGS. The hydraulic conductivity for the aquifer is estimated at 15.3 ft/day (0.0054 cm/sec).

Field investigations indicated that chlorinated VOC contamination in the surface and subsurface soil had impacted the groundwater to depths ranging from the upper boundaries of the unconfined aquifer to approximately 19-ft (5.8 m) BGS. Analysis of soil cores revealed the presence of PCE DNAPL. Visual inspection of soil cores also indicated the presence of DNAPL.

In 1998, a groundwater pump-and-treat system was installed to prevent migration of groundwater contaminants until a comprehensive remedial investigation could take place.

Laboratory experiments were first conducted to evaluate the extent of DNAPL destruction with the application of EZVI. It was found that treatment of dissolved phase TCE concentrations in the aqueous phase. The EZVI combined sequestration of the DNAPL with degradation of the VOCs by the nZVI. The EZVI also served to sequester potential untreated VOCs (Geo-Syntec 2006).

A field demonstration/validation was conducted at the site of the dry cleaning facility. The goal was to evaluate PCE and TCE degradation via biotic and biological processes as well as demonstrate the efficacy of EZVI at a scale large enough to generate full-scale design and cost information for technology application at DoD and other sites.

The EZVI used at the site was the same formulation used in laboratory treatability tests and was composed of nZVI, water, corn oil, and surfactant at 10, 51, 38, and 1 percent by weight, respectively. The EZVI was manufactured on-site, moved to a staging area and transferred to holding tanks. EZVI was injected into test plots in October 2006. The technology demonstration was designed to inject a maximum of 850 gal of EZVI into a pneumatic injection test plot and 50 gal into a direct injection test plot.

In some cases, so-called *daylighting* or *short-circuiting* of EZVI occurred during injection; in other words, some EZVI migrated upward into former soil borings to the ground surface. In the pneumatic injection test plot it was estimated that 576 gal of EZVI was injected into eight locations between 7- and 18.5-ft BGS, and an estimated 32 gal of EZVI returned to the surface. A total of 151 gal of EZVI was injected into four locations within the direct injection test plot between 6- and 12-ft BGS, and an estimated 5 gal of EZVI returned to the surface.

Reductions of approximately 93 percent in the mass of tetrachloroethene (PCE) DNAPL occurred in the pneumatic injection test plot following EZVI injection (table 14.1). In addition, significant reductions were measured in the mass flux (mass flow per unit area) of the parent compounds (~85% PCE reduction and ~86% TCE reduction), and of the degradation product cis-1,2-dichloroethene (cDCE) (~71% reduction). There were, however, significant increases in the mass flux of the degradation products vinyl chloride (VC) and ethene in the pneumatic injection test plot.

DNAPL was recovered from some wells where DNAPL had previously been absent, indicating that some of the DNAPL was mobile. However, the concurrent presence of daughter products (VC and ethene) indicated that the mass was being degraded and not simply displaced.

Although there were problems with short-circuiting of the EZVI to the surface during injection, this was believed to be site-specific, that is, due to the shallow nature of the treatment area and the presence of preexisting short-circuit pathways (old boreholes). The cost assessment of this project showed more than 62 percent cost savings compared to pump-and-treat.

Table 14.1 Estimates of contaminant mass in pneumatic injection plot before and after EZVI injection

Media	VOC	Pre-injection Mass (g)			Post-demonstration Mass (g)		
		Sorbed/Dissolved	DNAPL	Total	Sorbed/Dissolved	DNAPL	Total
Soil	Tetrachloroethene	2,760	29,028	31,788	730	2,137	2,867
	Trichloroethene	1,317	0	1,317	521	0	521
	cis-1,2-Dichloroethene	1,254	0	1,254	569	0	569
	Vinyl chloride	2,214	0	2,214	114	0	114
Groundwater	Tetrachloroethene	577	0	577	333	0	333
	Trichloroethene	267	0	267	182	0	182
	cis-1,2-Dichloroethene	588	0	588	819	0	819
	Vinyl chloride	12	0	12	45	0	45
Total Mass (g)		8,990	29,028	38,018	3,312	2,137	5,449
Reduction (%)					63%	93%	86%

14.5 HAZARDS ASSOCIATED WITH NANOTECHNOLOGY

Current understanding regarding fate and transport of nanoscale materials in the environment is limited. Likewise, the potential toxicological effects of nanoscale materials are largely unknown. Although the environment contains many natural particles at the nanoscale, some researchers suggest that manufactured NPs may behave differently (Handy et al. 2008). NPs are engineered to possess specific surface properties and chemistries that may not be found in natural particles.

A range of factors may influence both the toxicity of NPs and their behavior in the environment, including chemical composition, structure, molecular weight, water solubility, octanol-water partition coefficient, potential for particle aggregation/disaggregation, and presence of surface coatings (Nowack and Bucheli 2007; Mueller and Nowack 2010).

14.5.1 Mobility

Application of synthetic NPs for site remediation will inevitably result in their release into terrestrial, aquatic, and marine ecosystems. The mobility of NPs in the biosphere is strongly dependent upon whether the particles are dispersed, aggregate and settle, or form mobile nanoclusters (Karn et al. 2009).

Depending on groundwater chemistry and hydrogeologic conditions, some NPs are reported to travel great distances (Kersting et al. 1999; Novikov et al. 2006; Vilks et al. 1997; Karn et al. 2009). Some NPs form stable nanoclusters in groundwater that are likely to be highly mobile, possibly carrying with them surface-sorbed contaminants. These nanoclusters can migrate across new redox zones and inadvertently contaminate previously unaffected soil and groundwater (Waite et al. 1999; Karn et al. 2009). In addition to self-aggregation, NPs could associate with suspended colloidal solids in soil and water and possibly enter food chains.

14.5.2 Toxicity

A number of researchers have examined the potential for toxicity of NPs to humans and other receptors. Generally, findings indicate that smaller particles are more reactive and also more toxic than larger ones; substantial differences among different particle types exist, however (Oberdörster et al. 2007; Mueller and Nowack 2010).

NPs enter the body via inhalation, ingestion, or dermal absorption. Recent studies have shown that NPs can be taken up by a variety of mammalian cell types (Oberdörster et al. 2007; Mueller and Nowack 2010).

Studies are underway to determine the potential toxicity of synthetic nanoscale materials. Silver NPs are now known to be highly toxic to mammalian cells (Rana and Kalaichelvan 2013) and have been shown to damage brain cells, liver cells, and stem cells (Hussain et al. 2005, 2006). Lam et al. (2004) demonstrated that three carbon nanotube products induced dose-dependent lung lesions in mice. Warheit et al. (2004) reported that carbon nanomaterials induced pulmonary inflammation as well as altered lung cell proliferation and pathology. Hussain et al. (2005) found that high levels (100 to 250 µg/mL) of MoO_3 and TiO_2 NPs damaged rat liver cells in vitro. Lower doses (10 to 50 µg/mL) had no measured effect.

14.5.3 Ecotoxicology

Numerous ecotoxicity studies of engineered NPs are in progress. Findings from experiments under simplified conditions indicate that certain NPs are toxic to organisms even at low concentrations (Rana and Kalaichelvan 2013). The greater surface area and larger number of reactive sites of nanoscale materials may result in greater biological activity per unit mass compared to micro- or macroscale particles having the same composition.

Ecotoxicologic effects of synthetic nanomaterials have been reported on microorganisms, plants, invertebrates, and fish (Boxall et al. 2007; Rana and Kalaichelvan 2013). Laboratory studies have shown that fish, *Daphnia*, copepods, and other organisms can take up engineered NPs (Adams et al. 2006; Fortner et al. 2005; Lovern et al. 2007; Oberdörster et al. 2006; Karn et al. 2009). Single-walled carbon nanotubes induced significant hatching delay in Zebrafish embryos at concentrations greater than 120 mg/L (Cheng et al. 2007).

Phytotoxicity of NPs has been demonstrated for Zn and ZnO; effects on seed germination in various crops were reported by Lin and Xing (2007). Fifty percent inhibition of root growth was observed for nano-Zn and nano-ZnO at approximately 50 mg/L for radish, and about 20 mg/L for rape and ryegrass. Silver NPs were shown to impart toxic effects on nitrogen fixing and ammonifying bacteria and chemolithotrophic bacteria in soil communities (Rana and Kalaichelvan 2013). NPs of SiO_2, TiO_2, and ZnO were toxic to Gram-positive (*Bacillus subtilis*) and Gram-negative (*Escherichia coli*) bacteria in water suspensions.

Concerns about potential toxicity have limited the use of nanoscale materials for remediation by some companies. For example, DuPont has decided against the use of nZVI for remediation at its affected properties until issues concerning fate and transport have been more thoroughly researched. The company has stated concerns regarding of post-remediation persistence and potential human exposure to the particles (DuPont 2007).

Available data indicate that risks of engineered NPs to public health and the environment are probably low; however, knowledge of their potential impacts is still limited (Karn et al. 2009).

QUESTIONS

1. Provide three examples each of natural and incidental NPs.
2. What are the benefits of emulsified nanoscale ZVI for groundwater remediation?
3. What is the benefit of using a bimetallic NP (e.g., Pb-Fe) for site remediation, as compared to using nano-Fe alone?
4. If nanomaterials, for example, zerovalent iron, are to be used in a PRB, how can the site operator ensure that the NPs do not become entrained in groundwater? In other words, how can the NPs be 'fixed' in place and not migrate?
5. Check the published literature and determine which nanomaterials are appropriate for: (1) dechlorination of industrial solvents; (2) altering the oxidation states of arsenic and chromium; and (3) oxidizing aromatic contaminants such as dyes.
6. Explain the general structure and function of: dendrimers; ferritins; metalloporphyrinogens; and SOMS.
7. List and discuss several common drawbacks to the use of nanomaterials for in situ remediation of a contaminant plume. Consider mobility; potential toxicity; nontarget reactivity.

REFERENCES

ABS Materials. 2010a. *Osorb the VOC-Eater™ 10 Web Page*. Available at: http://www.absmaterials.com/ex-situ-remediation (Accessed March 22, 2018).

Adams, L. K., D. Y. Lyon, and P. J. J. Alvarez. 2006. Comparative ecotoxicity of nanoscale TiO_2, SiO_2 and ZnO water suspensions. *Water Research* 40(19):3527–32.

Arkas, M., R. Allabashi, D. Tsiourvas, E.-M. Mattausch, and R. Perfler. 2006. Organic/inorganic hybrid filters based on dendritic and cyclodextrin "nanosponges" for the removal of organic pollutants from water. *Environmental Science and Technology* 40(8):2771–77.

Barakat, M. A., M. H. Ramadan, M. A. Alghamdi, S. S. Algarny, H. T. Woodcock, and J. N. Kuhn. 2013. Remediation of Cu(II), Ni(II), and Cr(III) ions from simulated wastewater by dendrimer/titania composites. *Journal of Environmental Management* 117:50–7.

Boccieri, J. 2010. *Boccieri Announces $200,000 To Abs Materials Company, LLC*. Available at: http://bignews.biz/?id=899733&keys=Congressman-Roy-Blunt-Oil Spill (Accessed August 19, 2010).

Boxall, A. B. A., K. Tiede, and Q. Chaudhry. 2007. Engineered nanomaterials in soils and water: How do they behave and could they pose a risk to human health? *Nanomedicine* 2(6):919–27.

Cagin, T., G. Wang, R. Martin, and W. Goddard III. 2005. *Molecular Modeling of Dendrimers for Nanoscale Applications*. Seventh Foresight Conference on Molecular Nanotechnology, Foresight Institute.

Chen, Y., J. C. Crittenden, S. Hackney, L. Sutter, and D. W. Hand. 2005. Preparation of a novel TiO2-based P-N junction nanotube photocatalyst. *Environmental Science and Technology* 39(5):1201–08.

Cheng, J., E. Flahaut, and H. C. Shuk. 2007. Effect of carbon nanotubes on developing zebrafish (*Danio rerio*) embryos. *Environmental Toxicology and Chemistry* 26(4):708–16.

Diallo, M. S., L. Balogh, A. Shafagati, J. H. Johnson, W. A. Goddard, and D. A. Tomalia. 1999. Poly(amidoamine) dendrimers: A new class of high capacity chelating agents for Cu (II) ions. *Environmental Science and Technology* 22(5):820–24.

Dror, I., D. Baram, and B. Berkowit. 2005. Use of nanosized catalysts for transformation of Chloroorganic pollutants. *Environmental Science and Technology* 39(5):1283–90.

DuPont. 2007. *Nanomaterial Risk Assessment Worksheet, Zero Valent Nano Sized Iron Nanoparticles (nZVI) for Environmental Remediation*. Available at: http://www2.dupont.com/Media_Center/en_US/assets/downloads/pdf/NRAW_nZVI.pdf (Accessed May 17, 2018).

Edmiston, P. L. 2009. *Swellable Glass-Nano ZVI Composite Materials for the Remediation of TCE and PCE in Groundwater*. June 12, 2009. www.frtr.gov/meetings1.htm.

Edmiston, P. L. 2010. *Pilot Scale Testing of Swellable Organosilica-Nanoparticle Composite Materials for the In Situ and Ex Situ Remediation of Groundwater Contaminated with Chlorinated Organics*. May 13, 2010. Available at: www.frtr.gov/meetings1.htm.

ESTCP (Environmental Security Technology Certification Program). 2006. *Final Laboratory Treatability Report for: Emulsified Zero Valent Iron Treatment of Chlorinated Solvent DNAPL Source Areas*. ESTCP Project CU-0431. January 23, 2006.

ESTCP (Environmental Security Technology Certification Program). 2010. *Emulsified Zero-Valent Nano-Scale Iron Treatment of Chlorinated Solvent DNAPL Source Areas*. ER-200431. September 2010. US Department of Defense.

Fortner, J. D., D. Y. Lyon, C. M. Sayes, A. M. Boyd, J. C. Falkner, E. M. Hotze, et al. 2005. C60 in water: Nanocrystal formation and microbial response. *Environmental Science and Technology* 39(11):4307–16.

Fulekar, M. H., and B. Pathak. 2017. *Environmental Nanotechnology*. Boca Raton, FL: CRC Press.

Geitner, N. K., P. Bhattacharya, M. Steele, R. Chen, D. A. Ladner, and P. C. Ke. 2012. Understanding dendritic polymer-hydrocarbon interactions for oil dispersion. *RSC Advances* 2(25):9371–75.

Geosyntec Consultants, Inc. 2006. *Final Laboratory Treatability Report For: Emulsified Zero Valent Iron Treatment of Chlorinated Solvent DNAPL Source Areas*.

Report. Prepared for Environmental Security & Technology Certification Program (ESTCP), project CU-0431. January 17, 2006.

Gill, H. S. 2006. *Bimetallic Nanoscale Particles*. PARS Environmental Inc. PowerPoint Presentation (Accessed November 4, 2017).

Hussain, S. M., A. K. Javorina, A. M. Schrand, H. M. Duhart, S. F. Ali, and J. J. Schlager. 2006. The interaction of manganese nanoparticles with PC-12 cells induces dopamine depletion. *Toxicological Sciences* 92(2):456–63.

Hussain, S. M., K. L. Hess, J. M. Gearhart, K. T. Geiss, and J. J. Schlager. 2005. In vitro toxicity of nanoparticles in BRL 3A rat liver cells. *Toxicology in Vitro* 19(7):975–83.

Interstate Technology & Regulatory Council (ITRC). 2005. *Permeable Reactive Barriers: Lessons Learned/New Directions*. PRB-4. Available at: http://www.itrcweb.org/documents/prb-4.pdf (Accessed October 2, 2008).

Karn, B., T. Kuiken, and M. Otto. 2009. Nanotechnology and in situ remediation: A review of the benefits and potential risks. *Environmental Health Perspectives* 117(12):1823–31.

Kersting, A. B., D. W. Efurd, D. L. Finnegan, D. J. Rokop, D. K. Smith, and J. L. Thompson. 1999. Migration of plutonium in ground water at the Nevada Test Site [Letter]. *Nature* 397:56–9.

Klaine, S. J., P. J. J. Alvarez, G. E. Batley, T. E. Fernandes, R. D. Handy, D. Y. Lyon, S. Mahendra, M. J. McLaughlin, and J. R. Lead. 2008. Nanoparticles in the environment: Behavior, fate, bioavailability, and effects. *Environmental Toxicology and Chemistry* 27(9):1825–51.

Lam, C. W., J. T. James, R. McCcluskey, and R. L. Hunter. 2004. Pulmonary toxicity of singlewall carbon nanotubes in mice 7 and 90 days after intratracheal instillation. *Toxicological Science* 77:126–34.

Lin, D., and B. Xing. 2007. Phytotoxicity of nanoparticles: Inhibition of seed germination and root growth. *Environmental Pollution* 150(2):243–50.

Long, R. Q., and R. T. Lane. 2001. Carbon nanotubes as superior sorbent for dioxin removal. *Journal of the American Chemical Society* 123(9):2058–59.

Lovern, S. B., J. R. Strickler, and R. Klaper. 2007. Behavioral and physiological changes in Daphnia magna when exposed to nanoparticle suspensions (titanium dioxide, nano-C60 and C60HxC70Hx). *Environmental Science and Technology* 41(12):4465–70.

Mamadou, D., S. Christie, P. Swaminathan, J. Johnson, and W. Goddard. 2005. Dendrimer enhanced ultrafiltration. 1. Recovery of Cu(II) from aqueous solutions using PAMAM dendrimers with ethylene diamine core and terminal NH_2 groups. *Environmental Science and Technology* 39(5):1366–77.

Mansoori, G. A., T. R. Bastami, A. Ahmadpour, and Z. Eshaghi. 2008. Environmental application of nanotechnology. *Annual Review of Nano Research* 2:1–73.

Mauter, M. S., and M. Elimelech. 2008. Environmental applications of carbon-based nanomaterials. *Environmental Science & Technology* 42(16):5843–59.

Meridian Institute. 2006. *Overview and Comparison of Conventional Water Treatment Technologies and Nano-based Treatment Technologies*. Background Paper for the International Workshop on Nanotechnology, Water and Development, October 10–12, 2006, Chennai, India.

Mishra, A., and J. Clark. 2013. *Green Materials for Sustainable Water Remediation and Treatment*. RSC Green Chemistry Series. Cambridge, UK: Royal Society of Chemistry.

Nature. 2018. *Carbon Nanotubes and Fullerenes*. Available at: https://www.nature.com/subjects/carbon-nanotubes-and-fullerenes.

Novikov, A. P., S. N. Kalmykov, S. Utsunomiya, R. C. Ewing, F. Horreard, A. Merkulov, S. B. Clark, V. V. Tkachev, and B. F. Myasoedov. 2006. Colloid transport of plutonium in the far-field of the Mayak Production Association, Russia. *Science* 314(5799):638–41.

Nowack, B., and T. D. Bucheli. 2007. Occurrence, behavior and effects of nanoparticles in the environment. *Environmental Pollution* 150(1):5–22.

Nutt, M. O., J. B. Hughes, and M. S. Wong. 2005. Designing Pd-on-Au bimetallic nanoparticles for trichloroethylene hydrodechlorination. *Environmental Science and Technology* 39(5):1346–53.

Oberdörster, E., S. Zhu, T. M. Blickley, P. McClellan Green, and M. L. Haasch. 2006. Ecotoxicology of carbon-based engineered nanoparticles: Effects of fullerene (C60) on aquatic organisms. *Carbon* 44(6):1112–20.

O'Hara, S., T. Krug, J. Quinn, C. Clausen, and C. Geige. 2006. Field and laboratory evaluation of the treatment of DNAPL source zones using emulsified zero-valent iron. *Remediation Journal* Spring.

Rana, S., and P. T. Kalaichelvan. 2013. *Ecotoxicity of Nanoparticles*. ISRN Toxicology. Hindawi Pub. Available at: http://dx.doi.org/10.1155/2013/574648.

Savage, N., and M. S. Diallo. 2005. Nanomaterials and water purification: Opportunities and challenges. *Journal of Nanoparticle Research* 7(4–5):331–42.

Shipley, H. J., K. E. Engates, and A. M. Guettner. 2011. Study of iron oxide nanoparticle remediation of arsenic. *Journal of Nanoparticle Research* 13(6):2387–97.

Sivasubramanian, V. 2016. *Environmental Sustainability Using Green Technologies*. Boca Raton, FL: CRC Press.

Temple University. 2004. Researchers using proteins to develop nanoparticles to aid in environmental remediation. *ScienceDaily*. Available at: http://www.sciencedaily.com/releases/2004/09/040901090324.htm (Accessed February 2, 2018).

Theron, J., J. A. Walker, and T. E. Cloete. 2008. Nanotechnology and water treatment: Applications and emerging opportunities. *Critical Reviews in Microbiology* 34(1):43–69.

Tratnyek, P. G., and R. L. Johnson. 2006. Nanotechnologies for environmental cleanup. *Nanotoday* 1(2). Available at: http://dx.doi.org/10.1016/S1748-0132(06)70048-2 (Accessed May 20, 2015).

Triano, R. M., M. L. Paccagnini, and A. M. Balija. 2015. Effect of dendrimeric composition on the removal of pyrene from water. *SpringerPlus* 4:511.

U.S. EPA (U.S. Environmental Protection Agency). 2016. *Nanotechnology: Applications for Environmental Remediation*. Available at: https://clu-in.org/techfocus/default.focus/sec/Nanotechnology:_Applications_for_Environmental_Remediation/cat/Application/.

U.S. EPA. 2008a. *Nanotechnology for Site Remediation Fact Sheet. Solid Waste and Emergency Response*. EPA 542-F-08-009 (Accessed May 15, 2018).

U.S. EPA. 2008b. *Office of Superfund, Remediation and Technology Innovation. Nanotechnology: Practical Considerations for Use in Groundwater Remediation.* National Association of Remedial Project Managers Annual Training Conference, Portland, Oregon. July 7–11, 2008.

U.S. EPA. 2007. *Science Policy Council.* Nanotechnology White Paper. U.S. Environmental Protection Agency. EPA 100/B-07/001, Washington, DC. February 2007.

Vilks, P., L. H. Frost, and D. B. Bachinski. 1997. Field-scale colloid migration experiments in a granite fracture. *Journal of Contaminant Hydrology* 26(1–4):203–14.

Waite, T. D., A. I. Schafer, A. G. Fane, and A. Heuer. 1999. Colloidal fouling of ultrafiltration membranes: Impact of aggregate structure and size. *Journal of Colloid and Interface Science* 212(2):264–74.

Warheit, D. B., B. R. Laurence, K. L. Reed, D. H. Roach, G. A. Reynolds, and T. R. Webb. 2004. Comparative pulmonary toxicity assessment of single-wall carbon nanotubes in rats. *Toxicological Sciences* 77(1):117–25.

Watlington, K. 2005. *Emerging Nanotechnologies for Site Remediation and Wastewater Treatment.* U.S. Environmental Protection Agency. Office of Solid Waste and Emergency Response. Office of Superfund Remediation and Technology Innovation Technology Innovation and Field Services Division. Washington, DC.

Xu, Y., and D. Zhao. 2005. Removal of copper from contaminated soil by use of poly(amidoamine) dendrimers. *Environmental Science and Technology* 39(7):2369–75.

Xu, Y., and D. Zhao. 2006. Removal of lead from contaminated soils using poly(amidoamine) dendrimers. *Industrial and Engineering Chemistry Research* 45(5):1758–65.

Xu, J.-C., L. Mei, X.-Y. Guo, and H.-U. Li. 2005. Zinc ions surface-doped titanium dioxide nanotubes and its photocatalysis activity for degradation of methyl orange in water. *Journal of Molecular Catalysis, Part A: Chemical* 226(1):123–27.

Zhang, W.-X., and D. W. Elliot. 2006. Applications of iron nanoparticles for groundwater remediation. *Remediation Journal.* Available at: https://onlinelibrary.wiley.com/doi/abs/10.1002/rem.20078 (Accessed June 12, 2018).

Chapter 15

Technology Selection

My will shall shape the future. Whether I fail or succeed shall be no man's doing but my own . . . My choice; my responsibility.

—Elaine Maxwell

When choosing between two evils, I always like to try the one I've never tried before.

—Mae West

15.1 BACKGROUND

As discussed in chapter 1, the number of brownfields and other contaminated sites in the United States alone may number in the hundreds of thousands. The range of contamination includes crude and refined petroleum products, metallic wastes, inorganics, and radioactives. In addition to public health and environmental concerns related to soil and groundwater contamination, cleanups are notoriously expensive. The cost of remediating soil at a site contaminated from a leaking UST may range from $10,000 and well beyond. Costs for remediating sites with groundwater contamination can range from $100,000 to over $1 million depending on extent of contamination.

According to EPA a key factor in the high cost of site cleanups is the use of remediation technologies that are either inappropriately selected or not optimally designed and operated given specific conditions of the affected site. Excavation and landfilling, still used for managing some contaminated soils, does not truly remediate the site; rather, it simply transfers the hazard from one location to another. In addition to being costly, transporting contaminated soil off-site increases the risk to human health and the environment by dispersal of soil particles and vapors. Pump-and-treat continues to be the most popular method for remediating groundwater; however, the success of this method is often limited because either the source of contamination is not being adequately managed or the system is not optimized. Even when properly operated, pump-and-treat systems have inherent limitations: they may not work well in complex geologic settings or heterogeneous aquifers; they often stop reducing contamination long before reaching intended cleanup levels; and in some situations they can make sites more difficult to remediate by smearing contamination across the subsurface (see chapter 10).

With so many sites requiring remediation at such substantial cost, state and federal agencies are encouraging the use of faster, more effective, and less costly alternatives to conventional cleanup methods. EPA continues to encourage state and local governments to promote the use of the most appropriate cleanup technology for each affected site.

The purpose of this chapter is to focus on selection of appropriate technology for site cleanup, taking into consideration site-specific conditions and the nature and extent of contamination.

As stated in the preface, there is no standardized list of procedures for treatment or removal of hydrocarbon and metallic contaminants from surface soil, subsoil, and/or groundwater. The choice of technology is based upon a constellation of factors, technical and otherwise, for example the characteristics of the affected site, specific properties of the contaminant, details specific to the release itself, and the capabilities of the technology. Beyond these technical criteria issues such as cost, and even political considerations, must be addressed in choosing the appropriate remediation technology for a site.

This chapter focuses on engineering-related considerations for evaluating the technologies presented in this book. This chapter should be used alongside published technical sources, information from professional training courses, and scientific journals.

The discussion of each technology includes a brief overview plus checklists that provide the most important factors to evaluate for successful implementation of each technology.

15.2 ISOLATION (CONTAINMENT) (CHAPTER 6)

Isolation involves the physical segregation of subsurface contaminants from non-contaminated soil and groundwater. A number of systems are available to isolate the affected area so that contaminants are contained, either for permanent isolation or for removal or treatment at a later date. Specific systems which can effectively limit spread of the contaminant include diverting the flow of groundwater; subsurface barriers placed in the direction of flow to control lateral spread; and placement of an impermeable cap on the land surface to reduce infiltration of precipitation and run-on. These techniques have the common purpose of isolating the contaminant; they do not remove or destroy the contaminant.

Table 15.1 Soil and site properties for isolation (containment)

Parameter or Issue	Optimal	Not Preferred or Detrimental	Comments
Presence of sensitive environmental receptors (e.g., wetlands, wildlife refuges)	None occurring within a large radius	In proximity to affected site	If water table is drawn down, may affect hydrology of wetlands. May need to install buffers to protect sensitive sites. Regulatory concerns about excavation near sensitive sites.
Presence of subsurface structures	None	Fiber optic cables, electric, sewer, water, steam, and so on susceptible to damage by excavation.	May need to protect drinking water wells, utility lines, and so on. Document all buried structures in advance.
Depth to groundwater	Deep aquifer	Shallow aquifer	High water table makes containment difficult.
Soil particle size analysis	Soils high in clay	Soils high in sand and gravel (high porosity) may require containment	Dense (clayey) soils may be self-containing (i.e., limit contaminant migration).
Ks	$< 10^{-5}$ cm/sec	$> 10^{-3}$ cm/sec	Soils with high Ks may require containment.
Soil surface area	< 0.1 m^2/g	> 1 m^2/g	Higher adsorption hinders groundwater diversion.
Precipitation	Low to moderate with good infiltration	High precipitation and consequent runoff will make containment difficult	Additional engineering measures may be required, for example, berms, diversion ditches.

Source: FRTR 2018; IDEM 2017; NRC 2007; Quintal and Otero 2018.

Table 15.2 Contaminant properties for isolation (containment)

Parameter or Issue	Optimal	Not Desired or Detrimental	Comments
Contaminant types	VOCs, gasoline, other mobile compounds; soluble metals and inorganics	Technology not useful or necessary for heavy oils and some metals (e.g., lead)	
Contaminant phase (hydrocarbons only)	Liquid (free liquid or dissolved)	Vapor	
Contaminant viscosity	High (> 20 cPoise)	Low (< 2 cPoise)	
DNAPL contamination	Shallow depth to confining layer	Deep to aquiclude	Keyed-in wall necessary
Time since release	Recent	Long (> 1 year)	If too old (and plume too extensive), difficult to divert groundwater

Source: FRTR 2018; IDEM 2017; NRC 2007; Quintal and Otero 2018.

Table 15.3 System parameters for isolation (containment)

Parameter or Issue	Optimal	Not Desired or Detrimental	Comments
Managing the contaminant	Subsurface hydrology well defined.		The technology can be combined with pump-and-treat.
Runoff controls	If precipitation is moderate to heavy, install system for controlling runoff.	Inadequate runoff controls make containment difficult.	
Caps	Multilayers beneficial as they divert runoff. Also, vegetative layer protects surface layers.	Cap exposed to rainfall, sunlight, wind, and so on may be damaged.	Caps especially important when highly soluble contaminants present—prevents excess dispersion of plume.

Source: FRTR 2018; IDEM 2017; NRC 2007; Quintal and Otero 2018.

15.3 EXTRACTION PROCESSES (SOIL FLUSHING) (CHAPTER 7)

Soil flushing is an in situ technology that removes single or multiple contaminants from the subsurface. An appropriate extracting solution is either surface applied or injected below grade to solubilize contaminants. Extracted liquids are treated and collected for disposal, or reinjected to the subsurface.

Table 15.4 Soil and site properties for extraction processes (soil flushing)

Parameter or Issue	Optimal	Not Desired or Detrimental	Comments
Depth to groundwater	< 40 ft to contaminated aquifer	Deep and/or confined aquifers	The technology can be combined with pump-and-treat.
Soil particle size analysis	Sandy, gravelly, loamy soils	Dense, compacted clays	Rapid flow will enhance treatment and removal
Hydraulic conductivity	Faster than 1×10^{-3} cm/sec	Slower than 1×10^{-5} cm/sec	Rapid flow will enhance treatment and removal.
Soil chemical properties	Acidic or near-neutral pH (for metals removal), low organic matter content	High concentrations of Ca, Fe, or other metals (when using chelating agents and acids); high organic matter content.	Cationic soil metals (Ca^{2+}, Mg^{2+}, Fe^{2+}) will saturate chelating agents; organic matter may immobilize chelating agent.
Extent of plume	Small, confined contamination	Large, diffuse site (> 1 ha)	
Presence of subsurface structures	None	Fiber optic cables, electric, sewer, water, steam, and so on susceptible to damage by excavation.	Document all buried structures in advance.
Rock fractures	None	Present	
Precipitation	Low to moderate	High	If high precipitation may need to install berms and other run-on and runoff controls.
Location	Rural/industrial/suburban	Urban	NPDES permit may be required for disposal into storm drainage.

Source: Ahn et al. 2008; Bricker et al. 2001; U.S. EPA 1991; Wang et al 2007.

Table 15.5 Contaminant properties for extraction processes (soil flushing)

Parameter or Issue	Optimal	Not Desired or Detrimental	Comments
Number of contaminants	One	Multiple	Multiple contaminants may require several extracting solutions; technology may not work optimally.
Hydrocarbon contaminants	More soluble hydrocarbons, low molecular weight.	Heavy hydrocarbons (asphaltics, tars, etc.)	If nonaqueous (nonpolar) contaminants occur, may be necessary to obtain large volumes of surfactants; heavy hydrocarbons (e.g., coal tars) may be impossible to mobilize.
Contaminant phase	Dissolved	Vapor	
Aqueous contaminants	Aqueous-based extracting solutions (acids, chelating agents, other).	Oily hydrocarbons	Hydrocarbons and other nonpolar contaminants will hinder water movement.

Source: Ahn et al. 2008; Bricker et al. 2001; U.S. EPA 1991; Wang et al 2007.

Table 15.6 System parameters for extraction processes (soil flushing)

Parameter or Issue	Optimal	Not Desired or Detrimental	Comments
Nature of the extracting solution	Effectively solubilizes contaminant; must not itself be a hazardous compound.	May itself become a groundwater contaminant.	Some acids and surfactants may destroy soil structure; some may be toxic; acids will kill native microbes.
Extraction well installation	Wells installed within the radius of influence.	Wells too far apart; plume outside range of influence.	Radius of influence must be sufficiently large to recover all solubilized contaminant.

Source: Ahn et al. 2008; Bricker et al. 2001; U.S. EPA 1991; Wang et al 2007.

15.4 IN SITU SOLIDIFICATION/STABILIZATION (CHAPTER 8)

Solidification and stabilization (S/S) technologies are conducted by mixing contaminated soil with a binding agent to form a crystalline matrix which incorporates the contaminated material. Inorganic binders include cement, cement kiln dust, fly ash, and blast furnace slag. Certain organic wastes can be immobilized using organic binders such as bitumen (asphalt). During solidification, contaminants are immobilized within a solid matrix in the form of a monolithic block. Stabilization converts contaminants to a less- or a nonreactive form, typically by chemical processes.

Table 15.7 Soil and site properties for in situ S/S

Parameter or Issue	Optimal	Not Desired or Detrimental	Comments
Location	Remote (rural)	Urban, suburban	Regulatory approval may be easier in rural areas.
Depth to groundwater	Deep aquifer	Shallow potable aquifer	High groundwater levels make containment difficult.
Precipitation	Low to moderate with little runoff	High with excessive runoff	High precipitation will increase contaminant leaching.
Soil types	Mineral soils	High organic matter	Organic matter interferes with cement-based S/S fixative process.
Presence of subsurface structures	None	Fiber optic cables (electric, sewer, water, steam, etc.) are susceptible to damage by excavation.	Document subsurface structures in advance.

Source: Conner 1990; Donnelly and Webster 1996; Shi 2004; Stegemann 2004; U.S. EPA 1990a; 1995; 1997.

Table 15.8 Contaminant properties for in situ S/S

Parameter or Issue	Optimal	Not Preferred or Detrimental	Comments
Metallic contaminants	Similar chemistry of contaminants	Varying chemistry (e.g., Cr and As)	Varying chemistry may stabilize one contaminant while solubilizing a second. If Cr and As are the contaminants of concern, their valence states should be known.
Metals, radioactives are the primary inorganic contaminants.	Aqueous chemistry	Oily nonpolar hydrocarbons	Presence of hydrocarbons interferes with cement-based S/S fixative process.
Hydrocarbon contaminants	Low molecular weight	High molecular weight, viscous	A surfactant solution can mobilize lightweight hydrocarbons.

Source: Conner 1990; Donnelly and Webster 1996; Shi 2004; Stegemann 2004; U.S. EPA 1990a; 1995; 1997.

Table 15.9 System parameters for in situ S/S

Parameter or Issue	Optimal	Not Preferred or Detrimental	Comments
Portland cement or pozzolanic-based treatment	Metals, inorganic contaminants	Oily, nonpolar contaminants	Binder must be compatible with contaminant
Thermoplastic-based treatment	Oily wastes	Aqueous wastes	

Source: Conner 1990; Donnelly and Webster 1996; Shi 2004; Stegemann 2004; U.S. EPA 1990a; 1995; 1997.

15.5 SOIL VAPOR EXTRACTION (CHAPTER 9)

Soil vapor extraction (SVE) is an in situ technology that reduces concentrations of volatile constituents occurring in soil pores or adsorbed to soil in the vadose zone. A vacuum is applied to the soil to create a negative pressure gradient that stimulates movement of volatile vapors toward extraction wells. The extracted vapors are then treated, as necessary, and discharged to the atmosphere.

SVE technology has been proven effective in reducing concentrations of VOCs and certain SVOCs found in petroleum products at UST sites. SVE is generally more successful when applied to more volatile petroleum products such as gasoline. Diesel fuel, heating oil, lubricating oils, and kerosene, which are less volatile than gasoline, are not readily treated by SVE.

Table 15.10 Soil and site properties for soil vapor extraction

Parameter or Issue	Optimal	Not Preferred or Detrimental	Comments
Depth to groundwater	At least 15–20 ft	High water table	If high water table is encountered, sparging may be used.
Soil moisture content	Generally dry or moist. Working in the vadose zone.	Saturated. Water table fluctuates markedly within the affected zone during the year.	Consider biosparging when working below the water table.
Soil temperature	Warm soils (> 20°C)	Cool soils (< 20°C) limit volatilization of vapors.	Warm soils (> 20°C) favor SVE.
Soil particle size analysis	Sands and gravels may impart high Ks and therefore promote vapor extraction.	Dense, clayey soils	Dense soils will require many more extraction wells as they restrict the radius of influence. If soils are dense, install a surface seal. Install air injection or passive inlet wells.
Soil air conductivity	High (> 10^{-4} cm/sec)	Low (< 10^{-6} cm/sec)	
Site geology	Homogeneous	Heterogeneous	Undocumented layers of dense material will affect removal rate.
Precipitation	Low-to-moderate precipitation will promote the presence of air (and vapors) in pore spaces.	Heavy precipitation (and wet soils) will exclude vapors from soil voids.	Capping may be necessary to prevent excess water infiltration.
Presence of subsurface structures	None	Fiber optic cables, electric, sewer, water, steam, and so on are susceptible to damage by excavation.	Document all subsurface structures in advance.
Presence of surface structures	None	Many	If many structures, consider horizontal well installation.

Source: Byrnes et al. 2014; Cole 1994; Johnson et al. 1990; Jury 1986; Noonan and Curtis 1990; U.S. EPA 1989; 1991; 1995.

Table 15.11 Contaminant properties for soil vapor extraction

Parameter or Issue	Optimal	Not Preferred or Detrimental	Comments
Contaminant types	Light (low molecular weight) hydrocarbons (e.g., gasoline, jet fuel) are more amenable to SVE.	Heavy, high molecular weight hydrocarbons (heavy oils)	
Vapor pressure	> 100 mm Hg	< 10 mm Hg	If contaminant vapor pressure is < 0.5 mm Hg some type of enhancement (e.g., heated air injection) may be needed to increase volatility.
Boiling points of the contaminant constituents	< 300°C	> 300°C	A higher boiling point indicates higher molecular weight and hence less volatility.
Henry's law constant	> 100 atm	< 100 atm	
Water solubility	< 100 mg/l	> 1000 mg/l	Low solubility implies greater concentration of hydrocarbons.

Source: Byrnes et al. 2014; Cole 1994; Johnson et al. 1990; Jury 1986; Noonan and Curtis 1990; U.S. EPA 1989; 1991; 1995.

Table 15.12 System parameters for soil vapor extraction

Parameter or Issue	Optimal	Not Preferred or Detrimental	Comments
Water table	Deep (below base of extraction wells)	Shallow	May be necessary to lower water table via pumping. Alternatively, use air sparging.
Properties of extracted vapors	Relatively innocuous	Excessive vapor concentrations emitted, or concentrations of hazardous vapors above acceptable limits.	Install vapor treatment system. Many choices available depending on site factors and regulatory requirements.

Source: Byrnes et al. 2014; Cole 1994; Johnson et al. 1990; Jury 1986; Noonan and Curtis 1990; U.S. EPA 1989; 1991; 1995.

15.6 REACTIVE BARRIER WALLS (CHAPTER 10)

PRBs are subsurface structures installed downgradient of a contaminant plume to allow passage of groundwater while promoting degradation or removal of contaminants by specific chemical and physical reactions. PRBs have been used to successfully treat or remove metallic, radioactive, nonmetallic, and hydrocarbon contaminants from groundwater. Several variations of PRB configurations are available.

Table 15.13 Soil and site properties for permeable reactive barriers

Parameter or Issue	Optimal	Not Preferred or Detrimental	Comments
Location	Remote (rural)	Urban, suburban	Regulatory approval may be easier in rural areas.
Depth to groundwater	< 40 ft to contaminated aquifer	Excessively deep to contaminated aquifer	
Extent of plume	Small, confined contamination	Large, diffuse site (> 1 ha)	
Soil particle size analysis	Sandy, gravelly	Dense, heavy clays	
Hydraulic conductivity	Faster than 1×10^{-3} cm/sec	Slower than 1×10^{-5} cm/sec	Rapid flow will enhance treatment and removal.
Precipitation	Low to moderate	High	Run-on and runoff controls may be necessary.
Presence of subsurface structures	None	Fiber optics, water, sewer, steam and so on may interfere with installation of PRB.	Determine the location of all structures in advance.
Presence of surface structures	None	May interfere with installation of PRB.	

Source: Amos and Younger 2003; Benner et al. 1997; Blowes and Ptacek 1992; Cheng and Wu 2000; Longmire et al. 1991; Marsh et al. 2000; Sivavec et al. 1995; Thiruvenkatachari et al. 2008; U.S. EPA 1998.

The PRB is installed across the flow path of a contaminant plume. The reactive media is installed in order to be in intimate contact with the surrounding aquifer material. The reactive treatment zone inhibits contaminant movement via employing reactants and agents including zerovalent iron or other metals, zeolites, humic materials, chelating agents, sorbents, and active microbial cells.

Table 15.14 Contaminant properties for permeable reactive barriers

Parameter or Issue	Optimal	Not Preferred or Detrimental	Comments
Hydrocarbon contaminant	ZVI, reduced iron barrier		
Heavy metal contaminant	ZVI, reduced iron, limestone, apatite, microbial barrier		
Redox-sensitive elements (e.g., chromate, arsenate)	ZVI, reduced iron, limestone, apatite, microbial barrier		
Liquid viscosity	Low (< 2 cPoise)	High (> 20 cPoise)	
Time since release	Recent (< 1 month)	Long (> 12 months)	

Source: Amos and Younger 2003; Benner et al. 1997; Blowes and Ptacek 1992; Cheng and Wu 2000; Longmire et al. 1991; Marsh et al. 2000; Sivavec et al. 1995; Thiruvenkatachari et al. 2008; U.S. EPA 1998.

Table 15.15 System parameters for permeable reactive barriers

Parameter or Issue	Optimal	Not Preferred or Detrimental	Comments
Runoff controls	If precipitation is moderate to heavy, install controls for controlling runoff.		Inadequate runoff controls will make containment difficult.

Source: Amos and Younger 2003; Benner et al. 1997; Blowes and Ptacek 1992; Cheng and Wu 2000; Longmire et al. 1991; Marsh et al. 2000; Sivavec et al. 1995; Thiruvenkatachari et al. 2008; U.S. EPA 1998.

15.7 IN SITU MICROBIAL REMEDIATION (CHAPTER 11)

In situ bioremediation encourages the growth and proliferation of indigenous microorganisms for biodegradation of organic contaminants in either the saturated or the vadose zone. In situ bioremediation degrades hydrocarbons which are dissolved in groundwater and adsorbed to aquifer solids.

Bioremediation requires a delivery system for providing an electron acceptor, nutrients, and an energy source (carbon). In a typical in situ bioremediation system, non-contaminated groundwater is extracted using wells and, if necessary, treated to remove residual dissolved constituents such as iron. The treated groundwater is then mixed with an electron acceptor and nutrients,

Table 15.16 Soil and site properties for in situ bioremediation

Parameter or Issue	Optimal	Not Preferred or Detrimental	Comments
Hydraulic conductivity	$> 10^{-3}$ cm/sec	$< 10^{-5}$ cm/sec	
Impermeable layers	Present	Absent	
Groundwater chemistry	Dissolved Fe $<$ 10 mg/L	Dissolved Fe $>$ 10 mg/L	High Fe concentrations will cause clogging of well screens.
Soil moisture content	$>$ 30% by vol.	$<$ 10% by vol.	
Soil pH	6–8	$<$ 5.5 or $>$ 8.0	Extremes in pH will denature microbial enzymes.
Soil and groundwater temperature	10°C–45°C	$<$ 10°C or $>$ 45°C	Extremes in temperatures will denature microbial enzymes and inactivate microbial populations.
Total heterotrophic bacteria count	$> 10^6$ CFU/g dry soil		
C:N:P ratio	Between 100:10:1 and 100:1:0.5	$<$ 100:10:1 or $>$ 100:1:0.5	Low N will inhibit microbial reproduction.
Soil moisture content	40–85%	$<$ 40% or $>$ 85%	

Source: Aggarwal et al. 1991; Chikere et al. 2011; Das and Chandran 2011; Eweis et al. 1998; Thomas et al. 2006; U.S. EPA 1990c; 1992b; 1994; 1995b; Vira and Fogel 1991.

and reinjected upgradient of the contaminant source. Infiltration galleries or injection wells may be used to reinject treated water. Extracted water that is not reinjected is discharged, typically to surface water or to a POTW. In situ bioremediation can be implemented in a number of treatment modes, including aerobic, anaerobic, and co-metabolic.

Table 15.17 Contaminant properties for in situ bioremediation

Parameter or Issue	Optimal	Not Preferred or Detrimental	Comments
Degradability of constituents	Biorefractory index > 0.1	Biorefractory index < 0.01	
Dominant phase	Dissolved	Free liquid (NAPL)	
Contaminant solubility	High (> 1000 mg/l)	Low (< 100 mg/l)	
Total petroleum constituent concentrations	< 50,000 mg/l		Excessively high hydrocarbon concentrations, even of a biodegradable contaminant, can be toxic.
Total heavy metals	< 2,500 mg/l		Excessive metals will inactivate microbial enzymes.
Time since release	Long (> 12 months)	Short (< 1 month)	Longer times will allow more time for hydrocarbon transformation to dissolved forms.

Source: Aggarwal et al. 1991; Chikere et al. 2011; Das and Chandran 2011; Eweis et al. 1998; Thomas et al. 2006; U.S. EPA 1990c; 1992b; 1994; 1995b; Vira and Fogel 1991.

Table 15.18 System parameters for in situ bioremediation

Parameter or Issue	Optimal	Not Preferred or Detrimental	Comments
Presence of free product	Free product recovery system needed.		
Determination of area of in fluence	Appropriate well placement, given total area to be cleaned up and area of influence of each injection/extraction well.		
Subsurface soil and groundwater sampling for tracking constituent reduction and biodegradation conditions	Set up schedule for tracking constituent reduction. Control nutrient addition on a periodic or continuous basis.		Install appropriate nutrient delivery systems.

Source: Aggarwal et al. 1991; Chikere et al. 2011; Das and Chandran 2011; Eweis et al. 1998; Thomas et al. 2006; U.S. EPA 1990c; 1992b; 1994; 1995b; Vira and Fogel 1991.

15.7.1 Landfarming

Landfarming, also known as land application, is an aboveground remediation technology for soil that reduces concentrations of petroleum constituents through biodegradation and aeration (venting). This technology involves spreading excavated contaminated soils in a thin layer on the ground surface and stimulating aerobic microbial activity via aeration and/or addition of

Table 15.19 Soil and site properties for land treatment

Parameter or Issue	Optimal	Not Preferred or Detrimental	Comments
Total heterotrophic bacteria count	> 10^6 CFU/g dry soil	< 1,000 CFU/g dry soil	
Soil pH	6–8	< 5.5 or > 8.0	
Soil moisture content	40–85% (vol/vol)		
Soil temperature	10°C–45°C	< 10°C or > 45°C	
C:N:P ratio	100:10:1 to 100-1:0.5		
Precipitation rate	30–45 in/y	Soil too wet will impair aerobic microorganisms.	Irrigation may be needed for excessively dry soils.

Source: Pope and Matthews 1993; U.S. EPA 1984; 1995a.

Table 15.20 Contaminant properties for land treatment

Parameter or Issue	Optimal	Not Preferred or Detrimental	Comments
Contaminants to be treated	Lightweight hydrocarbons (gasoline, jet fuel, diesel fuel, kerosene).	Heavy (high MW) hydrocarbons; PAHs; heavy metals	Air emissions may need to be monitored and, if necessary, controlled.
Degradability of constituents	Biorefractory index > 0.1	Biorefractory index < 0.01	
Concentration of total petroleum constituents	< 50,000 mg/kg		Excessively high hydrocarbon concentrations, even of a biodegradable contaminant, can be toxic.
Total heavy metals	< 2,500 mg/kg	> 3,000 mg/kg	

Source: Pope and Matthews 1993; U.S. EPA 1984; 1995a.

nutrients and moisture. The enhanced microbial activity results in degradation of adsorbed petroleum constituents. If contaminated soils are shallow (i.e., < 3 ft below ground surface), it may be possible to effectively stimulate microbial activity without excavating soil. If contaminated soil is deeper than 5 ft, it should be excavated and placed on to the ground surface.

Table 15.21 System parameters for phytoremediation

Parameter or Issue	Optimal	Not Preferred or Detrimental	Comments
Technical feasibility	A treatability study has been conducted; biodegradation has been demonstrated, nutrient application defined, and potential toxic conditions checked.		
Available land area	Large (> 10 acres)	Small (< 1 acre)	Consider landfarm depth and additional space for berms and access.
Treatment of heavy hydrocarbons	Frequent mixing of soil material		
Daily operations	Run-on and runoff controlled; erosion control measures specified; frequency of application and composition of nutrients and pH adjustment materials specified; moisture addition; other suboptimal natural site conditions addressed in landfarm design.		
Operation plan	Anticipate frequency of aeration, nutrient addition, moisture addition.		
Monitoring progress of bioremediation	Conduct quarterly monitoring for soil pH, moisture content, bacterial populations, nutrient levels, contaminant concentrations. Monitor contaminant reduction and biodegradation conditions in LTU soils.		

Source: Baker et al. 2000; Banks and Schwab; 1993; Blaylock et al. 1997; Chaney et al. 2000; Fuhrmann et al. 2002; Hinchman et al. 1997; Pichtel et al. 2000; Smith et al. 1995; U.S. EPA 2000; 2005; Xia et al. 2003.

15.8 PHYTOREMEDIATION (CHAPTER 12); PHYTOEXTRACTION OF METALS

Phytoextraction involves the use of hyperaccumulating plants to transport metals from soil to concentrate them into roots and aboveground shoots. Following harvest of the extracting crop, the metal-rich plant biomass can be processed to recover the contaminant (e.g., valuable heavy metals, radionuclides). If recycling the metal is not economically feasible, the small amount of ash (compared to the original plant biomass or the large volume of contaminated soil) can be disposed appropriately.

Table 15.22 Soil and site properties for phytoremediation

Parameter or Issue	Optimal	Not Preferred or Detrimental	Comments
Location	Brownfield, urban		Phytoremediation zones can provide 'green belts.'
Depth of contamination	Well within rooting zone of plants	Beyond root zone	If beyond rooting zone, affected soil can be excavated and placed in layers on surface for treatment.
Soil fertility status	Adequate concentrations of N, P, K, micronutrients	Low concentrations of N, P, K, micronutrients	Commercial and organic fertilizer materials are usually readily available.
Soil organic matter	Low (< 0.1%)	Moderate to high (> 1%)	Organic matter can be added in the form of manures, composts, and so on.
Soil pH	5.5–7.5	Strongly acidic or alkaline	
Soil salinity	Low (< 1 dS/m)	High (> 4 dS/m)	
Presence of toxins	None	Presence of excess heavy metals, salts, and so on.	

Source: Baker et al. 2000; Banks and Schwab; 1993; Blaylock et al. 1997; Chaney et al. 2000; Fuhrmann et al. 2002; Hinchman et al. 1997; Pichtel et al. 2000; Smith et al. 1995; U.S. EPA 2000; 2005; Xia et al. 2003.

Table 15.23 Contaminant properties for phytoremediation

Parameter or Issue	Optimal	Not Preferred or Detrimental	Comments
Contaminants	One	Multiple	Multiple contaminants may create toxicity issues, require more frequent crop rotation.
Contaminant types	aqueous	Hydrophobic compounds (e.g., oily wastes, hydrocarbon solvents)	Hydrophobic materials will inhibit water movement and may cause plant toxicity.
Contaminant form (metals)	Soluble, exchangeable	Crystalline, insoluble	Site-specific metal chemistry must be determined.

Source: Baker et al. 2000; Banks and Schwab; 1993; Blaylock et al. 1997; Chaney et al. 2000; Fuhrmann et al. 2002; Hinchman et al. 1997; Pichtel et al. 2000; Smith et al. 1995; U.S. EPA 2000; 2005; Xia et al. 2003.

Table 15.24 System parameters for nanoremediation

Parameter or Issue	Optimal	Not Preferred or Detrimental	Comments
Soil management	Removal of weeds (tillage, herbicides); adequate moisture.		Plant growth must be optimized to achieve both maximal root and shoot growth, and thus maximum access to soil metals, and maximum extraction.
Crop management	Optimal planting density	Plants excessively crowded together or spaced too far apart.	Maximize soil material covered by healthy extracting plants.
Chelate application	Chelate not toxic, readily biodegradable	Potentially toxic chelating agent applied in single large dose.	May be beneficial to apply at multiple intervals instead of as one large dose.

Source: Baker et al. 2000; Banks and Schwab; 1993; Blaylock et al. 1997; Chaney et al. 2000; Fuhrmann et al. 2002; Hinchman et al. 1997; Pichtel et al. 2000; Smith et al. 1995; U.S. EPA 2000; 2005; Xia et al. 2003.

15.9 NANOREMEDIATION (CHAPTER 14)

This technology involves the application of reactive nano-sized materials for immobilization, destruction, and/or detoxification of selected contaminants in soil, groundwater, and surface water. Nanomaterials possess properties that enable chemical reactions, including catalysis, to transform pollutants to nonhazardous forms. In many cases, reactions catalyzed by nanoparticles are significantly more rapid than those by micro-sized particles. Nanomaterials can be introduced to the soil by installation of a PRB, direct injection, and other methods.

Table 15.25 Soil and site properties for nanoremediation

Parameter or Issue	Optimal	Not Preferred or Detrimental	Comments
Presence of subsurface structures	None	Fiber optic cables, electric, sewer, water, steam, and so on susceptible to damage by excavation.	May need to protect drinking water wells, utility lines, and so on. Document all buried structures in advance.
Site geology	Homogeneous	Heterogeneous	Undocumented layers of dense material will affect reaction and removal rates.
Soil moisture content	Generally moist or wet	Dry	Water movement will allow for contaminants to reach nanoparticles.
Hydraulic conductivity	$> 10^{-3}$ cm/sec	$< 10^{-5}$ cm/sec	
Groundwater chemistry	Low dissolved Fe (< 10 mg/L)	Dissolved Fe > 10 mg/L	High Fe concentrations will cause clogging of injection well screens.
Soil pH	6–8	< 5.5 or > 8.0	Extremes in pH may alter properties and behavior of nanoparticles.

Source: Boxall et al. 2007; ESTCP 2006; Karn et al. 2009; Mansoori et al. 2008; Nowack and Bucheli 2007; Rana and Kalaichelvan; 2013; U.S. EPA. 2008a; 2008b; 2007; Watlington 2005; Zhang and Elliot 2006.

Table 15.26 Contaminant properties for nanoremediation

Parameter or Issue	Optimal	Not Preferred or Detrimental	Comments
Number of contaminants	One	Multiple	Technology may not work optimally for multiple contaminants.
Hydrocarbon contaminant	nZVI, TiO_2, bimetallic, others		Concerns about passivization of ZVI. Many nanoparticles are under study.
Heavy metal contaminant	nZVI, limestone, others		Many nanoparticles are under study.
Redox-sensitive elements (e.g., chromate, arsenate)	nZVI, limestone, others		Many nanoparticles are under study.
Liquid viscosity	Low (< 2 cPoise)	High (> 20 cPoise)	
Time since release	Recent (< 1 month)	Long (> 12 months)	
DNAPL contamination	Shallow depth to confining layer	Deep to aquiclude	Keyed-in wall may be necessary.

Source: Boxall et al. 2007; ESTCP 2006; Karn et al. 2009; Mansoori et al. 2008; Nowack and Bucheli 2007; Rana and Kalaichelvan; 2013; U.S. EPA. 2008a; 2008b; 2007; Watlington 2005; Zhang and Elliot 2006.

Table 15.27 System parameters for nanoremediation

Parameter or Issue	Optimal	Not Preferred or Detrimental	Comments
Managing the contaminant	Subsurface hydrology well defined.		This technology can be combined with pump-and-treat.
Cap	Layer is beneficial as it diverts runoff.		Cap is especially important when highly soluble contaminants present—prevents excess dispersion of plume.
Nature of the nanoparticle	Effectively immobilizes and/or destroys contaminant; must not itself be a hazardous compound.	May itself become a groundwater contaminant.	Nanoparticles may agglomerate and migrate in groundwater.
Injection well installation	Wells installed within radius of influence.	Wells too far apart; plume is outside range of influence.	Radius of influence must be sufficiently large to react with all contaminant.

Source: Boxall et al. 2007; ESTCP 2006; Karn et al. 2009; Mansoori et al. 2008; Nowack and Bucheli 2007; Rana and Kalaichelvan; 2013; U.S. EPA. 2008a; 2008b; 2007; Watlington 2005; Zhang and Elliot 2006.

REFERENCES

American Society for Testing and Materials. 1997. *ASTM Standards Relating to Environmental Site Characterization*. West Conshohocken, PA: ASTM.

———. 2005. *Standard Practice for Environmental Site Assessments: Phase I Environmental Site Assessment Process*. E1527-00. West Conshohocken, PA: ASTM.

Cole, G. M. 1994. *Assessment and Remediation of Petroleum Contaminated Sites*. Boca Raton, FL: CRC Press.

Hess-Kosa, K. 2007. *Environmental Site Assessments, Phase I: A Basic Guide*, 3rd ed. Boca Raton, FL: CRC Press.

Reible, D. 2013. *Processes, Assessment and Remediation of Contaminated Sediments (SERDP ESTCP Environmental Remediation Technology)*. New York: Springer.

U.S. Environmental Protection Agency. 2004. *How to Evaluate Alternative Cleanup Technologies for Underground Storage Tank Sites: A Guide for Corrective Action Plan Reviewers*. EPA 510-B-94-003; EPA 510-B-95-007; and EPA 510-R-04-002. http://www.epa.gov/OUST/pubs/tums.htm.

U.S. Environmental Protection Agency. 2012. *Sampling and Analysis Plan. Guidance and Template*. Version 3, Brownfields Assessment Projects. R9QA/008.1. http://www.epa.gov.

Chapter 16

Revegetation of the Completed Site

He plants trees to benefit another generation.

—Caecilius Statius

16.1 INTRODUCTION

To date, thousands of acres of land on brownfields, Superfund, and RCRA sites have been treated for removal of metals and hydrocarbons. Many restored sites occur on active commercial and industrial properties and can be returned their original use or for other purposes: the affected area is simply graded, paved, and provided with the required drainage and runoff control. Other sites occur in remote areas or in locations of low commercial value. Regardless, however, the site manager may be required to revegetate the restored land. Some increasingly common end-uses include parks, hiking trails, wildlife habitat, sports fields, and golf courses (U.S. EPA 2015).

Sites that have been treated to remove hazardous contaminants may still present challenges to plant growth and establishment, however, including:

- The soil surface may be heavily compacted due to truck traffic and heavy equipment operation. Soil crusting may have occurred as a result of flooding the site with chemical reagents. Changes in physical properties such as density, aggregation, and texture reduce water and air infiltration and moisture-holding capacity and hinder efforts to revegetate a site.
- Surface soil may be very infertile due to incorporating and mixing material from significant depth. Deficiencies in N, P, and K are common. These are

significant as they are key nutrients for plant and microbial growth. Soil may likewise be deficient in organic matter.
- Microbial populations and activity may be limited if subsurface material is mixed with surface soil.
- Certain treatments, for example soil washing (chapter 7) or electrokinetics (chapter 13), may alter soil pH. An excessively low pH range can solubilize metals to the point that concentrations are toxic. In contrast, a very high pH can result in infertility, as nutrient elements are rendered unavailable to plants.

Many of the above challenges are addressed by installation of quality topsoil cover or application of soil substitutes (amendments). This chapter will address strategies for installation of a productive soil cover followed by establishment of a viable plant community.

16.2 USE OF SOIL SUBSTITUTES AND AMENDMENTS

In certain situations it may be feasible to stockpile surface soil material prior to initiating a remediation activity. At completion of the project the stockpiled soil layer is replaced. Grading, fertilization, and irrigation are provided as needed. Alternatively, so-called 'borrow topsoil' can possibly be obtained from nearby locations. It may be necessary to determine the chemical properties of borrow soil to anticipate potential problems with infertility, trace chemicals, or salts. Borrow soils may require amendments; infertile soils will require augmentation with fertilizers.

In many cases, the affected site occurs where local topsoil is unavailable and alternate materials are necessary to establish and support permanent vegetative cover. A number of materials are available that are considered by-products or residuals from commercial and industrial processes, but possess beneficial properties when applied to soil. Commonly used amendments include composted wastewater biosolids, animal manures, coal combustion products such as fly ash, as well as conventional agricultural fertilizers. These materials are often readily available. Sources include POTWs, CAFOs, coal-fired power plants, and pulp and paper mills, as well as retail sources. A partial list of soil amendment materials appears in table 16.1.

Applied properly and in appropriate quantities, amendments reduce exposure of soil to the effects of water and wind. Addition of amendments restores soil quality by balancing pH, adding nutrients and organic matter, increasing water-holding capacity, reestablishing microbial communities, and alleviating compaction. Many materials also immobilize contaminants and thus limit their availability to biota.

Table 16.1 Types and characteristics of selected soil amendments

Amendment	Availability	Uses	Public Acceptance	Cost	Advantages	Disadvantages
Biosolids	Sustainable supply; high quantities in urban areas.	Nutrient source; organic matter source; sorbent properties increase with increasing iron content.	Concerns about odors and pathogens; concerns largely driven by perception.	Materials are typically free; municipalities may pay for transport and use.	Multiple benefits as a soil amendment; highly cost-effective; EPA regulated; well characterized; consistent quality.	Public concern/public perceptions; high nutrient loadings in some settings; Some have high moisture content.
Manures	Sustainable supply; higher quantities near CAFOs	Nutrient source; organic matter source.	Concerns about odors and pathogens.	Materials generally free; Transport and application fee.	Widespread and readily available.	Not consistently regulated; variable quality; not routinely treated for pathogen reduction.
Compost	Location-dependent; volumes limited.	Nutrient source; organic matter source.	Readily accepted.	Product and transport costs can be high.	Readily accepted; stable product.	High cost; limited availability; N quantity usually significantly lower than non-composted materials.
Papermill sludges	Material available locally (northwest and east)	Organic matter source; slope stabilizer.	Possible odor problems; may contain chlorinated dioxins; primary sludges deficient in N.	Materials generally free; transport and application costs.	High C content; very dense texture is suitable for slope stabilization; large volumes; locally available.	Highly variable quality; may contain other residuals, for example, fly ash, waste lime, clay, which can be beneficial or detrimental for intended use. Low nutrient value.

(Continued)

Table 16.1 (Continued)

Amendment	Availability	Uses	Public Acceptance	Cost	Advantages	Disadvantages
Ethanol production by-products	New material; very location dependent.	Nutrient source; organic matter source.	Possible odor problems.	To be determined; transport and application fee.	High C content; large volumes; locally available.	Variable quality; not routinely treated for pathogen reduction; generally uncharacterized.
Limestone	Widespread.	Increase pH; Increase Ca content.	Highly accepted.	Product, transport, and application are $8–30/ton based on transport distances.	Regulated; well characterized; very uniform; enhances soil aggregation.	Agricultural limestone has low solubility and can become coated and ineffective in strongly acidic soil. Can be source of fugitive dust.
Gypsum	Large quantities locally available.	Good for sodic soil; good for low pH soil; improves soil structure.	Variable.	Materials generally free; transport fee.	Improves aggregation; offsets aluminum toxicity.	Different sources of waste gypsum and wide range of potential contaminants, many of which are regulated.
Coal combustion products	Most available in eastern United States.	Increase pH; Source of mineral nutrients (e.g., Ca).	Variable.	Materials generally free; transport and application fee.	Regulated; well characterized; enhances soil aggregation; light color reduces surface temperature for seedlings; increases soil moisture-holding capacity; reduces odor of organic amendments.	Varies plant to plant; can be high in B and salt content; can leach Se and As.

Cement kiln dust; lime kiln dust	Locally available	Increase pH; High Ca.	Variable.	Transport and application fee.	Highly soluble and reactive.	Potential fugitive dust; highly caustic; variable nutrient content; may contain contaminants.
Red mud	Locally available in TX and AR in U.S.	Increases pH; sorbent properties.	Variable.	Commercial product from a residual under development.	Demonstrated effective in limited testing in Australia and other sites at moderating pH and sorbing metals.	Potentially costly, high salt content; variable CCE.
Lime-stabilized biosolids	Locally available	Increase pH; organic matter and nutrient source; potential sorbent.	See biosolids.	See biosolids.	See biosolids; Potential multipurpose soil amendment.	Can have strong odor; lower N content than conventional biosolids; variable lime content.
Foundry sand	Large quantities locally available	Modifies texture; sorbent.	Variable.	Materials generally free; transport and handling fee.	Good filler; sand replacement.	n/a
Steel slag	Locally available	CCE, sorbent, and Mn fertilizer.	Accepted.	Materials generally free; transport and grinding fee.	Combination of CCE and sorbent, including Mn.	May volatilize ammonia.

(Continued)

Table 16.1 (Continued)

Amendment	Availability	Uses	Public Acceptance	Cost	Advantages	Disadvantages
Dredged material	Large quantities locally available	Modifies texture.	Variable.	Materials generally free; transport may be paid by generator.	Ideal for blending with other residuals.	Needs dewatering; can contain wide range of contaminants.
Water treatment residuals	Available wherever water is treated	Good for binding P; Potential sorbent.	Accepted.	Materials generally free; Transport costs may be covered by generator.	Moderates P availability when mixed with high P soil amendments.	Different materials have variable reactivities; May contain As and radioactive isotopes.

Source: U.S. EPA (2007).

16.2.1 Organic Soil Amendments

Organic amendments are most frequently used to provide essential nutrients (e.g., N and P), to rebuild soil organic matter content and reestablish microbial populations. Other benefits associated with increased organic matter content include:

- enhanced water infiltration
- enhanced water-holding capacity
- increased aggregation
- improved aeration

Selected organic amendments are discussed below.

16.2.1.1 Biosolids

These are nutrient-rich organic materials resulting from the treatment of domestic wastewater at a municipal facility. Over 7 million tons of biosolids are generated annually by municipal wastewater treatment plants in the United States, and about 55 percent of this material is land applied, primarily to agricultural land (Northeast 2007).

The chemical characteristics of biosolids vary between sources; however, composition is typically rather predictable from any single source. The nitrogen content of biosolids is generally of the 'slow-release' type: N occurs in plant and animal proteins, amino sugars, nucleic acids, and so on. As microorganisms decompose the solids, N becomes available to vegetation over several years following application. Biosolids often possess significant liming and sorbent properties as well as organic matter and nutrients.

Use of biosolids may be limited by excessive nutrient loading at higher application rates. In particular, there are concerns about excess N losses via runoff and leaching. Odors may also cause issues of public acceptance.

16.2.1.2 Manures

Over 335 million tons of animal manures are generated annually in the United States (USDA 2005).

Manures vary widely in terms of moisture content—some contain over 80 percent water, which could result in excessive costs if it must be transported a significant distance. The nutrient content of manures is highly variable, depending on type and age of animal, feed source, housing type, handling method, temperature, and moisture content. Because of this variability, land application decisions should be based on the nutrient content of the manure to be applied (Clemson 1996).

The nitrogen content of manures is typically readily available to vegetation and does not persist in soil as long as that in biosolids (U.S. EPA 2007). As is the case with application of biosolids, excessive nutrient loading may occur at higher application rates of manures; likewise, public opposition may occur due to concerns about odors, pathogens, and potential for surface water and groundwater contamination.

Some manures are dewatered or otherwise stabilized for beneficial use but most are applied 'as is' to soil.

16.2.1.3 Paper mill Sludges

Pulp and paper mills generate several types of sludges from treatment of wastes derived from virgin wood fiber sources, recycled paper products, and non-wood fibers. The nutrient, heavy metal, and organic composition of paper mill sludges is determined by the fiber source and treatment process, and is quite varied (Camberato et al. 2006). So-called primary sludges are mainly composed of short lignin fibers which are high in C but low in plant nutrients. In a survey of fifty-four paper mills, median N and P were reported to be 2.7 and 1.6 g kg^{-1}, respectively (Thacker and Vriesman 1984). Additional treatment of primary sludge is accomplished by addition of N and P to facilitate microbial degradation, which results in higher nutrient contents of the remaining solids, which are termed secondary sludge.

Many paper mills combine other wastes such as fly ash or kaolin with pulp sludges, which may enhance their soil amendment potential (U.S. EPA 2007). Both primary and secondary paper mill sludges provide large quantities of organic matter to soil.

16.2.2 Inorganic Soil Amendments

16.2.2.1 Fly Ash

Fly ash is the finely divided residue resulting from combustion of coal. It is an amorphous ferro-alumino-silicate mineral which contains most soil elements with the exception of N. Fly ash tends to be quite alkaline, sometimes with pH values exceeding 10 or 11. The calcium carbonate equivalent of fly ash can vary from 0 to > 50 percent. It is, therefore, a useful and low-cost liming material. In rare instances, fly ash is acidic. Fly ash particles are often clay- and silt-sized; therefore, addition to soil could improve moisture retention capacity.

16.2.2.2 Flue gas desulfurization (FGD) sludge

FGD sludge is produced when SO_2 is removed from coal combustion flue gas. The chemical composition of FGD depends on the type of scrubber

Table 16.2 Macronutrient composition of selected amendment materials in g kg⁻¹

Nutrient	Poultry Manure	Livestock Manure	Biosolids	Compost	Fly Ash	Paper mill Sludge
N	25–30	20–30	30–50	5–30	0.3–3	0.07–2
P	20–25	4–10	10–40	1–20	0.04–8	0.01–0.4
K	11–20	15–20	1–6	2–10	20–70	0.01–1
Ca	40–45	5–20	10–50	8–35	100–200	0.03–21
Mg	6–8	3–4	3–20	3–6	10–80	0.02–1.9
S	5–15	4–50	10–40	1–20	20–60	0.5

Sources: Schöneggera et al. (2018); Sullivan et al. (2015); Kuokkanen et al. (2008); Barker et al. (2000); Miller et al. (2000); Page et al. (1979).

technology used. Sulfur dioxide may be reacted with alkaline materials such as $CaCO_3$ and form calcium sulfite, $CaSO_3$, or calcium sulfate, $CaSO_4$ (gypsum). The resulting product has a high water content and must be dewatered for conversion to a slurry.

More than 100 million tons of coal fly ash and flue gas desulfurization (FGD) lime sludge are produced annually in the United States (U.S. EPA 2007). Both fly ash and FGD sludge are commonly comingled at generating facilities.

High levels of soluble salts and boron (B) in FGD sludge may limit application rates to soil. FGD materials commonly are higher in CCE than fly ash. Heavy metal concentrations should be determined prior to use of fly ash and FGD sludge. Concentrations of metals can vary considerably.

The plant macronutrient composition of several organic and inorganic amendments is shown in table 16.2.

16.2.3 A Living Amendment—Mycorrhizal Fungi

Arbuscular mycorrhizal fungi (AMF) refers to a symbiotic association between a fungus and the root cortical cells of vascular plants (Brundrett 2002). AMF are ubiquitous soil microbes which occur in almost all habitats. Taxonomically, all AMF are affiliated to the Glomeromycota phylum of fungi (Schüβler et al. 2001).

AMF can interact symbiotically with almost all plants—they are found in the roots of about 80 to 90 percent of plant species (primarily grasses, agricultural crops, and herbs) (Wang and Qui 2006). Mycorrhizal symbiosis maintains and often stimulates plant growth. The fungal mycelium acts as an extension of the plant root system, thus allowing the plant to optimize use of soil water and mineral nutrients from a much greater volume of soil (figure 16.1). Mycorrhizal associations are essential for plant uptake of nutrients such as N and P, especially in P-deficient soils. This association ultimately serves to reduce fertilizer requirements.

Figure 16.1 Photograph showing the effect of increased volume of soil in contact with a plant root via association with mycorrhizal fungi.

AM fungi further act to improve soil quality and ultimately plant health by increasing the diversity and abundance of soil microflora. Healthier plants are better equipped to resist environmental stresses such as contamination, drought, and temperature extremes, and have an improved capacity to survive certain bacterial and fungal pathogen attacks. AM fungi reduce soil erosion by increasing plant rooting capacity (Canada 2017).

Application of AMF to disturbed and infertile soils is highly recommended. As stated earlier in this chapter, soils which have undergone remediation by chemical or physical means may be drastically altered from the original material—soil may be deficient in organic matter and available nutrients, and lack the necessary microbial populations that support plant establishment and growth. AMF are available for purchase from a wide range of suppliers.

16.2.4 Application

Numerous approaches can assist in determining the appropriate application rate for amendments at a completed site. One method is to replicate the rates used for similar reclamation scenarios. For example, barren coal mine spoils have been restored with biosolids at rates from 20 to > 100 dry tons per acre (Hearing et al. 2001; Pichtel et al. 1994). Metal-contaminated sites have been restored with mixtures of biosolids and lime, with biosolids added at rates of 25 to 100 tons per acre and higher (U.S. EPA 2007).

Published research is an extremely valuable resource as regards types and application rates of amendments. Data are available on nutrient levels, microbiological diversity, and presence of possible interferences and toxins in various materials. Information on amendment use may focus on geographic location, climate, previous land use (e.g., industrial, brownfield, mine), soil type, compatibility of plant type with amendment, economics, and other factors.

Another approach is to conduct chemical analyses of both the amendment(s) of interest and the recipient soil. Once data have been compiled and integrated the site manager can calculate optimal amendment application rates. Of particular importance is ensuring adequate levels of nitrogen, phosphorus, and potassium, as they are key macronutrients required in plant nutrition. It is also appropriate to measure soil organic carbon level, pH, and lime requirement. Beyond chemical factors, it is suggested that the manager determine soil bulk density, particle size, and moisture content. It is strongly recommended that soil chemical and physical data be shared with a county extension agent or similar expert.

In reestablishing quality soil cover, it is essential to include a mixture of N-rich materials with C-rich materials. Added carbon will provide soil organic matter and promote microbial growth while reducing the potential for N leaching. In general, a bulk amendment C:N ratio between 20:1 and 40:1 is recommended, but higher C additions may be effective (U.S. EPA 2007). It also may be appropriate to include a mineral amendment such as foundry sand, steel slag, or gypsum as part of the mixture as a source of inorganic plant nutrients.

The moisture content of the amendment, commonly reported as % solids, is a key characteristic that directs application procedures and timing. In the case of biosolids application the range of solids content includes: liquid sludge at 2 to 8 percent solids, which can be pumped easily; semisolid biosolids at 8 to 18 percent solids, which also can be pumped; and solid biosolids cake at 20 to 40 percent solids, which may be dispersed from a manure-type spreader or end-dumped (Brown and Henry n.d.).

Application rates typically are calculated on a dry weight basis. This means that, for an average dewatered biosolids load (20% solids), an application of 90 dry tons per acre would involve applying 450 tons per acre of raw material. This is a significant volume and may pose complications with incorporation into soil.

Many types of equipment are available to perform direct spreading, including farm manure wagons, all-terrain vehicles with rear tanks, and dump trucks (figure 16.2). Heavy applications can be accomplished using several straightforward techniques which are simple and inexpensive.

Figure 16.2 Application of composted biosolids to a metal-contaminated site.

16.2.4.1 Single application

The simplest and most cost-effective application method is to distribute the entire calculated load in a single 'lift.' Depending upon application rate and percent solids, the lift may range from as little as 1 to 36 in in depth. Amendments can be allowed to dry on the surface prior to incorporation. Conventional farm disks or chisel plows are used to incorporate the mixture into the soil.

16.2.4.2 Multiple lifts

Amendment applications can be made in smaller or partial lifts. Some states require incorporation of biosolids over a certain timeframe. When multiple heavy applications are needed within a short period, working the soil becomes difficult because repeated applications without drying will make the soil excessively wet (U.S. EPA 2007).

16.2.4.3 Blending

Individual amendments can be combined with other materials to create characteristics optimal for site revegetation. For example, the goal may be to create a blend containing a full suite of plant nutrients while simultaneously increasing soil pH. In such a scenario, composted biosolids may be mixed with a desired ratio of power plant fly ash or cement kiln dust.

16.2.5 Concerns with Amendment Use

Potential soil amendments must be characterized for all relevant physical, chemical, and biological properties prior to acquisition and application. Revegetation plans should address potential adverse impacts from amendment use. For example, excessive nitrate-N leaching may occur if nutrient-rich materials are applied during winter or early spring.

The amendment material should be tested for the presence of potential toxins. In the past, concerns have arisen about the potential for Cd and/or Pb contamination of sewage sludges from certain sources. Sludges and manures may also have a high salt content. Paper mill sludges may contain some quantity of chlorinated hydrocarbons, a by-product from the paper bleaching process (Thacker et al. 2007). Fly ash may be contaminated with heavy metals.

16.2.6 Permitting and Regulations

A number of regulatory requirements may need to be addressed if amendments are applied for site cover. For example, biosolids are regulated under 40 CFR Part 503, under the combined authority of the CWA, RCRA, and CAA. Amendments such as foundry sand may be regulated as hazardous waste under RCRA.

Site managers must be aware of regulatory requirements when two or more soil amendments are blended. For example, when mixing biosolids with fly ash, the biosolids are regulated under the Clean Water Act Part 503 as well as state water or solid waste programs, and the fly ash is regulated as a solid waste under RCRA. If such blends are being planned, regulatory issues should be identified early in the project.

16.3 SELECTION AND MANAGEMENT OF PLANTS

The vegetation planted on the final soil cover serves a number of functions. As a protective soil stabilizing agent, vegetation absorbs the energy of rainfall, thus reducing so-called impact erosion on the soil layer, and improving moisture capture. Vegetation also controls soil moisture via uptake and evapotranspiration. The plant community is, additionally, an aesthetic and an ecological mitigation strategy as it provides a reconstructed habitat for plant and animal species.

Several approaches are available for creation of a permanent plant cover at the site. This section distinguishes between revegetation and environmental restoration as follows (CAEPA 1999):

Revegetation is defined as the placement of plants, horticultural or native, to the affected site. Few other environmental restoration techniques are included. The selected plants can be an arbitrary choice of the project manager, with no consideration for native species, their distribution, or plant community design. A restored brownfield which has been graded to engineering specifications and planted with non-native grasses in regulatory compliance is an example of simple revegetation.

Environmental restoration includes revegetation but also embraces elaborate design and return to original site contours, soil properties, and vegetative communities. The ultimate goal is for the site to be assimilated back into the surrounding environment. Environmental restoration typically takes into account: a detailed reconstruction of site topography; soil types conducive to local native plants; surface hydrology; and native plant species, including their diversity and distribution (CAEPA 1999).

This section focuses on revegetation only.

16.3.1 Practical Considerations

Some of the initial practical issues to be addressed when planning to revegetate a site include (EPA 2007):

- Seedbed preparation. This practice is essential for improving probability of seeding success. Preparation includes leveling the surface, breaking up large clods, and reducing the presence of competitive plants, including weeds.
- Including legumes in the mix. Legumes are recommended when grasses are to be grown. Legumes can serve as an excellent source of soil N; when grown together with dense-rooted grasses, a self-sustaining ecosystem can be established.
- Use of mulch. Mulch can stabilize seeded areas prior to plant establishment—it serves to decrease water erosion, reduce wind velocity, reduce soil crusting, decrease rainfall impact, limit evaporation from the surface, and decrease soil surface temperature. A variety of mulches are available, from natural (e.g., straw and wood chips) to synthetic (plastic sheeting).
- Irrigation. Watering may be necessary for revegetation in some regions to ensure successful plant establishment and avoid the potential for replanting in case of drought.

- Weed control. Weed species pose one of the greatest threats to long-term success of revegetation efforts. Careful monitoring of the site during establishment and control of invasive species is critical, as weeds and other invasive species can quickly disperse and invade open land. Weeds may out-compete native plants that help control erosion. Certified weed-free plant seed should be used to avoid introduction of invasive species that are difficult to eliminate once they become established.
- Grazing management. Activities of animals—whether wildlife such as deer or cattle and other livestock—should be monitored and controlled. Wildlife and ruminants can overbrowse a newly planted site and leave it vulnerable to invasive species.

16.3.1.1 Plant Types

Prior to initiating the revegetation activity, several criteria must be considered when selecting the plants that are desired in the final vegetative community.

The plant community must be compatible with the soil cover. As indicated earlier in this chapter, many soils that are used for vegetative cover are not the original native material and may not be quality topsoil, lacking the primary nutrients found in natural soil (CAEPA 1999).

An emphasis on plant diversity in vegetation selection is important for long-term vigor and resilience of cover. The site manager should avoid monoculture, or planting of only one type of tree or shrub. Species heterogeneity helps reduce disease dispersal or blights and encourages wider environmental diversity in the restored habitat (CAEPA 1999).

16.3.1.2. Water Management

Once the remediation project is complete, the final cover may include a clay or geomembrane layer directly beneath the soil (see chapter 6). When such layers are used, irrigation or routine watering must be included to maintain required soil moisture levels to safeguard against desiccation.

Watering should be scheduled based upon annual precipitation levels and local climate including rainfall volumes and periods of occurrences. If irrigation is to be used, the source water should be analyzed for possible presence of contaminants. Control of drainage off-site must be considered in final project design (CAEPA 1999).

16.4 GRASSES

Grasses are common to open environments (i.e., exposed to bright sunlight), but are also adapted to other locations including the shaded

Table 16.3 Common grasses suitable for revegetation at a completed site

Common Name (Scientific name)	Drought Resistance	Cold Tolerance	Acid Tolerance	Salt Tolerance	High Water Tolerance	Lower pH Limit	Comments
Kentucky bluegrass (*Poa pratensis*)	Poor	Good	Fair	Fair	Fair	5.5	Shallow-rooted sod former.
Perennial ryegrass (*Lolium perenne*)	Poor	Good	Poor	Fair	Fair	4.5	Short-lived perennial. Dominates stands for two years.
Timothy (*Phleum pratense*)	Fair	Poor	Good	Good	Fair	4.5	Good quality hay and pasture. Does not tolerate heavy grazing. Fertility demanding.
Smooth brome (*Bromus inermis*)	Good	Good	Poor	Good	Fair	5.0	Forms dense sod. Good erosion control.
Tall fescue (*Festuca arundinacea*)	Good	Good	Good	Good	Fair	4.5	Drought resistant. Common in reclamation of mined lands.
Weeping lovegrass (*Eragrostis curvula*)	Good	Fair	Good	Fair	Fair	4.0	Tolerant of acid soils and dry conditions. Short-lived perennial.
Redtop (*Agrostis gigantea*)	Good	Good	Good	Good	Good	4.0	Sod former. Adapts to a wide variety of soils. Short lived if not managed.
Switchgrass (*Panicum virgatum*)	Good	Good	Good	Fair	Good	4.0	Rhizomatous, acid tolerant, tall. Slow to establish.

Source: Skousen, J., and C.E. Zipper n.d.

understory of woodlands. Grasses occur as annuals or perennials depending on species.

Grasses are commonly seeded in revegetation programs for several reasons: (1) a wide range of species is available for seeding; (2) seed of species adapted to disturbed areas can be obtained readily and at reasonable costs; and (3) the grass family as a whole is tolerant of a variety of environmental and soil conditions (Skousen and Zipper n.d.).

Several grasses are well adapted to the infertile and/or droughty soil often associated with disturbed sites. Many species are capable of producing substantial biomass in a short time and responding quickly to fertilizer and management. Other species may be slower growing but persist on the site for long periods without management. For these reasons, revegetation programs frequently contain grasses as a major component of the seed mix. Grass species suggested for site revegetation are shown in table 16.3.

16.5 LEGUMES

The family of plants termed legumes (Leguminosae) comprises approximately 20,000 species (Skousen and Zipper n.d.). A healthy and persistent legume component is important to nitrogen fertility on borrow and replacement soils. Legumes fix (capture) nitrogen from the atmosphere to support their own nutrition; however, they also increase the concentration of soil nitrogen, making it available to companion species.

Legumes fix atmospheric N_2 by virtue of a symbiotic association with bacteria of the genus Rhizobium. Once the symbiotic bacteria become established in root nodules, bacterial enzymes allow the nodules to fix N_2 gas which is subsequently incorporated into plant proteins. As a result, legumes are able to grow in soils having little available soil nitrogen.

Legumes are often adapted to specific climates and soil conditions, so regional and soil-specific characteristics should be considered in selecting the appropriate species. It is important that legumes be inoculated with their specific Rhizobium symbiont prior to planting. Rhizobia inocula are available commercially.

Some common legumes suggested for revegetation purposes are listed in table 16.4.

16.6 NATIVE PLANTS

Native plant communities are the optimal choice for ensuring ecological diversity and long-term sustainability of the landscape. Native plants adapted to the conditions of the project site can reduce irrigation requirements as well

Table 16.4 Legumes suitable for revegetation at a completed site

Common Name (Scientific Name)	Drought Resistance	Cold Tolerance	Acid Tolerance	Salt Tolerance	High Water Tolerance	Precipitation Range (inches)	Lower pH Limit	Comments
Alfalfa (*Medicago sativa*)	Good	Good	Poor	Fair	Fair	15–20	6.0	Soil pH must be maintained above 6.0. P needed in high quantities. Good drainage required.
Crimson clover (*Trifolium incarnatum*)	Poor	Good	Fair	Fair	Fair	14–50	5.0	Winter annual legume. Reseeds itself. Tolerates pH to 4.0.
Red clover (*Trifolium pratense*)	Poor	Good	Fair	Fair	Fair	20–50	5.0	Used for erosion control. Short-lived perennial, but reseeds itself. Requires high P levels.
White (ladino) clover (*Trifolium repens*)	Poor	Fair	Fair	Fair	Good	18–45	5.5	Sod former. P and Ca are critical.
Alsike clover (*Trifolium hybridum*)	Fair	Poor	Good	Good	Good		5.0	More tolerant of moist, acidic soils than other clovers.
Flat pea (*Lathyrus sylvestris*)	Good	Good	Good	Fair	Fair	20–50	4.5	Slow establishment but has hardy rhizomes. Drought and acid resistant.

Common lespedeza (*Lespedeza striata*)	Fair	Fair	Good	Poor	Fair	25–45	4.5	Establishes quickly, reseeds. Tolerates acid soils.
White sweetclover (*Melilotus alba*)	Good	Good	Poor	Fair	Fair	14–40	5.5	Grows in early spring. Has large taproot.
Yellow sweetclover (*Melilotus officinalis*)	Good	Good	Poor	Fair	Fair	14–45	5.5	More drought tolerant and competitive than white.
Birdsfoot trefoil (*Lotus corniculatus*)	Fair	Fair	Good	Good	Fair	18–45	4.5	Grows well in mixtures. Rhizomatous.
Vetch, hairy (*Vicia villosa*)	Fair	Good	Good	Good	Fair	20–50	5.5	Fall plant for good winter cover. Important to inoculate.

Source: Skousen, J., and C.E. Zipper n.d.

as maintenance and pest control. Native plants are adapted to defend against many pests. Using native plants in final cover planting should therefore increase chances of vegetative survival, which reduces costs of replacement of plants lost to pest infestations (CAEPA 1999).

The USDA NRCS operates twenty-five Plant Materials Centers (PMCs) which are dedicated to the evaluation and improvement of plants and vegetative technologies. PMCs identify vegetative solutions to reduce soil erosion, increase soil health and plant productivity, produce forage and biomass, improve air quality, and stabilize coastal areas, among other benefits.

16.7 TREES

The trees most likely to produce healthy stands on restored sites are those well suited to local growing conditions. Most native tree species grow well in moderately acidic soils, that is, a pH range of 5.0 to 6.5. Alkaline soils (> pH 7.0) may limit tree selection. Bur oak and Shumard oak can tolerate soil pH greater than 7.5. A few species, including pin oak, can tolerate soil pH less than 4.0 (Davis et al. 2017).

Soil compaction limits species selection. If the site surface is compacted, future tree productivity will be restricted. A limited number of species such as green ash and American sycamore can survive in compacted soils, but most do not grow well. Where equipment traffic causes soil compaction, soils should be ripped to produce loose conditions before planting (figure 16.3).

Figure 16.3 Soil ripper. *Source*: Wikimedia Commons.

Many hardwood species such as northern red oak and white oak occur throughout the eastern United States and can be planted widely, but some should be restricted only to certain climatic conditions. Species such as sugar maple, bigtooth aspen, and red spruce are adapted to cool climates and are more successful in northern areas and at elevations above 3,000 ft. In contrast, southern red oak and others are adapted to the warmer climates of southern United States and lower elevations (table 16.5) (Davis et al. 2017).

16.8 THE VALUE OF FOREST HETEROCULTURE

Native forests, particularly those east of the Mississippi, are typically diverse in species distribution. The presence of multiple species helps a plant community to endure if a pest or pathogen damages one or more species. Planting a mixture of native trees and shrubs is also optimal for biodiversity of wildlife as well as plants (Cunningham et al. 2015). For these reasons, it is recommended that multiple species be planted.

16.9 BENEFITS OF CROP TREES, WILDLIFE TREES, AND NITROGEN-FIXING TREES

For recently restored sites, the site manager may consider three types of tree species for revegetation (Davis et al. 2017):

- crop trees that form a forest canopy;
- species selected for wildlife benefits;
- species that fix atmospheric nitrogen.

Crop trees include black cherry, tulip tree, sugar maple, and oaks that can produce economic value for the landowner and form a rich canopy. Many crop tree species provide wildlife benefits; however, some tree and shrub species are of low commercial value but are important to wildlife. It is recommended to prescribe other tree and shrub species in addition to crop trees for improving wildlife habitat in the revegetation program. Species such as flowering dogwood and eastern redbud become established and grow rapidly, generating a canopy used by birds for cover and nesting, and fruits and seeds that serve as wildlife food. Attracting wildlife aids natural succession and forest development. Mammals and birds consume fruits and seeds in natural habitats and then move through the restored area, where they deposit the seeds in their waste.

Table 16.5 Woody species appropriate for site revegetation

Species	Scientific name	Leaf Type[1]	Site Type	Potential Crop Tree	Growth Rate	N Fixer	pH Range[3]	Climate[4]
boxelder	Acer negundo	d	wet		rapid		M-H	
red maple	Acer rubrum	d	all		rapid		L-M-H	
sugar maple	Acer saccharum	d	moist, flat	yes	slow		L-M-H	C
gray alder	Alnus incana	d	wet		rapid	yes	M	
speckled alder	Alnus incana ssp. rugosa	d	wet		mod.[2]	yes	L-M-H	
hazel alder	Alnus serrulata	d	wet		rapid	yes	M	
mountain alder	Alnus viridis ssp. crispa	d	wet		mod.	yes	L-M-H	
Allegheny serviceberry	Amelanchier laevis	d	moist, flat		mod.		L-M-H	
false indigo bush	Amorpha fruticosa	d	moist		slow	yes	L-M-H	
yellow birch	Betula alleghaniensis	d	moist, flat		slow		L-M-H	C
sweet birch	Betula lenta	d	moist, flat		mod.		L-M	
river birch	Betula nigra	d	wet	yes	rapid		L-M	W
bitternut hickory	Carya cordiformis	d	moist, flat		slow		L-M-H	
pignut hickory	Carya glabra	d	dry		slow		L-M-H	
shellbark hickory	Carya laciniosa	d	moist, flat		slow		M	
shagbark hickory	Carya ovata	d	moist, flat		slow		L-M-H	
mockernut hickory	Carya tomentosa	d	dry		slow		L-M	
American chestnut	Castanea dentata	d	dry, moist		rapid		L	
northern catalpa	Catalpa speciosa	d	moist, flat		rapid		L M	
New Jersey tea	Ceanothus americanus	d	dry, moist		slow		L L-M	
common hackberry	Celtis occidentalis	d	moist, flat		rapid		M-H	
common buttonbush	Cephalanthus occidentalis	d	moist, wet		mod.		L-M-H	
eastern redbud	Cercis canadensis	d	moist, flat		slow		M-H	
silky dogwood	Cornus amomum	d	moist, flat		mod.		M	
flowering dogwood	Cornus florida	d	moist, flat		mod.		L-M-H	
gray dogwood	Cornus racemosa	d	all		mod.		L-M	

Common name	Scientific name		Site conditions		Growth rate	Shade tolerance	
American hazelnut	Corylus americana	d	moist, flat		mod.	M	
green hawthorn	Crataegus viridis	d	moist, flat, wet		mod.	L-M-H	
common persimmon	Diospyros virginiana	d	moist, wet		slow	L-M-H	
American beech	Fagus grandifolia	d	moist, flat		slow	L-M-H	
white ash	Fraxinus americana	d	moist, flat		mod.	L-M-H	
green ash	Fraxinus pennsylvanica	d	moist, flat, wet		rapid	L-M-H	
water locust	Gleditsia aquatica	d	wet		mod.	L-M-H	
honeylocust	Gleditsia triacanthos	d	moist, wet		rapid	L-M-H	
Kentucky coffeetree	Gymnocladus dioicus	d	moist, flat		slow	L-M-H	
American witchhazel	Hamamelis virginiana	d	moist, flat		slow	L-M	
American holly	Ilex opaca	e	all		slow	L-M-H	
common winterberry	Ilex verticillata	d	moist, flat		mod.	L-M-H	
black walnut	Juglans nigra	d	moist, flat		rapid	L-M-H	
eastern redcedar	Juniperus virginiana	e	moist, flat		slow	L-M-H	
sweetgum	Liquidambar styraciflua	d	moist, wet	yes	rapid	L-M-H	
Yellow poplar (tuliptree)	Liriodendron tulipifera	d	moist, flat, wet	yes	rapid	L-M	
sweet crab apple	Malus coronaria	d	moist, flat		slow	M	
red mulberry	Morus rubra	d	moist, flat		mod.	M	
hophornbeam	Ostrya virginiana	d	moist, flat		slow	L-M-H	
sourwood	Oxydendrum arboreum	d	dry, flat		slow	L-M	
red spruce	Picea rubens	e	moist, flat	yes	mod.	L-M	C
shortleaf pine	Pinus echinata	e	moist, flat	yes	rapid	L-M	W
pitch pine	Pinus rigida	e	dry		rapid	L	
eastern white pine	Pinus strobus	e	moist, flat	yes	rapid	L-M	
loblolly pine	Pinus taeda	e	dry	yes	rapid	L-M-H	W
Virginia pine	Pinus virginiana	e	dry		rapid	L-M-H	
American sycamore	Platanus occidentalis	d	moist, flat, wet	yes	rapid	L-M	
eastern cottonwood	Populus deltoides	d	moist, wet	yes	rapid	L-M	
bigtooth aspen	Populus grandidentata	d	moist, flat, wet		rapid	L-M	C
American plum	Prunus americana	d	moist, flat		mod.	M	
pin cherry	Prunus pensylvanica	d	moist, flat		rapid	L-M-H	

(Continued)

Table 16.5 (Continued)

Species	Scientific name	Leaf Type[1]	Site Type	Potential Crop Tree	Growth Rate	N Fixer	pH Range[3]	Climate[4]
black cherry	Prunus serotina	d	moist, flat	yes	rapid		L-M-H	C
white oak	Quercus alba	d	dry, moist, flat	yes	slow		L-M	
scarlet oak	Quercus coccinea	d	dry	yes	rapid		L-M	
southern red oak	Quercus falcata	d	dry, flat	yes	mod.		L-M-H	W
bur oak	Quercus macrocarpa	d	dry, moist, flat	yes	mod.			
chestnut oak	Quercus montana	d	dry	yes	slow		L-M	
chinkapin oak	Quercus muehlenbergii	d	dry	yes	mod.		M-H	
pin oak	Quercus palustris	d	moist, wet	yes	rapid		L-M	
northern red oak	Quercus rubra	d	moist, flat	yes	mod.		L-M-H	
Shumard oak	Quercus shumardii	d	dry, flat	yes	mod.		M-H	W
post oak	Quercus stellata	d	dry	yes	slow		L-M	
black oak	Quercus velutina	d	dry	yes	mod.		L-M	
bristly locust	Robinia hispida	d	dry, moist, flat		rapid	yes	L-M-H	
black locust	Robinia pseudoacacia	d	all		rapid	yes	L-M-H	
black willow	Salix nigra	d	wet		rapid		L-M-H	
American black elderberry	Sambucus nigra ssp. canadensis	d	moist, flat, wet		rapid		L-M-H	
sassafras	Sassafras albidum	d	moist, flat		mod.		L-M-H	
American basswood	Tilia americana	d	moist, flat	yes	mod.		L-M-H	
American elm	Ulmus americana	d	moist, flat		rapid		M-H	
slippery elm	Ulmus rubra	d	moist, flat		rapid		M-H	
highbush blueberry	Vaccinium corymbosum	d	wet		mod.		L-M-H	
southern arrowwood	Viburnum dentatum	d	all		slow		L-M	
blackhaw	Viburnum prunifolium	d	dry, moist		slow		L-M-H	

[1] Leaf type: d = deciduous; e = evergreen.
[2] mod. = moderate.
[3] Soil pH range: L = low (pH < 5); M = medium (pH 5–7); H = high (pH > 7).
[4] Climate suitability. C = does well in cool climates and at higher elevations (> 3,000 ft); W = does well in warm climates.
Source: Davis et al. (2017).

Some tree species occurring in natural forests at relatively low densities, such as common persimmon and black walnut, produce large fruits and seeds. The large seeds make them valuable as wildlife food sources and furthermore limit their spread out of the restored landscape by wind and animals.

Nitrogen-fixing trees remove N_2 from the air, transforming it to organic forms that enrich the soil. Examples include black locust, mimosa, alder, redbud, autumn olive, Kentucky coffee tree, golden chain tree, acacia, mesquite, and others. As mentioned earlier, recently restored soils are often deficient in N, an essential plant nutrient. If not taken up by plants, any N applied as fertilizer may remain in soil for a few years only. Therefore, planting at least one tree species that is capable of fixing atmospheric N is recommended.

QUESTIONS

1. List and discuss several challenges to plant growth and establishment at a site that was recently treated for metals removal by flushing with a dilute acid.
2. Organic amendments are often used to provide essential nutrients to recently treated sites. List five benefits of organic matter additions to a barren brownfield.
3. Use of biosolids as a soil substitute may be limited by what factors? Include environmental concerns and public considerations.
4. Sodium may occur in certain commercial and industrial wastes. How does excess Na adversely affect soil properties?
5. Before fly ash is to be used as an amendment on a highly acidic soil, what attributes should first be tested for in the ash?
6. A very high pH can cause soil infertility, as the concentrations of available nutrient elements to plants:
    (a) are excessive and potentially toxic; (b) are very low; (c) fluctuate; (d) create massive soil structure.
7. Biosolids often possess significant liming properties (True/false).
8. Paper mill sludges are typically much lower in N content than are biosolids (True/false).
9. A significant hazard with land application of animal manures is the potential for release of pathogenic microorganisms—over 100 diseases can be transmitted from manures to humans. (True/false) Check the published literature.
10. Legumes are recommended when grasses are to be grown at a reclaimed site. What are the benefits of including legumes at an affected site?
11. If legumes are being introduced to a recently restored site, the soil should be inoculated with the correct Rhizobia (True/false).

12. List and discuss the benefits of mycorrhizal fungi for revegetation of a barren brownfield.
13. The nitrogen content of biosolids is primarily of the slow-release form. Why is this beneficial in terms of both plant growth and environmental considerations?
14. When revegetating a brownfield or Superfund site, heteroculture is recommended. What are the benefits of heteroculture over monoculture plantings at these sites?
15. Why is it important to consider soil microbial populations and activities after remediation is complete? What can be inferred from measurements of microbial biomass, microbial respiration, or enzyme production?

REFERENCES

Barker, A. V., M. L. Stratton, and J. E. Rechcigl. 2000. Soil and by-product characteristics that impact the beneficial use of by-products. In: *Land Application of Agricultural, Industrial, and Municipal By-Products*, edited by J. M. Bartels. Madison, WI: Soil Science Society of America.

Brown, S. L., and C. L. Henry. n.d. *Using Biosolids for Reclamation/Remediation of Disturbed Soils (White Paper)*. Seattle, WA: University of Washington.

Brundrett, M. C. 2002. Coevolution of roots and mycorrhizas of land plants. *New Phytologist* 154(2): 275–304.

California Environmental Protection Agency (CAEPA). 1999. *A Guide to the Revegetation and Environmental Restoration of Closed Landfills*. Integrated Waste Management Board. October 1999.

Camberato, J. J., B. Gagnon, D. A. Angers, M. H. Chantigny, and W. L. Pan. 2006. Pulp and paper mill by-products as soil amendments and plant nutrient sources. *Canadian Journal of Soil Science* 86(4): 641–53.

Canada Agriculture and Agri-Food Canada. 2017. *Glomeromycota in Vitro Collection (GINCO)*. http://www.agr.gc.ca/eng/science-and-innovation/research-centres-and-collections/glomeromycota-in-vitro-collection-ginco/?id=1236786816381 (Accessed May 31, 2018).

Clemson University. 1996. *Land Application of Animal Manure*. https://www.clemson.edu/public/regulatory/ag-srvc-lab/animal-waste/manure.pdf.

Corker, A. 2006. *Industry Residuals: How They Are Collected, Treated and Applied*. Intern Paper. Prepared for U.S. EPA Office of Superfund Remediation and Technology Innovation. 52 pp.

Cunningham, S. C., R. McNally, P. J. Baker, T. R. Cavagnaro, J. Beringer, J. R. Thomson, and R. M. Thompson. 2015. Balancing the environmental benefits of reforestation in agricultural regions. *Perspectives in Plant Ecology, Evolution and Systematics* 17(4): 301–17.

Davis, V., J. A. Burger, R. Rathfon, and C. E. Zipper. 2017. Selecting tree species for reforestation of Appalachian mined lands. Chapter 7. In: *The Forestry Reclamation Approach: Guide to Successful Reforestation of Mined Lands,* edited by M. B. Adams. Gen. Tech. Rep. NRS-169. Newtown Square, PA: U.S. Department of Agriculture, Forest Service, Northern Research Station.

Haering, K. C., W. L. Daniels, and S. E. Feagley. 2000. Reclaiming mined land with biosolids, manures and papermill sludge, pp. 615–44. In: *Reclamation of Drastically Disturbed Lands,* edited by R. I. Barnhisel et al. American Soc. of Agron. Monograph #41, Madison, WI, 1082 pp.

Kuokkanen, T., H. Nurmesniemi, R. Pöykiö, K. Kujala, J. Kaakinen, and M. Kuokkanen. 2008. Chemical and leaching properties of paper mill sludge. *Chemical Speciation and Bioavailability* 20(2): 111–22.

Miller, D. M., W. P. Miler, S. Dudka, and M. E. Sumner. 2000. Characterization of industrial by-products. In: *Land Application of Agricultural: Industrial, and Municipal By-Products*, edited by J. M. Bartels. Madison, WI: Soil Science Society of America.

North East Biosolids and Residuals Association. 2007. Biosolids management trends in the U.S. *BioCycle* 48(5): 47.

Page, A. L., A. A. Elseewi, and I. R. Straughan. 1979. Physical and chemical properties of fly ash from coal-fired power plants with special reference to environmental impacts. *Residue Reviews* 71: 83–120.

Pichtel, J., W. A. Dick, and P. Sutton. 1994. Comparison of amendments and management practices for long-term reclamation of abandoned mine lands. *Journal of Environmental Quality* 23(4): 766–72.

Schöneggera, D., M. Gómez-Brandón, T. Mazzier, H. Insam, R. Hermanns, E. Leijenhorst, T. Bardelli, and M. Fernández-Delgado Juárez. 2018. Phosphorus fertilising potential of fly ash and effects on soil microbiota and Crop. *Resources, Conservation and Recycling* 134: 262–70.

Schüßler, A., D. Schwarzott, and C. Walker. 2001. A new fungal phylum, the Glomeromycota: Phylogeny and evolution. *Mycological Research* 105(12): 1413–21.

Sheoran, V., A. S. Sheoran, and P. Poonia. 2010. Soil reclamation of abandoned mine land by revegetation: A review. *International Journal of Soil, Sediment and Water* 3(2): 1–20.

Skousen, J., and C. E. Zipper. n.d. *Revegetation Species and Practices.* Powell River Project. Publication 460-122. Blacksburg, VA: Virginia Cooperative Extension.

Sullivan, D. M., C. G. Cogger, and A. I. Bary. 2015. *Fertilizing with Biosolids.* Pacific Northwest Extension. Oregon State University. https://catalog.extension.oregonstate.edu/sites/catalog/files/project/pdf/pnw508_0.pdf.

Thacker, N. P., V. C. Nitnaware, S. K. Das, and S. Devotta. 2007. Dioxin formation in pulp and paper mills of India. *Environmental Science and Pollution Research International* 14(4): 225–26.

U.S. Department of Agriculture, Agricultural Research Service. 2005. *National program 206: Manure and Byproduct Utilization* (FY-2005 Annual Report). Retrieved August 28, 2012.

U.S. Environmental Protection Agency. 2007. *The Use of Soil Amendments for Remediation, Revitalization, and Reuse.* EPA 542-R-07-013. Washington, DC.

———. 2015. *Revegetating Landfills and Waste Containment Areas. Fact Sheet.* https://www.epa.gov/sites/production/files/2015-08/.../revegetating_fact_sheet.pdf.

Wang, B., and Y.-L. Qui. 2006. Phylogenetic distribution and evolution of mycorrhizas in land plants. *Mycorrhiza* 16(5): 299–363.

Acronyms and Abbreviations

AAI	All-Appropriate-Inquiry
ASTM	American Society for Testing and Materials
AST	aboveground storage tank.
BDL	below detection limit
BGS	below ground surface
BNP	bimetallic nanoscale particle
BOD	biochemical oxygen demand
BTEX	benzene, toluene, ethylbenzene, and xylene
Btu	British thermal unit
C	carbon
°C	degrees Celsius
CAA	Clean Air Act
CAFO	concentrated animal feeding operation
CCl_4	carbon tetrachloride
cDCE	*cis*-dichloroethene
CEC	cation exchange capacity.
CERCLA	Comprehensive Environmental Response, Compensation, and Liability Act of 1980
CFR	Code of Federal Regulations
CO	carbon monoxide
CO_2	carbon dioxide
CSPE	chlorosulfonated polyethylene
CWA	Clean Water Act
DNAPL	dense nonaqueous phase liquid
DOD	Department of Defense
DOE	Department of Energy
DOT	Department of Transportation

DRE	destruction and removal efficiency
EDB	ethylene dibromide
EDC	ethylene dichloride
EPA	Environmental Protection Agency
°F	degrees Fahrenheit
Fe°	zerovalent iron
FID	flame ionization detector
FWPCA	Federal Water Pollution Control Act
GAC	granular activated carbon
GPR	ground-penetrating radar
GC	gas chromatography
HCl	hydrochloric acid (or hydrogen chloride)
HDPE	high-density polyethylene
HSWA	Hazardous and Solid Waste Amendments to RCRA
H_2O_2	hydrogen peroxide
ICP	inductively coupled plasma
kg	kilogram
KPa	kilopascal
L	liter
LBP	lead-based paint
LDPE	low-density polyethylene
LDR	Land Disposal Restrictions
LEL	lower explosive limit
LFL	lower flammable limit
LNAPL	light nonaqueous phase liquid
MCL	maximum contaminant level
MS	mass spectrometry
MSW	municipal solid waste
MTBE	methyl tert-butyl ether
NAAQS	National Ambient Air Quality Standards
NAD	nicotinamide adenine dinucleotide
NAPL	nonaqueous phase liquid
NIMBY	Not in My Backyard
NEPA	National Environmental Policy Act
ng	nanogram (1 gram x 10^{-9})
NOx	nitrogen oxides
NP	nanoparticle
NPL	National Priorities List
OSHA	Occupational Safety and Health Administration
PAH	polycyclic aromatic hydrocarbon
PCB	polychlorinated biphenyl
PCDD	polychlorinated debenzodioxin

PCDF	polychlorinated dibenzofuran
PCE	perchloroethylene (tetrachloroethene)
PCP	pentachlorophenol
PID	photoionization detector
PAH	polynuclear aromatic hydrocarbon
PLM	polarized light microscopy
POTW	Publicly-Owned Treatment Works
ppb	parts per billion
ppm	parts per million
ppt	parts per trillion
PRB	permeable reactive barrier
PRP	potentially responsible party
psi	pounds per square inch
PVC	polyvinyl chloride
RCRA	Resource Conservation and Recovery Act
RFH	radio-frequency heating
SBOD	soluble biochemical oxygen demand
SEM	scanning electron microscopy
S/S	solidification/stabilization
SG	specific gravity
SOMS	swellable organically modified silica
SOx	sulfur oxides
SVE	soil vapor extraction
TCE	trichloroethylene
TCLP	Toxicity Characteristic Leaching Procedure
TDS	total dissolved solids
TEM	transmission electron microscopy
TOC	total organic carbon
TOCP	tri-o-cresylphosphate
TPH	total petroleum hydrocarbons
TSCA	Toxic Substances Control Act
UEL	upper explosive limit
UFL	upper flammable limit
µg	microgram
µm	micrometer
UST	underground storage tank
VC	vinyl chloride
VLDPE	very low-density polyethylene
VOC	volatile organic compound
VLDPE	very low-density polyethylene
ZVI	zero-valent iron

Glossary

Accumulator Green plant that absorbs high concentrations of an element or compound into tissue with no apparent detrimental effect.

Acid A compound that donates a proton (H^+) to another compound. A substance that causes destruction to skin tissue at the site of contact or that corrodes steel. Liquids possess a pH of less than 7.0.

Actinomycetes A group of heterotrophic, mostly filamentous aerobic microorganisms.

Activated carbon Pyrolyzed carbonaceous material used to remove potentially toxic substances from gaseous or aqueous media.

Activated sludge A process of removing BOD from wastewater. Microbial cells are introduced into a reaction vessel and allowed to decompose organic compounds. Newly produced microbial biomass is collected and reintroduced into the process.

Acute effect An adverse effect on an organism, generally after a single exposure, with severe symptoms developing rapidly.

Acute toxicity Detrimental effects of a chemical that occur within a relatively short time frame (hours to months).

Adsorption Attraction of solid, liquid, or gas molecules, ions, or atoms to particle surfaces by physiochemical forces.

Advection Unidirectional bulk movement, such as water or a dissolved ion, under the influence of a hydraulic gradient.

Aerobic System or process in which oxygen, O_2, is required or is present. The biological state of living in the presence of oxygen.

Aliphatic hydrocarbon Class of hydrocarbons that contain no aromatic rings. The class includes alkanes, alkenes, alkynes, and cyclic alkane hydrocarbons.

Alkali A liquid or solid substance that is caustic. Strong alkalis in solution are corrosive to the skin and mucous membranes; substances with a pH greater than 7.0.

Alkalinity A solution having a pH value greater than 7.0. A measure of the capacity of liquids to neutralize strong acids. Alkalinity results from the presence of bicarbonates, carbonates, hydroxides, silicates, phosphates, and others.

Alkene Hydrocarbons composed of molecules that contain one or more carbon-carbon double bonds. Also known as *olefins*.

Alkylaromatic Aromatic compounds containing alkyl substituents.

Alkynes Hydrocarbons composed of molecules that contain one or more carbon-carbon triple bonds.

Anaerobic System or process in which oxygen is not required or is absent.

Anhydrous Free from water.

Anion An ion that is negatively charged.

Anoxic Conditions lacking molecular oxygen.

Anthropogenic Man-made.

Aquiclude An impermeable layer of geologic strata occurring beneath the surface. Will not permit groundwater to flow through.

Aquifer Underground formation of porous geologic strata such as sand, rock, gravel, and so on, that can store and supply groundwater to wells or springs.

Aquifer, confined Aquifer possessing a confining layer between the zone of saturation and the surface.

Aquifer, unconfined Aquifer that has no confining layers between the zone of saturation and the surface.

Aquitard See *aquiclude*.

Aromatic hydrocarbons Hydrocarbons composed of six-membered rings, with alternating double and single carbon-carbon bonds.

Artesian An aquifer situated between two impermeable layers and subject to greater than atmospheric pressure.

Asphalt A black, bituminous material composed of hydrocarbons having a high boiling point. Found in nature and can be prepared by pyrolysis of coal, tar, petroleum, and lignite tar. Melts on heating and is insoluble in water.

AST Aboveground storage tank.

ASTM American Society for Testing and Materials.

Bacteria Single-celled microscopic organisms. Aerobic, anaerobic, facultative anaerobes exist.

Bacteria, aerobic Bacteria that require the presence of dissolved or molecular oxygen to carry out metabolic processes.

Bacteria, anaerobic Bacteria that do not require oxygen for metabolism; growth may be hindered by the presence of oxygen.

Bacteria, facultative Bacteria that can exist under either aerobic or anaerobic conditions.

Bentonite A 2:1 aluminosilicate clay formed from weathering of feldspars and composed mainly of montmorillonite and beidellite. Characterized by high swelling upon wetting. Bentonite is commonly used as a landfill liner and as fill around well casings.

Benzene C_6H_6. An aromatic hydrocarbon characterized by a six-carbon ring, with alternating double and single bonds.

Berm A constructed ridge of soil.

Binder A cement-like material or resin used to hold particles together.

Bioaccumulation The increase in concentration of a certain substance as it moves up a food chain. An important mechanism responsible for concentrating pesticides and heavy metals in animals at the top of food chains.

Biocide A substance that, when absorbed, ingested, inhaled, or otherwise consumed in small quantities, causes illness or death, or retardation of growth.

Biodegradability Degree to which a substance may be decomposed by the enzymatic activities of microorganisms.

Biodegradation Decomposition of a substance into simpler compounds by the action of microorganisms.

Biohazard Biological hazard. Infectious agents presenting a risk to the well-being of humans or other biota.

Biological treatment Process by which hazardous waste is rendered nonhazardous or is reduced in volume by the action of microorganisms.

Biomass Living plant, animal, or microbial tissue.

Bioremediation The use of biological processes to degrade organic contaminants in soil, sediments, strata, or water.

Biosolids Solids derived from the treatment of municipal wastewaters. Also known as sewage sludge.

Biosphere The thin sphere of life surrounding the earth. Embraces parts of the atmosphere, the lithosphere, and the hydrosphere.

Bitumen Naturally occurring or pyrolytically obtained dark, tarry hydrocarbons consisting almost entirely of carbon and hydrogen, with little oxygen, nitrogen, or sulfur. These hydrocarbons possess a very high boiling point.

Brownfield An abandoned or underutilized industrial site. Has the potential to contain a hazardous condition from wastes and other sources.

BTEX Benzene, toluene, ethylbenzene, and xylene. Added to automotive gasoline to improve combustibility. All are hazardous per RCRA.

Buffer A solution that resists changes in pH.

Capping system An impermeable system designed to reduce surface water infiltration, control gas and odor emissions, improve aesthetics, and provide a stable surface over a site.

Carcinogen Substance capable of causing cancer.

Cation A positively charged ion.

Cation exchange capacity A measure of the number of equivalents of negative charge on a colloidal surface such as clay or organic matter. Often measured in units of milliequivalents per 100 grams solids, or cmol/kg.

CEC Cation exchange capacity.

Cement A mixture of calcium aluminates and silicates made by combining lime and clay while heating.

Centigrade (Celsius) A scale for measuring temperature, in which 100° is the boiling point of water at sea level (one atmosphere) and 0° is the freezing point.

CERCLA Comprehensive Environmental Response, Compensation and Liability Act of 1980. CERCLA sets liability standards for environmental damage and authorizes identification and remediation of abandoned waste sites.

CFR Code of Federal Regulations. The U.S. government document in which all federal regulations are published. Each title focuses on a different federal department or agency.

Chalcophile Minerals that crystallized in a reducing environment to form sulfide minerals.

Chelate The bonding of a multidentate organic molecule with a metal via more than one bond. Typically a strong association.

Chemical oxidation Process that increases the oxidation state of an atom through loss of electrons.

Chemical precipitation The use of chemicals to convert dissolved and suspended matter to a solid form.

Chemical reduction Process that decreases the oxidation state of an atom through the acquisition of electrons.

Chlorinated dibenzodioxins A group of polychlorinated compounds characterized by two benzene rings linked by two oxygen bridges.

Chlorinated dibenzofurans A group of polychlorinated compounds characterized by two benzene rings linked by one oxygen bridge.

Chronic effect Adverse effects resulting from repeated doses of, or exposures to, a substance over a prolonged period.

Clay Finest-grained portion of soil. Particles that exhibit plasticity within a range of water contents and that exhibit considerable strength when air-dry. The USDA definition includes all particles less than 2 μm in diameter.

Colloid A particle measuring less than 1 μm across. Colloids tend to remain suspended in water due to Brownian movement.

Comprehensive Environmental Response, Compensation, and Liability Act Also known as the Superfund Law. Provides a mechanism for the cleanup of the most dangerous, abandoned, and uncontrolled hazardous waste sites in the United States. See *CERCLA*.

Confined aquifer Aquifer bounded above and below by impermeable strata; an aquifer containing confined groundwater.
Containment Technologies that reduce the mobility of a contaminant plume in the subsurface via construction of physical barriers. Also utilized to reduce the flow of groundwater through contaminated media.
Contaminant An undesirable minor constituent that renders another substance impure.
Diffusion Movement of molecules toward an equilibrium driven by concentration gradients (i.e., mass transfer).
Dioxins See *chlorinated dibenzodioxins*.
Disposal drum Drum used to overpack damaged or leaking containers of hazardous materials for shipment.
DNAPL Dense nonaqueous phase liquid. A nonpolar (i.e., hydrophobic) liquid that is more dense than water and will sink to the bottom of an aquifer if released to the subsurface.
Electrokinetics A technology that removes metals and other contaminants from soil and groundwater by applying an electric field in the subsurface.
Endophytic Within a green plant.
EPA See *United States Environmental Protection Agency*.
Evapotranspiration Return of water to the atmosphere by the combined action of evaporation and release by vegetation through the stomata.
Excluder Green plant that survives on contaminated soil by excluding particular toxins from entering the root.
Explosive limits The minimum and maximum concentration of a substance in air which can be detonated by spark, shock, fire, and so on. See *flammable limits*.
Exothermic reaction Chemical reaction that releases energy in the form of heat.
Exposure Contact with a toxic substance or harmful chemical or physical agent through inhalation, ingestion, puncture, or absorption.
Ex situ External to the system. For example, excavation of soil from a site followed by treatment.
FID Flame ionization detector.
Fermentation Microbial process in which organic compounds serve as both electron donors and electron acceptors.
Flammable limits The minimum and maximum concentration of flammable hydrocarbon vapors in air that will support combustion. The lowest concentration is the lower flammable limit (LFL) and the highest concentration is the upper flammable limit (UFL).

Flashpoint The lowest temperature at which a liquid gives off enough vapor to form an ignitable mixture with air and support a flame when a source of ignition is present.

Fly ash The finely divided residue from combustion of coal or other solids (e.g., MSW), which is transported out of the firebox by flue gas.

Fume Cloud of fine solid particles arising from the heating of a solid material such as lead.

Fungi Nonphotosynthetic unicellular and multicellular microorganisms that require organic compounds for growth.

Furans See *chlorinated dibenzofurans*.

Gas A state of matter in which a material has very low density, can expand and contract greatly in response to changes in temperature and pressure, easily diffuses into other gases, and uniformly distributes itself throughout a container.

GC/MS Gas chromatography/mass spectrometry. Analytical method and apparatus used for determination and quantification of organic compounds.

Groundwater Water occurring beneath the earth's surface that fills the pores between solids such as sand, soil, or gravel.

Grout Material injected into a soil or rock formation to change the physical characteristics of the formation. In solidification/stabilization applications, *grout* is a synonym for *binder*.

Grout curtains Containment barrier formed by grout injection.

Halogenated organic compounds Organic compounds that contain halogens such as chlorine, bromine, or fluorine within their structure.

Hazardous and Solid Waste Amendments Enacted in 1984, a set of sweeping amendments to RCRA that include specifications for hazardous waste incineration systems, hazardous waste landfills, and bans on land disposal of hazardous wastes.

Hazardous material Any substance or mixture having properties capable of producing adverse effects on public health or the environment if improperly managed.

Hazardous waste Any material listed as such in Title 40 CFR 261, or that possesses any of the characteristics of corrosivity, ignitability, reactivity, or toxicity as defined in Title 40 CFR 261, or that is contaminated by or mixed with any of the previously mentioned materials (40 CFR 261.3).

Heavy metals Metals of high atomic weight and density, such as lead and cadmium, that are toxic to living organisms.

HSWA Hazardous and Solid Waste Amendments.

Humus Stabilized organic material that remains after microbial degradation of plant and animal matter.

Hydraulic conductivity Measure of the amount of water that can move through a cross-section of material, for example soil, per unit time.

Glossary

Hydrophobic Literally, *water fearing*. A compound that is insoluble in water and soluble in hydrocarbons.

Hydrophilic Literally, *water loving*. A compound that is soluble in water and insoluble in hydrocarbons.

Hygroscopic Property of adsorbing moisture from the air.

Hyperaccumulator Plants that take up toxic elements and accumulate them in aboveground biomass at levels many times the expected concentrations, with little or no adverse effect to the plant.

Immobilization The reduced ability of a contaminant to move through or escape from soil or waste.

Impermeability A state in which fluids, particularly water, cannot penetrate in significant quantities through soil or other media.

Incineration An engineered process using controlled combustion to thermally degrade hazardous waste. Devices commonly used for incineration include rotary kilns and liquid injectors. Incineration is used primarily for the destruction of organic wastes.

Infiltration The entry of fluid into a medium through pores or small openings. Commonly used to denote the entry of water into soil.

Inorganic compounds Chemical compounds that do not contain carbon.

Ion An atom or molecule that has acquired a net electric charge by the loss or gain of electrons.

In situ In place. For example, within the intact soil at a site.

Kaolinite A common 1:1-type clay mineral having the general formula $Al_2(Si_2O_5)(OH_4)$.

Kiln dust Fine particulate by-product of cement production or lime calcination.

Land treatment facility A facility where hazardous waste is applied or incorporated into the soil surface. Such facilities are designated *disposal facilities* if the waste remains after closure.

Landfill An engineered waste disposal facility. Used for disposal of MSW, hazardous waste, or special wastes such as fly ash. Modern landfills are required per the HSWA to possess impermeable liners and systems for leachate collection and removal.

Leachate Any liquid, including any dissolved or suspended components in the liquid, that has percolated through or drained from a material.

LEL Lower explosive limit. See *LFL*.

LFL Lower flammable limit. The lowest concentration of a hydrocarbon vapor in air that can support ignition from a spark or flame.

Light nonaqueous phase liquid Contaminant that is not soluble in water and is less dense than water. LNAPLs float on groundwater.

Liner A protective layer, manufactured of natural or synthetic materials, installed along the bottom and sides of a landfill. The purpose of a liner is to reduce migration of leachate into groundwater beneath the site or laterally away from the site.

Lipophilic Literally, *fat loving*. A nonpolar molecule that dissolves readily in hydrocarbons.

LNAPL See *light nonaqueous phase liquid*.

LTU Land treatment unit or treatment cell used in landfarming of contaminated soil.

LUST Leaking underground storage tank.

Maximum contaminant level The maximum amount of a contaminant in water detectable by standard analytical methods. The Safe Drinking Water Act requires EPA to establish MCLs for water delivered to users of public water systems.

MCL See *maximum contaminant level*.

Metalliferous Metal-enriched.

Metalloid An element possessing properties of both metals and nonmetals. A semimetal, for example, arsenic or selenium.

Microorganisms Microscopic organisms including bacteria, actinomycetes, fungi, some algae, slime molds, protozoa, and some multicellular organisms.

Miscible Soluble in water.

Monolith A freestanding solid.

Montmorillonite A group of 2:1 aluminosilicate clay minerals characterized by a sheet-like structure. These clays swell on wetting, shrink on drying, and possess high cation exchange capacity.

MSW See *municipal solid waste*.

Municipal solid waste (MSW) Solid waste generated at residences, commercial establishments, and institutions. Also known as *domestic solid waste*.

Mutagen An agent that permanently damages genetic material. A substance capable of causing genetic damage.

NAPL Nonaqueous phase liquid.

National Priorities List (NPL) List of CERCLA sites (40 CFR Part 300 Appendix B). Sites that pose the greatest overall hazard to public health and the environment and are given highest priority for funding for cleanup.

Neutralization Process by which the acid or alkaline properties of a solution are reduced by addition of reagents to make hydrogen (H^+) and hydroxide (OH^-) concentrations approximately equal.

Nonaqueous phase liquid (NAPL) Organic liquid that will partition to a distinct organic phase, that is, will not dissolve into water.

Nonpolar An uncharged molecule. A compound that is lipophilic and hydrophobic.

NPL See *National Priorities List*.

Organic Compounds that contain carbon in combination with one or more elements; typically derived from living organisms.

OSHA Occupational Safety and Health Administration. Federal agency established by the Occupational Safety and Health Act of 1970.

Oxidation Chemical reaction that involves removal of electron(s) from an atom, thus resulting in an increase in the atom's oxidation state.

Oxidation/reduction Change in oxidation state of an element resulting from the transfer of electrons.

Oxidizer A chemical that initiates or promotes combustion of other materials via the release of oxygen.

Ozone O_3, a highly reactive form of oxygen.

PAH Polycyclic aromatic hydrocarbon. Petroleum hydrocarbon containing multiple, fused benzene rings. Several PAHs are hazardous to human health and the environment.

Paraffin hydrocarbons Hydrocarbons that contain no carbon-carbon multiple bonds. Paraffin hydrocarbons are also known as *saturated hydrocarbons* or *alkanes*.

Partitioning Distribution of a solute between two or more phases.

Pathogen Any microorganism capable of causing disease.

PCBs Polychlorinated biphenyls. Molecule consisting of two benzene (phenyl) rings connected by a carbon-carbon bridge, and bound to several chlorine atoms.

PCB transformer Any transformer that contains 500 mg/L PCBs or greater (40 CFR 761.3).

PCDF Polychlorinated dibenzofurans. Class of toxic chemical compounds occurring as a thermal degradation product of PCBs. A group of polychlorinated compounds characterized by two benzene rings linked by one oxygen bridge.

PCP Pentachlorophenol, a chlorinated phenol used as a wood preservative.

Percolation Movement of water under hydrostatic pressure or gravity through the interstices of rock, soil, or wastes. Typically involves deep movement into subsurface aquifers.

Permeability A measure of flow of a liquid through the pore structure of soil, strata, or waste. A function of both the fluid and solid media. If the permeating fluid is water, the permeability is termed *hydraulic conductivity*.

Petroleum A naturally occurring mixture of several hundred hydrocarbons. Crude petroleum is refined to produce gasoline, diesel, jet fuel, and other products.

pH A measure of the hydrogen ion concentration of an aqueous solution. A pH of 7.0 is neutral. Higher values indicate alkalinity and lower values indicate acidity.

PID Photoionization detector.

Plume The amorphous and mobile volume of a contaminant in soil and/or groundwater.

Polar A charged molecule; hydrophilic.

Polychlorinated biphenyls (PCBs) Any of 209 compounds or isomers of the biphenyl molecule that have been chlorinated to various degrees. Regulated under 40 CFR 761.3.

Polymerization Chemical reaction in which a large number of simple molecules combine to form a chain-like macromolecule. This reaction can occur with release of a significant amount of heat.

Pore A cavity or void in a solid.

Pore size distribution The total range of pore sizes in solids. Soils and geologic strata have a unique pore size distribution and associated permeability.

Porosity The ratio of the volume of voids in rock, soil, etc., to the total volume of the material.

Portland cement A cement produced by pulverizing clinker consisting of calcium silicates ($CaO \cdot SiO_2$) and aluminum- and iron-containing compounds.

Potentially Responsible Party PRP, the individual or organization that is potentially liable for the contamination and cleanup costs of a CERCLA site.

POTW Publicly owned treatment works. A municipal wastewater treatment plant or a sewage treatment plant.

Pozzolan A siliceous or siliceous/aluminous material which possesses little or no cementitious value but will, in finely divided form and in the presence of moisture, chemically react with calcium hydroxide to form cementitious compounds.

ppb Parts per billion. A unit for measuring concentration. Also µg/kg on a solids basis, or µg/L on a liquid basis.

ppm Parts per million. A unit for measuring concentration. Also mg/kg on a solids basis, or mg/L on a liquid basis.

Precipitation Process in which dissolved or suspended matter in water aggregates to form solids that separate from the liquid phase by gravity.

Pyrophoric Chemical that will ignite spontaneously in air.

RCRA See *Resource Conservation and Recovery Act*.

Recalcitrant Difficult to degrade, whether in the context of chemical or biological degradation.

Redox Oxidation-reduction.

Reduction Chemical reaction in which an atom gains electrons, thereby decreasing its oxidation state. Opposite of an oxidation reaction.

Resource Conservation and Recovery Act (RCRA) Comprehensive set of federal regulations enacted in 1976 that address the proper management of solid and hazardous waste.

Rhizosphere Zone directly adjacent to the plant root. An area active biologically and chemically due to the release of compounds from the root and by the presence of microbial biomass.

Rotating biological contactor Mechanical unit in which contaminated water is treated aerobically via promoting the formation of microbial films on thin plastic disks that are continuously rotated through the contaminated water and air.

Semimetal. See *metalloid*.

Sheet piles Vertical groundwater barriers constructed by driving piling, often steel or concrete, into the subsurface.

Sludge A solid, semisolid, or liquid waste generated from a municipal, commercial, or industrial wastewater treatment plant, water supply treatment plant, or air pollution control facility, with certain exclusions (40 CFR 260.10).

Slurry Fluid mixture of water and fine insoluble particles.

Slurry wall Vertical barrier composed of slurry material and constructed in a trench.

Smoke An air suspension (aerosol) of particles, often originating from combustion. Smoke generally contains liquid droplets and solid particles.

Soil flushing Process involving extraction and injection of aqueous solutions to remove contaminants from the subsurface in situ, that is, without excavation of the contaminated material.

Soil permeability Ease with which water can pass through soil.

Soil washing System of reacting contaminated soil with a selected extraction solution in order to remove a contaminant. A physical or chemical separation, often carried out in a reactor vessel, that is, ex situ.

Solidification Process in which a binder is added to contaminated soil or hazardous waste to convert it to a solid and/or to improve its handling and physical properties. The process may or may not involve chemical bonding between the soil, contaminants, and binder.

Solidification/stabilization (S/S) A treatment process which inhibits mobility or interaction in the environment through chemical reactions and/or physical interactions to retain or stabilize a contaminant.

Solubility The maximum concentration of a substance dissolved in a solvent at a given temperature.

Solubility product An equilibrium constant defined for equilibria between solids and their respective ions in solution.

Sorption The processes by which an element, ion, or compound attaches to the surface of a particle by physicochemical processes.

S/S-treated waste A waste liquid, slurry, or sludge that has been converted to a stable solid (granular or monolithic) by an S/S treatment process.

Stabilization Process by which a waste is converted to a more chemically stable form. The term may include solidification, but emphasizes chemical changes to reduce contaminant mobility.

Storage tank Any manufactured, non-portable, covered device used for containing pumpable hazardous waste.

Superfund The Comprehensive Environmental Response, Compensation and Liability Act (CERCLA). Also refers to sites listed on the National Priorities List (NPL).

Surface impoundment Any natural depression or excavated and/or diked area built on the land which is fixed, uncovered, and lined with soil or synthetic material, and is used for treating, storing, or disposing of wastes. Examples include holding ponds and aeration ponds.

Surfactant Surface-active agent, a soluble compound that reduces the surface tension of liquids or reduces interfacial tension between two liquids or a liquid and a solid.

SW-846 Test Methods for Evaluating Solid Waste, Physical/Chemical Methods. SW-846 is a compendium of approved test methods, sampling, and monitoring guidance for use in solid waste analysis.

Synthetic liner Landfill or lagoon liner manufactured of polymeric materials, for example, polyvinyl chloride.

TCLP See *Toxicity Characteristic Leaching Procedure*.

Teratogen A physical or chemical agent that causes nonhereditary birth defects.

Toxic Substances Control Act Law passed by the U.S. Congress in 1976 and administered by EPA that regulates the introduction of new or already existing chemicals to the market. Polychlorinated biphenyls (PCBs) are regulated under this Act.

Toxicity Capacity of a substance to produce injury or illness through ingestion, inhalation, or absorption through any body surface.

Toxicity Characteristic Leaching Procedure (TCLP) An analytical extraction and test to predict the leaching potential of landfilled hazardous contaminants in solid waste.

TPH Total Petroleum Hydrocarbons. Refers to U.S. EPA Method 418.1 or 8015, which describe the procedures for quantifying the petroleum hydrocarbon content of a sample.

TSCA See *Toxic Substances Control Act*.

Transpiration Release of water and gases from a green plant for eventual return to the atmosphere.

UEL Upper explosive limit. See *UFL*.

UFL Upper flammable limit. The highest concentration of a hydrocarbon vapor in air that will support combustion.

Underground storage tank A tank regulated under RCRA Subtitle I to store petroleum products or hazardous materials.

UST See *underground storage tank*.

United States Environmental Protection Agency The main federal agency charged with setting regulations to protect the environment.

Vapor Dispersion of a liquid or solid in air at standard temperature and pressure.

Vapor density The ratio of the vapor weight of a substance compared to that of air. If the ratio is greater than 1, the vapors are heavier than air and may settle to the ground. If less than 1, the vapors will rise.

Vapor pressure The pressure of a vapor in equilibrium with the liquid phase at a specified temperature. High vapor pressure indicates high volatility.

Vegetative cover Plant growth occurring on soil, spoils, landfill covers, and so on.

Vegetative uptake Metals are taken up through the root systems of plants and, in some cases, translocated.

Vertical barrier A rigid structure placed at the perimeter of a contaminated site. Reduces movement of contaminated groundwater off-site or limits the flow of uncontaminated groundwater through the site.

Virus Small particle typically composed of a strand of ribonucleic acid in a protein coat.

Viscosity Resistance of a material to flow.

Vitrification Technology that utilizes high-temperature treatment for reducing the mobility of metals and other contaminants in soil by incorporation within a vitreous (glasslike) monolith.

Volatile organic compound (VOC) Organic compound with a low boiling point. Converts readily from the liquid to the gaseous phase at ambient conditions.

Volatile matter Material capable of being vaporized or evaporated quickly.

Wastewater Contaminated process water from the treatment of wastewater, soil, sediment, and sludges.

Xenobiotic Anthropogenic compound considered foreign to the environment. From the Greek *xenos* (strange). Often recalcitrant and hazardous.

Index

Abd-Elfattah, A., 34
Aberdeen Proving Grounds,
    phytoremediation for, 332–36, *335*
aboveground storage tanks (ASTs):
    aerial photographs of, 114;
    nanoremediation for, 374–76;
    phase I ESAs for, 119
accumulators, 321
acetone, 236
acetylene (ethyne), 56
ACI. *See* American Concrete Institute
Acid Extraction Treatment System (AETS), 174
acid rain, 80
ACM. *See* asbestos-containing material
actinomycetes, in soils, 92–93, *93*
activated sludge, for liquid phase bioremediation, 290, *290*
ADA. *See* N-(acetamido)iminodiacetic acid
adsorption:
    of GAC, 167, 234–36;
    of lead, 32–35;
    of PAHs, 57–58;
    PRBs for, 251–52
aeration, in slurry bioremediation, 288–89, *289*, 303–4

aerial photographs, for phase I ESAs, *114*, 114–15, *115*
aerobic respiration, 266, *268*
AETS. *See* Acid Extraction Treatment System
Ahn, C. K., 179
Ahnstrom, Z., 182
Air Quality Act, 4
air sparging, *227*
air stripping, 166–67
Al. *See* aluminum
aliphatics, 52–56;
    in lubricating oils, 67–68
alkanes, 53–54, 60, *69–70*;
    bioremediation of, 272
alkenes, 55–56
alkylaromatics, 57
alkynes, 56
allophane, 85
aluminosilicates:
    arsenic and, 26;
    PRBs for, 251;
    selenium and, 39
aluminum (Al):
    in clays, 82–83, *83*;
    soil pH and, 80–81;
    for SOMS, 371
aluminum oxides:
    arsenic and, 26;

chromium and, 31;
    in hydrous oxides, 84–85
American Concrete Institute (ACI), 213–14
American Society for Testing and Materials (ASTM):
    on cement based-S/S, 202;
    on ESAs, 109;
    on S/S, 213–14
AMF. *See* arbuscular mycorrhizal fungi
ammonia:
    anaerobic respiration and, 267;
    in fertilizers, 91;
    RBCs for, 295
anaerobic respiration, 266–67, *268*
aniline, 58, *70*
anthracene, 57
antimony, 247
apatite, 250
APS. *See* ATP-sulfurylase
aquicludes, 98, 101;
    SERP and, 353
aquifers, 99, *99*;
    bioremediation for, 397;
    SVE for, 225
aquitards, 98
arbuscular mycorrhizal fungi (AMF), 415–16, *416*
aromatic hydrocarbons, 56–57, 61;
    bioremediation of, 272;
    in gasoline, 64–65;
    for petrochemicals, *69–70*;
    PRBs for, *245*;
    solubility of, 71;
    SVE for, 221.
    See also polycyclic aromatic hydrocarbons
arsenate:
    cadmium and, 29;
    PRBs for, 249;
    soil flushing of, 173
arsenic (As), *25*, 25–27, *26*;
    cement-based S/S for, 202;
    in lubricating oils, 68;
    organic S/S for, 205;
    PBRs for, 251;
    S/S for, 210
arsenite, 173
arsine, 25–26
As. *See* arsenic
asbestos-containing material (ACM), 17;
    in phase I ESAs, 123–26, *124*, *125*
Asel Gzar, H., 174, 176
asphalts, 63;
    bioremediation of, 272;
    organic S/S of, 204
ASTM. *See* American Society for Testing and Materials
ASTs. *See* aboveground storage tanks
ATP-sulfurylase (APS), 323
autotrophs, 263

B. *See* boron
Ba. *See* barium
backfill:
    bioremediation and, 305;
    GPR for, 129;
    for S/S, 211
bacteria:
    for bioremediation, 264, 274, *276*;
    hydrocarbons and, 52;
    mercury and, 36;
    nanoremediation and, 366;
    in soils, 88–90, *89*;
    tillage and, 90
Baker, A. J. M., 320
barite, 27
barium (Ba), 27–28;
    in lubricating oils, 68
base, neutral, and acid extractables (BNAs), 224
batteries:
    lead from, 32, *32*, *33*;
    zinc in, 40.
    See also nickel-cadmium batteries
Baziar, M., 176, 179
beidellite, 83
bench-scale tests, for electrokinetic remediation, 348–49

Index

bentonite, 226
benzene, 52, *70*;
   aromatic hydrocarbons and, 56, 57–58;
   bioremediation of, 270;
   in gasoline, 64;
   in groundwater, 71;
   PAHs and, 57;
   redox for, 350;
   from soil flushing, 178;
   SVE for, 221, 231
benzene, toluene, ethylbenzene, and xylene (BTEX):
   bioremediation of, 284;
   in gasoline, 64–66;
   phase II ESAs and, 134;
   SVE for, 224
benzo[a]pyrene, 52, 57
benzoic acid, 58
Bhopal, India, 7
bichromate, 30
bimetallic nanoscale particles (BNPs), 367
bioaugmentation, 273
biochemical oxygen demand (BOD):
   RBCs and, 294;
   soil flushing and, 186;
   trickling filters and, 294
biofilters, for SVE, 231
BioGenesis™, 179–80, *180*
bioreactors:
   for groundwater, 166;
   for liquid phase bioremediation, 289–90
bioremediation, 263–311, 395–97, *396*, *397*;
   bacteria for, 264, 274, *277*;
   at Burlington Northern Superfund site, 306–9, *308*, *309*;
   carbon dioxide for, 266;
   clogging in, 283–84;
   contaminant properties for, 267–68;
   exogenous microorganisms for, 273;
   field technologies for, 273–301;
   forced injection for, 280;
   of free product, 277;
   at French Limited Superfund Site, 302–6, *303*, *305*;
   fungi for, 264;
   groundwater for, 277–79, 284, 300;
   hydraulic conductivity and, 284;
   of hydrocarbons, 268–72, *269–71*;
   indigenous microorganisms for, 272–73;
   liquid phase, 291–97, *292–94*, *296*;
   metals and, 284–85;
   of NAPLs, 277;
   nutrients for, 266–69, *265*, 277, *277*, 297–98;
   off-gases from, 284;
   oxygen for, 265, 281;
   pH and, 263, 266, 267, 279, 283, 289;
   principles of, 264;
   reagents for, 280, 281–83;
   recovery systems for, 284;
   requirements for, 264–70;
   in situ processes for, 273–84, *276*;
   slurry for, 286–91, *286*, 302–7, *303*, *305*;
   subsurface application for, 280–81;
   surface application for, 280;
   trace metals for, 267;
   water for, 265–66;
   yeasts for, 275.
   See also landfarming
biosolids:
   as soil amendments, *409*, 413;
   zinc from, 40, 41
Bipp, H. P., 176
BNAs. *See* base, neutral, and acid extractables
BNPs. *See* bimetallic nanoscale particles
BOD. *See* biochemical oxygen demand
boron (B), FGD and, 415
Brown, G. A., 177
brownfields, 10–12, *11*;
   phytoextraction of, 322;
   revegetation of, 420
BTEX. *See* benzene, toluene, ethylbenzene, and xylene
buffering capacity, of soils, 81

bulk density, of soils, 97
Burlington Northern Superfund site,
    bioremediation at, 306–9, *307–9*
butadiene, *69*
butane, 221
butanol, *70*

CAA. *See* Clean Air Act
cadmium (Cd), 28–30;
    cement-based S/S for, 200–2;
    electrokinetic remediation of, 347;
    phytoextraction of, 322, 325;
    soil amendments and, 419.
    See also nickel-cadmium batteries
CAFOs. *See* concentrated animal
    feeding operations
calcium:
    for landfarming, 296;
    zinc and, 40–41
calcium carbonate (limestone):
    for arsenic toxicity, 27;
    barium and, 27–28;
    for bioremediation, 277;
    cadmium and, 29;
    FGD and, 415;
    lead and, 34;
    for PRBs, 248;
    for soil amendments, *410*;
    soil flushing and, 185;
    soil pH and, 81
capillary fringe, *98*, 98–99;
    SVE for, 223
capping systems, for isolation, 160–64,
    162, *164*
carbon:
    alkane and, 53–54;
    autotrophs and, 263;
    in humus, 85;
    from manures, 296;
    for nanotubes, 370–71, *371*;
    in paper mill sludge, 414;
    for PBRs, 252;
    for SVE, 228
carbonates:
    cement-based S/S for, 201;
    of lead, 32;
    lead and, 33, 35;
    metal precipitation from, 168;
    phytoextraction and, 322;
    zinc and, 40–41.
    See also calcium carbonate
carbon dioxide:
    for bioremediation, 265;
    FIDs and, 145
carbonic acid, 174
carbon monoxide:
    MTBE and, 66;
    NAAQS and, 4
carbon tetrachloride:
    metalloporphyrinogens for, 371;
    PRBs for, 246;
    SVE for, *234*, 234–35
Carter, Jimmy, 8
catalytic oxidation units, for SVE, 230
cation exchange capacity (CEC):
    barium and, 27–28;
    cadmium and, 29;
    GPR and, 129;
    in humus, 85, 86;
    metal from soil flushing and, 174;
    soil flushing and, 184;
    soil pH and, 81
CCA. *See* chromated-copper-arsenate
Cd. *See* cadmium
cDCE. *See cis*-dichloroethene
CEC. *See* cation exchange capacity
cement:
    for S/S, 200–2, *203*;
    for SVE, 226–27
CERCLA. *See* Comprehensive
    Environmental Response,
    Compensation, and Liability Act
    of 1980
CERCLIS. *See* Comprehensive
    Environmental Response,
    Compensation, and Liability
    Information System
cetane, in diesel fuel, 67
chelating agents:
    chromium and, 32;

lead and, 35;
nickel and, 38;
PRBs and, 242, 394;
for soil flushing, 173, 186;
for soil flushing of metals, 173–77, *175*
chemical weapons, 2
chemotrophs, 263
Chen, T. C., 176
Chen, Y., 369
chloride (Cl):
  aliphatics and, 54–55;
  cadmium and, 29;
  electrokinetic remediation of, 349;
  organic S/S for, 203
chloroform, 246
chlorosis, 38
chlorosulfonated polyethylene (CSPE), 161
chromate, 30;
  cadmium and, 29;
  PRBs for, 248–49;
  in soils, 31;
  S/S for, 209
chromated-copper-arsenate (CCA), 211
chromium (Cr), *24*, 30–32;
  cement based-S/S for, 202;
  dendrimers for, 370;
  electrokinetic remediation of, 349;
  ferritin for, 371;
  in groundwater, 249;
  in situ S/S for, 205–6;
  phytostabilization for, 331;
  for plants, 319
Churchill, Winston, 2
cis-benzene dihydrodiol, 270
cis-dichloroethene (cDCE), 246;
  nanoremediation for, 377;
  phytoremediation of, *335*
Cl. *See* chloride
clays, *84*;
  anaerobic respiration in, 266–67;
  arsenic and, 26;
  barium and, 28;
  cadmium and, 29;

colloidal fraction of, 82–83, *83*;
GPR for, 129;
hydraulic conductivity of, 101–4;
lead and, 32;
liners, for SVE, 229;
nickel from, 38;
PAHs in, 58;
for PBRs, 251–52;
SVE in, 231;
texture of, 94–95;
zinc and, 40–41
Clean Air Act (CAA), 3–4
Clean Water Act (CWA), 5;
  soil amendments and, 419
Co. *See* cobalt
coal tars, 57
cobalt (Co), 319
colloidal fraction, of soils, 82–86
colorimetric detector tubes, 143–45, *144*
co-metabolism, 330
Comprehensive Environmental Response, Compensation, and Liability Act of 1980 (CERCLA), 12–16;
  ESAs and, 109–10;
  on gasoline, 66;
  groundwater and, 165;
  SVE and, 233–34
Comprehensive Environmental Response, Compensation, and Liability Information System (CERCLIS), 117, *117*
Conca, J., 251
concentrated animal feeding operations (CAFOs), for soil amendments, 408, *409*
condensation, SVE and, 229–30
confined aquifers, 99, *99*
confining layers, 98;
  for SERP, 353
constructed wetlands, 318;
  for groundwater, 166, *167*
containment. *See* isolation
continuous reactive barriers, for PRBs, 242–43, *243*

copper (Cu):
  cement based-S/S for, 200;
  dendrimers for, 370;
  for plants, 319
Cr. *See* chromium
crude petroleum:
  composition of, *60*, 60–62;
  refining of, *62*, 62–64, *63*;
  SVE for, 221
CSPE. *See* chlorosulfonated polyethylene
cumene, 57;
  in gasoline, 66
CWA. *See* Clean Water Act
cyanide:
  organic S/S for, 203;
  RBCs for, 293;
  soil flushing of, 173
cycloalkanes, 61;
  bioremediation of, 270, *270*, 272
cyclohexane, 54;
  in dendrimers, 370

Darcy, Henri, 102–3
Da Vinci, Leonardo, 79
daylighting EZVI, 375
DDT, 2
dendrimers, 369–71, *370*
dense nonaqueous phase liquids (DNAPLs), 71;
  EZVI for, 368;
  in groundwater, 100;
  nanoremediation for, 374–75;
  phase II ESAs and, 132;
  SVE for, 223
Department of Defense (DOD), 10;
  phytoremediation and, 322
Department of Energy (DOE), 10;
  phytoremediation and, 322
Dewey, John, 345
Diamond Vogel Paint Company, 18
dichromate, 30
dickite, 83
diesel fuel, 51–52, 63, 67;
  landfarming for, 301;
  PAHs from, 57;
  SVE for, 221, 394;
  viscosity of, 73
diethylene triamine pentaacetic acid (DTPA), 174–77, *175*
diffuser stacks, for SVE, *229*, 231
Dillard, Annie, 199
dimethylarsine, 25
DNAPLs. *See* dense nonaqueous phase liquids
DOD. *See* Department of Defense
DOE. *See* Department of Energy
Doyle, Arthur Conan, 107
DTPA. *See* diethylene triamine pentaacetic acid
due diligence, CERCLA and, 13–14
Dynamic Underground Stripping, 354

Earth Day, 4
EDB. *See* ethylene dibromide
EDC. *See* ethylene dichloride
EDDS. *See* [S,S] ethylenediaminedisuccinic acid
EDTA. *See* ethylenedinitrilotetraacetic acid
EDX. *See* energy dispersive X-ray
electrical resistivity:
  for phase II ESAs, 131;
  for SVE, 221–23
electrokinetic remediation, *246*, 345–50, *346*, *347*
electromagnetics (EM), for phase II ESAs, 130–31, *130*
electron transport chain, 266, *266*
Elliott, H. A., 177
EM. *See* electromagnetics
Emergency Planning and Community Right-to-Know Act (EPCRA), 7
Emerson, Ralph Waldo, 317
emulsified zerovalent iron (EZVI), 367–68, *368*, 374–75, *376*
energy dispersive X-ray (EDX), 126;
  for S/S, 209
Energy Policy Act of 2005, 66

Environmental Protection Agency
    (EPA), 4–5;
  ACM and, 17;
  AETS of, 174;
  on benzene in groundwater, 71;
  on bioremediation, 300;
  on cadmium, 30;
  CERCLA and, 13;
  on cost of remediation, 385–86;
  Health Hazard Evaluation by, 17;
  HRS of, 211;
  on lead in gasoline, 64;
  on MTBE, 66;
  on radon, 121;
  RCRA and, 18;
  TSCA and, 7;
  Used Oil Management Standards of, 68;
  USTs and, 6–7
environmental site assessments (ESAs), 107–49, *109*;
  CERCLA and, 109–10. *See also* phase I ESAs;
  phase II ESAs
EPA. *See* Environmental Protection Agency
EPCRA. *See* Emergency Planning and Community Right-to-Know Act
ESAs. *See* environmental site assessments
ethane:
  as alkane, 53;
  BNPs for, 367;
  PRBs for, *245*, 246
ethanol, 66;
  for soil amendments, *410*
ethene:
  nanoremediation for, 375;
  PRBs for, *245*, 246
ethylbenzene, 57, 58;
  in gasoline, 64;
  from soil flushing, 178;
  SVE for, 221.
  See also benzene, toluene, ethylbenzene, and xylene

ethylene, *69*
ethylene dibromide (EDB), 64
ethylene dichloride (EDC), 64
ethylenedinitrilotetraacetic acid (EDTA):
  phytoextraction and, 322;
  for soil flushing, 185;
  for soil flushing of metals, 174–77, *175*
ethyne (acetylene), 56
Evangelou, M., 176
excluders, 319
Exide Technologies, 18
exogenous microorganisms, for bioremediation, 272
explosive limits, of hydrocarbons, 73–74, *75*
extinctions, 1
extraction processes. *See* soil flushing
Exxon Valdez oil spill, 5
EZVI. *See* emulsified zerovalent iron

Fairchild Superfund site, SVE at, 235–38, *237*, *238*
Farrah, H., 33
Federal Communications Commission (FCC), 357
Federal Water Pollution Control Act (FWPCA), 5
fermentation, 267, *268*;
  redox and, 81
ferritin, 371
ferrous sulfate, 27
fertilizers:
  ammonia in, 91;
  bacteria and, 90;
  fungi and, 91;
  nickel from, 38;
  soil pH and, 80;
  zinc from, 40–41
FGD. *See* flue gas desulfurization sludge
FIDs. *See* flame ionization detectors
Fischer, K., 176
fish, mercury in, 35

flame ionization detectors (FIDs), 145
flashpoint, of hydrocarbons, 74
flue gas desulfurization sludge (FGD), 414–15
fly ash:
  selenium from, 39;
  as soil amendment, 414–15
forced injection, for bioremediation, 280
formaldehyde, *69*;
  for organic S/S, 203
Forster, E. M., 151
fractional distillation, of crude petroleum, 62–64, *63*
free product, 132;
  bioremediation of, 275
French Limited Superfund Site, bioremediation at, 301–6, *303*, *305*
friability, of ACM, 124
fuel oil, 67;
  landfarming for, 300–301;
  phase II ESAs and, 134;
  specific gravity of, 71;
  SVE for, 221–40;
  viscosity of, 73
fungi:
  for bioremediation, 262;
  for soil amendments, 415–16, *416*;
  in soils, 90–92, *91*
fungicides, mercury from, 35, 38
funnel-and-gate system, for PRBs, 242, *243*
FWPCA. *See* Federal Water Pollution Control Act

GAC. *See* granulated activated carbon
galena, 32
gas chromatograph (GC), *139*, 145–46, *146*
gasoline, 51–52, 64–66, *65*;
  alkylaromatics in, 57;
  cyclohexane in, 54;
  landfarming for, 301;
  octane number for, 64;
  PAHs from, 57;

phase II ESAs and, 134;
from soil flushing, 178;
solubility of, 71;
specific gravity of, 71;
SVE for, 221, 232, 394;
viscosity of, 73;
weathering of, *135*
Gatea, I. M., 176
GC. *See* gas chromatograph
genetic engineering, for phytoextraction, 323–24
geobotanical indicators, 319
geomembranes, for SVE, 228–29
Giannis, A., 176
gibbsite, 85
Gidarakos, E., 176
goethite, 85
gold:
  mining, mercury from, 35;
  for SOMS, 371–72
GPR. *See* ground-penetrating radar
granulated activated carbon (GAC):
  adsorption, 167, 233–34;
  with in situ bioremediation, 276;
  pneumatic fracturing and, 355–56;
  for SVE, 233–35
grasses, revegetation of, *422*, 421–23
ground-penetrating radar (GPR), 128–29, *129*
groundwater:
  air sparging for, 221;
  air stripping for, 166–67;
  benzene in, 71;
  bioreactors for, 166;
  for bioremediation, 275–77, 282, 298;
  bioremediation for, 395;
  constructed wetlands for, 166, *167*;
  electrokinetic remediation for, 346;
  flow of, 100–101, *100*, 103;
  GAC adsorption for, 167;
  ion exchange for, 167–68;
  isolation and, 152–54, *153*, *154*, 165–68;
  landfarming and, 300–301;

manures and, 413–14;
mercury in, 36;
metalloporphyrinogens for, 371;
metal precipitation for, 168;
MTBE in, 134;
PAHs in, 58;
phase II ESAs and, 133;
PRBs and, 243;
pump-and-treat technology for, 165–68;
recharge of, 101;
sand and, 102;
SERP for, 351–54, *353*;
soils and, 98–104;
sprinkler irrigation for, 168;
SVE for, 222, 225–26.
See also specific applicable topics
grout curtains, for isolation, 158–59, *159*
Gull, William Withey, 365

halides:
aliphatics and, 54–55;
alkenes and, 56
halloysite, 83
halogens:
aliphatics and, 54–55;
alkenes and, 56;
aromatic hydrocarbons and, 58–59;
RBCs for, 294
Hanford Site, SVE at, 233–35
Hazardous and Solid Waste Amendments of 1984 (HSWA), 6
hazardous waste:
by Diamond Vogel Paint Company, 18;
hydrocarbons in, 52;
by Wal-Mart, 19;
zinc from, 40–41
Hazard Ranking System (HRS), of EPA, 211
HCl. *See* hydrochloric acid
HDPE. *See* high-density polyethylene
Health Hazard Evaluation, by EPA, 17
heavy metals, 4;

phytoextraction of, 322;
PRBs for, 247–49;
SDWA on, 5;
slurry bioremediation and, 284–87
Henry's law constant, 231
heptamethylnoname, 67
Hessling, J. L., 176
heterotrophs, 265
hexane, 221
hexavalent chromium, 30
Hg. *See* mercury
high-density polyethylene (HDPE), 161;
for landfarming, 300–301;
for S/S, 209;
for SVE, 229
Hong, A., 176
Hooker Chemical Company, 7–8, *8*
hot gas injection, 356
HRS. *See* Hazard Ranking System
HSWA. *See* Hazardous and Solid Waste Amendments of 1984
humus, 85–86, *86*, *87*;
arsenic and, 26;
barium and, 28;
mercury and, 36
hydraulic conductivity:
bioremediation and, 284;
for hydrocarbons, *296*;
of soils, 101–4, *102*, 138, 230;
SVE and, 232
hydrocarbons, *69–70*;
bioremediation of, 267–72, *267–69*, 396;
chemistry and properties of, 51–77;
ESAs for, 109;
explosive limits of, 73–74, *75*;
ferritin for, 371;
flashpoint of, 74;
fuels from, 59–68, *60*, *62*, *63*;
landfarming for, 295–301, *296*, *297*;
NAPLs and, 121;
phytoremediation of, 323–24;
PRBs for, 395;
products from, *61*;
properties of, *72*;

RBCs for, 293;
    from soil flushing, 178–79;
    solubility of, 71;
    specific gravity of, 71;
    S/S for, 208, 209;
    structure and nomenclature of, 52–59;
    SVE for, 229–30, 355–56;
    vapor pressure of, 73, *74*;
    vapors from, phase II ESAs and, 133;
    viscosity of, 73.
    See also specific types
hydrochloric acid (HCl), 182, 354;
    for soil flushing of metals, 174
hydrogen peroxide, 230;
    for bioremediation, *276*, 280
hydroxides:
    arsenic and, 26;
    cement-based S/S for, 201;
    lead and, 32;
    metal precipitation from, 168;
    PRBs for, 248;
    S/S for, 210;
    zinc and, 40–41;
    ZVI and, 254
hyperaccumulators, 319–30, *320*;
    genetic engineering and, 323–24

iminodisuccinic acid (IDSA), 185
incinerators. *See* municipal solid wastes incinerators
indicator plants, 319
indigenous microorganisms, for bioremediation, 272–73
infiltrate, 98
injection grouting, for in situ S/S, 207
in-place mixing, for in situ S/S, *206*
insecticides, 2
in situ processes:
    for bioremediation, 273–84, *276*;
    for nanoremediation, 372–74, *373*
in situ vitrification (ISV), 358–62, *359*, *361*, *362*
International Society of Soil Science (ISSS), 94

International Union of Soil Sciences (IUSS), 94
ion exchange, for groundwater, 167–68
iron:
    anaerobic respiration and, 267;
    for BNPs, 367;
    for PBRs, 250;
    for plants, 319;
    soil flushing and, 186.
    See also zerovalent iron
iron oxides:
    arsenic and, 26;
    chromium and, 31;
    in hydrous oxides, 84–85;
    lead and, 34;
    selenium and, 39–40
iso-alkanes, 54
isobutane, 54
isolation (containment), 151–69, 386–88, *387*, *388*;
    capping systems for, 160–63, *162*, *164*;
    equipment for, 164–65;
    general design for, 164;
    groundwater and, 152–54, *153*, *154*, 165–68;
    grout curtains for, 158–59, *159*;
    monitoring of, 165;
    PRBs for, 241, 245–46;
    sheet piling for, 159–60, *160*;
    slurry walls for, 155–58, *155–58*;
    subsurface barriers to, 154–60, *155–60*
iso-octane, 64
ISSS. *See* International Society of Soil Science
IUSS. *See* International Union of Soil Sciences

jet fuel, 63;
    from soil flushing, 178;
    SVE for, 221;
    viscosity of, 73
Jung, C. G., 241

K. *See* potassium
Kabata-Pendias, A., 27
kaolinite, 82, 83
Keon, N. H., 182
kerosene, 63, 67;
    landfarming for, 301;
    SVE for, 394
Kin-Buc Landfill, New Jersey, 8, *9*
King of Prussia Technical Corporation, soil flushing at, 187–91, *189, 190*

lactone, 270
Lam, C. W., 378
land disposal restrictions (LDRs):
    in situ S/S and, 205;
    landfarming and, 295
landfarming, *296*, 399–400;
    for gasoline, 301;
    groundwater and, 300–301;
    for hydrocarbons, 297–301, *294*;
    for methylene chloride, 309, *309*;
    for PAHs, 301, 306–9, *307, 309*;
    PIDs for, 300;
    tillage for, 299, 307–8
landfills:
    capping systems for, 161;
    electrokinetic remediation for, 346;
    lead from, 32–35;
    nickel from, 38–39;
    radioactive waste in, 18
land treatment unit (LTU), 295–301, *296, 297*, 307–9, *308, 309*
LBP. *See* lead-based paint
LDRs. *See* land disposal restrictions
lead (Pb), *32*, 32–35, *33*;
    cement-based S/S for, 200–2;
    dendrimers for, 370;
    EDTA for, 177;
    in gasoline, 64;
    from incinerators, 3;
    in lubricating oils, 68;
    NAAQS and, 4;
    in phase I ESAs, 122–23, *123*;
    phytoextraction of, 326;
    phytostabilization for, 331;
    Romans and, 1;
    soil amendments and, 419;
    S/S for, 211
lead-based paint (LBP), in phase I ESAs, 121–22, *122*
legumes, revegetation of, 423, 424–25
LEL. *See* lower explosive limit
Leštan, D, 177
light nonaqueous phase liquids (LNAPLs), 71;
    in groundwater, 100;
    isolation and, 151;
    SERP for, 353;
    slurry walls for, 155;
    SVE for, 223
lime:
    for bioremediation, 277;
    for nickel toxicity, 39;
    for soil amendments, *411*;
    soil pH and, 81;
    for zinc toxicity, 41
limestone. *See* calcium carbonate
Lin, D., 380
liquid phase bioremediation, 289–95, *290–92, 294*
Liu, L., 174
Liu, R., 330
LNAPLs. *See* light nonaqueous phase liquids
Longmire, P. A., 251
Love Canal, 7–8, *8*
lower explosive limit (LEL), 73–74, *75*
LTU. *See* land treatment unit
lubricating oils, 67–68;
    landfarming for, 300–301;
    phase II ESAs and, 134;
    SVE for, 394

Macauley, E., 176
macropores, 97
Madadian, E., 179
magnesium, 82–83
maleic anhydride, *70*
manganese (Mn):
    barium and, 27;
    lead and, 34;
    for plants, 319

manifest system, of RCRA, 6
manures:
    actinomycetes and, 92–93;
    carbon from, 298;
    as soil amendments, *409*, 413–14
maximum concentration limits (MCLs), *73*
Maxwell, Elaine, 385
MCLs. *See* maximum concentration limits
McRae, C. W., 249
mercury (Hg), 35–38, *37*;
    Minamata disease from, 2–3;
    S/S for, 209
mesophiles, 90, 92
metallic chromium, 30
metalloporphyrinogens, 371
metals:
    bioremediation and, 282–83;
    chemistry of, 23–49, *25*;
    ISV for, 358;
    phytoremediation of, 319–28, *320*, *321*, *323*;
    phytostabilization and, 331;
    RFH and, 356–58;
    soil flushing for, 173–77, *175*.
    See also heavy metals; trace metals
methane:
    as alkane, 53–54, 60;
    anaerobic respiration and, 265;
    PRBs for, *245*
methylarsinic acid, 25
methylation, of mercury, 36–38
methylbenzene, 57
methylene chloride:
    landfarming for, 308–9, *309*;
    PRBs for, 248
methylglycinediacetic acid (MGDA), 185
methylhexane, 61
methyl tert-butyl ether (MTBE), 66;
    phase II ESAs and, 134;
    SVE for, 232
MGDA. *See* methylglycinediacetic acid
microencapsulation, 202

micropores, 97
middle distillates:
    of crude petroleum, 63;
    phase II ESAs and, 134
Minamata disease, from mercury, 2–3
MixFlo, 303–4
Mn. *See* manganese
molybdenum (Mo):
    PBRs for, 251;
    for plants, 319
monoaromatics, 68
montmorillonite, 83;
    selenium and, 39–40;
    zinc and, 40–41
MSWs. *See* municipal solid wastes
MTBE. *See* methyl tert-butyl ether
multilayer caps, 163, *164*
municipal solid wastes (MSWs), 4;
    cadmium from, 28;
    CERCLA and, 13;
    chromium from, 28;
    HSWA for, 6;
    incinerators, 3;
    Resource Recovery Act of 1970 for, 5;
    selenium from, 39–40;
    zinc from, 40–41.
    See also landfills
mycorrhiza, 35

N. *See* nitrogen
NAAQS. *See* national ambient air quality standards
N-(acetamido)iminodiacetic acid (ADA), 174, 175
nacrite, 83
NADP. *See* nicotinamide adenine dinucleotide phosphate
n-alkanes, 54;
    bioremediation of, 271–72
nanoparticles (NPs), 366–72
nanoremediation, 366–72, *366*, 400, *402*;
    hazards of, 377–79;
    in situ processes for, 372–74, *373*;

NPs for, 366–72;
  at Parris Island Marine Corps
    facility, 374–75;
  ZVI for, 367
nanotubes, 368–69, *369*
naphthalene, 57, *70*;
  in lubricating oils, 68;
  S/S for, 209;
  SVE for, 232
NAPLs. *See* nonaqueous phase liquids
national ambient air quality standards
    (NAAQS), 3
National Pollutant Discharge
    Elimination System (NPDES), 5;
  phase I ESAs and, 118
National Priorities List (NPL,
    Superfund), 10, *15*;
  brownfields and, 12;
  CERCLA and, 13–14;
  for phase I ESAs, 117;
  on revegetation, 407;
  S/S and, 209
native plants, revegetation of, 423–26
natural gas, as alkane, 53
Naval Air Weapons Station,
    electrokinetic remediation for,
    349
n-heptane, 64
n-hexadecane, in diesel fuel, 67
nickel (Ni), 38–39;
  cement based-S/S for, 200;
  dendrimers for, 370;
  for plants, 321
nickel-cadmium batteries:
  cadmium from, 28;
  organic S/S for, 205
nicotinamide adenine dinucleotide
    phosphate (NADP), 266, *267*
nitrate:
  for bioremediation, 281;
  soil amendments and, 419;
  soil flushing of, 173
nitric acid, 174
nitrilotriacetic acid (NTA), 174–77, *175*
nitrobenzene, *70*

nitrogen (N):
  AMF and, 415;
  anaerobic respiration and, 266–67;
  in humus, 85;
  legumes and, 423;
  in manures, 414;
  for paper mill sludge, 414;
  revegetation and, 407;
  selenium and, 39–40
nitrogen oxides, 4
Nixon, Richard, 5
nonaqueous phase liquids (NAPLs):
  bioremediation of, 275;
  isolation and, 151;
  phase II ESAs and, 132;
  phytoremediation of, 323;
  slurry walls for, 156;
  soil phase I ESAs and, 120–21;
  soil water-holding capacity and, 104;
  SVE for, 223, 230–31.
  See also dense nonaqueous phase
    liquids; light nonaqueous phase
    liquids
nontronite, 83
NPDES. *See* National Pollutant
    Discharge Elimination System
NPL. *See* National Priorities List
NPs. *See* nanoparticles
NTA. *See* nitrilotriacetic acid

ocean dumping, 4
octane number, for gasoline, 64
off-gases, 208;
  from bioremediation, 282;
  from hydrocarbons, 185;
  from ISV, 361;
  from RBCs, 293
Oil Pollution Act (OPA), 5, 51
oil spills, 51;
  phase II ESAs for, 134–35
OPA. *See* Oil Pollution Act
organics:
  ISV for, 360;
  phytoremediation of, *328*, 328–30;
  PRBs for, 245–47;

soil flushing for, 178–81, *179*, *180*;
S/S of, 202–3, 210
osmotic potential, 264
Osorb™, 371–72
oxidants, for SVE, 230–31
oxidation reduction (redox):
cadmium and, 29;
cement-based S/S and, 201;
chromium and, 31;
mercury and, 35;
PRBs and, 247;
for remediation, 348–49, *352*;
selenium and, 39;
of soils, 81;
ZVI and, 254
oxides:
arsenic and, 26;
cadmium and, 29;
cement-based S/S for, 201;
of lead, 32;
selenium and, 40;
ZVI and, 254
oxygen:
for bioremediation, 263, 269, 279;
in plants, *318*
oxyhydroxides, ZVI and, 254
ozone:
hydrogen peroxide and, 280;
NAAQS and, 4

P. *See* phosphorus
PAHs. *See* polycyclic aromatic hydrocarbons
palladium:
for BNPs, 367;
for SOMS, 371
PAMAM. *See* poly(amidoamine)
paper mill sludge, as soil amendment, *409*, 414
paraffins, 63;
in diesel fuel, 67;
for organic S/S, 203
Parker, R., 182
Parris Island Marine Corps facility, nanoremediation at, 374–76

particle density, of soils, 96
particulate matter, 3
Pb. *See* lead
PCBs. *See* polychlorinated biphenyls
PCDD. *See* polychlorinated dibenzo-p-dioxins
PCDF. *See* polychlorinated dibenzofurans
PCE. *See* perchloroethylene
PCP. *See* pentachlorophenol
PEI. *See* polyethylenimine
pentachlorophenol (PCP), 212–13;
fungi and, 92;
S/S for, 211, 213–15, *214*
pentane, 221
perchloroethylene (PCE), 56;
metalloporphyrinogens for, 371;
nanoremediation for, 374–76;
phytoremediation of, *335*;
PRBs for, 245–47
percolation, 98
permeability, of soils, 101, *101*;
phytoremediation and, 323;
SERP and, 351;
SVE and, 232, 356–57
permeable reactive barriers (PRBs), 241–60, *242*, *243*, 395, *395*;
for adsorption, 251–52;
advantages of, 254–55;
disadvantages of, 255;
groundwater and, 243;
installation and configuration, 242–44;
for organics, 245–47;
reactive media for, 252;
suitable contaminants for, 245–52, *245*;
treatment mechanisms, 244
pesticides, 2;
arsenic from, 25, 27;
dendrimers for, 370;
RBCs and, 293;
from soil flushing, 178
petrochemicals, 68–70, *69–70*
petroleum. *See* hydrocarbons

pH:
- actinomycetes and, 92–93;
- arsenic and, 26;
- for bacteria, 91;
- bioremediation and, 261, 264, 265, 277, 281, 287;
- cadmium and, 29;
- calcium carbonate and, 248;
- cement-based S/S and, 201;
- chromium and, 31;
- electrokinetic remediation and, 347–48;
- for fungi, 91;
- of humus, 86;
- of hydrous oxides, 85;
- for in situ S/S, 208;
- lead and, 34, 35;
- mercury and, 37–38;
- metal from soil flushing and, 174;
- metal precipitation and, 168;
- for PRBs, 252;
- phytoremediation and, 327;
- phytostabilization and, 331;
- revegetation and, 408;
- slurry bioremediation and, 287;
- of soils, 80–81;
- for S/S, 213;
- zinc from, 40–41

phase I ESAs, 110–27;
- ACM in, 123–26, *124*, *125*;
- aerial photographs for, *114*, 114–15, *115*;
- for hazardous materials, 118–19, *119*;
- interviews for, 126;
- LBP in, 121–22, *122*;
- lead in, 122–23, *123*;
- radon in, 121;
- report of, 127;
- Sanborn Fire Insurance Maps for, 115–16, *116*;
- site history for, 113, *113*;
- site reconnaissance for, 118–26;
- USDA soils map and, 112, *112*;
- USGS topographic map for, 110–11, *111*, *117*, 126–27;
- for USTs, 114, 119, *120*

phase II ESAs, 128–46;
- contaminant considerations for, 133–36;
- electrical resistivity for, 131;
- EM for, 130–31, *130*;
- field sampling for, 140–43, *141–42*;
- field testing for, 141–46;
- GPR for, 128–29, *129*;
- noninvasive technologies for, 128–31;
- release considerations for, 131–33;
- site considerations of, 136–40, *137–39*

phenanthrene:
- dendrimers for, 370;
- in lubricating oils, 68

phenol, 58, *70*;
- in humus, 86;
- organic S/S for, 203;
- RBCs for, 293;
- S/S for, 209

phosphate:
- arsenic and, 26;
- bioremediation and, 279;
- cadmium and, 29;
- lead and, 33, 34, 35;
- mercury and, 38–39;
- for nickel toxicity, 38–39;
- PRBs for, 248

phosphoric acid, 174

phosphorus (P):
- AMF and, 415;
- barium and, 27;
- in humus, 85;
- lead and, 34–35;
- for paper mill sludge, 414;
- revegetation and, 407;
- selenium and, 39–40;
- zinc and, 40–41

photochemical smog, 3, *3*

photoionization detectors (PIDs), 143, 144–45, *145*;
- for landfarming, 300–301

photosynthesis:
  mercury and, 37–38;
  nickel and, 38–39
phototrophs, 263
phthalic, 70
phyllosilicates, 82
phytoremediation (phytoextraction, phytodegradation), 317–43, 397–98, *398–400*;
  for Aberdeen Proving Grounds, 332–36, *334, 335*;
  advantages and disadvantages of, 334, *335*;
  agronomic practices for, 326–27;
  benefits of, 327;
  disadvantages of, 327–28;
  genetic engineering for, 323–24;
  of metals, 319–28, *323, 328*;
  of organics, *328*, 328–30;
  site preparation for, 325–26;
  soils and, 325
phytostabilization, 331
Pichtel, J., 174
Pickering, W. F., 33
PIDs. *See* photoionization detectors
Pilon-Smits, E. A. H., 323
Plant Materials Centers (PMCs), of USDA, 426
plants:
  arsenic and, 26–27;
  barium in, 28;
  cadmium in, 29–30;
  chromium in, 31;
  lead in, 34–35;
  mercury in, 36–38;
  nickel in, 38–39;
  oxygen in, *318*;
  selenium in, 39;
  water in, *318*;
  zinc in, 40–41.
  See also phytoremediation; revegetation
plastics, 2;
  cement-based S/S for, 202;
  for SVE, 224
PLM. *See* polarized light microscopy

PMCs. *See* Plant Materials Centers
pneumatic fracturing, *355*, 355–56
Pociecha, M., 177
polarized light microscopy (PLM), 125
poly(amidoamine) (PAMAM), 370
polychlorinated biphenyls (PCBs), 7;
  bioremediation for, 262, 302;
  cement based-S/S for, 200;
  fungi and, 92;
  from soil flushing, 178;
  in South Korea, 17
polychlorinated dibenzofurans (PCDF), 211
polychlorinated dibenzo-p-dioxins (PCDD), 211
polycyclic aromatic hydrocarbons (PAHs), 57–58;
  dendrimers for, 370;
  fungi and, 92;
  landfarming for, 300–301, 306–9, *308, 311*;
  in lubricating oils, 67–68;
  phytoremediation of, 332;
  from soil flushing, 178;
  S/S for, 211
polyethylene, 177;
  for organic S/S, 202, 203.
  See also high-density polyethylene
polyethylenimine (PEI), 177, 370
polymers:
  for organic S/S, 205;
  for soil flushing of metals, 177
polypropylene, 161
polyvinyl chloride (PVC), 161;
  for landfarming, 296;
  for SVE, 224, *225*, 226
population growth, 2
porosity, of soils, 97, 232;
  electrokinetic remediation and, 346;
  SVE and, 232
porous-media flow, 103
potassium (K):
  barium and, 27;
  revegetation and, 407
potentially responsible parties (PRPs), CERCLA and, 12–13

POTWs. *See* Publicly Owned Treatment Works
pozzolanic reactions, for cement-based S/S, 201
PRBs. *See* permeable reactive barriers
propane:
   as alkane, 53;
   PRBs for, *245*
PRPs. *See* potentially responsible parties
psychrophiles, 89
Publicly Owned Treatment Works (POTWs), 5, 398;
   for soil amendments, 408
pump-and-treat technology:
   electrokinetic remediation and, 349;
   for groundwater, 165–68;
   for SVE, 223, 233–35.
   See also permeable reactive barriers
PVC. *See* polyvinyl chloride
pyrene, 211
pyrolysis, 360

radioactive waste:
   in landfills, 18;
   in ocean dumping, 4;
   organic S/S of, 202;
   PRBs for, 396–97
radio-frequency heating (RFH), 356–58, *357*
radionuclides, electrokinetic remediation for, *246*, 345–50, *347*, *350*
radius of influence, 225
radon, 121
Raghu, V., 28
Rainbow Disposal site, SERP at, 354
rainfall infiltration rate, 138;
   SVE and, 233
RBCs. *See* rotating biological contactors
RCRA. *See* Resource Conservation Recovery Act
reactors, for slurry bioremediation, 289–90
reagents:
   for bioremediation, 278, 279–81;
   for PRBs, 395;
   for S/S, 209

recharge, of groundwater, 101
Record of Decision (ROD):
   for bioremediation, 302;
   for S/S, 209;
   for SVE, 231–33
recycling, 4–5
redox. *See* oxidation reduction
remedial investigation/feasibility study (RI/FS), for S/S, 209
removal action, CERCLA and, 14
resonance, of aromatic hydrocarbons, 57
Resource Conservation Recovery Act (RCRA), 5–6, 18;
   on bioremediation, 266, *267*;
   capping systems for, 161;
   Corrective Action, 9;
   groundwater and, 165;
   for phase I ESAs, 117, *117*;
   on phytoremediation, 322;
   on revegetation, 407;
   on slurry bioremediation, 285;
   soil amendments and, 419;
   TSD of, *117*
Resource Recovery Act of 1970, 5
revegetation, 407–34;
   of grasses, 421–23, *422*;
   of legumes, 423, *424*, 425;
   of native plants, 423–26;
   plant selection for, 419–21;
   soil amendments for, 408–19, *409–12, 416, 418*;
   of trees, *426*, 426–27, *428–30*
RFH. *See* radio-frequency heating
RI/FS. *See* remedial investigation/feasibility study
Rivers and Harbors Act of 1899, 5
ROD. *See* Record of Decision
rotating biological contactors (RBCs), *291*, 291–94, *294*
Rugh, C. L., 325

S. *See* sulfur
Safe Drinking Water Act (SDWA), 5
Sanborn Fire Insurance Maps, for phase I ESAs, 115–16, *116*

sand:
    groundwater and, 102;
    for soil amendments, *411*
saponite, 83
SARA. *See* Superfund Amendments and Reauthorization Act
saturated zone, *98*, 99, 335
SDWA. *See* Safe Drinking Water Act
selenate, 39;
    cadmium and, 29
selenides, 39–40
selenite, 39;
    cadmium and, 29;
    soil flushing of, 173
selenium (Se), 39–40;
    in groundwater, 249;
    PBRs for, 249
Selma Pressure Treating (SPT), S/S at, 210–15, *213–15*
semivolatile organic compounds (SVOCs):
    RFH for, 356–57;
    SVE for, 394
SERP. *See* Steam Enhanced Recovery Process
sewage sludge:
    cadmium in, 30;
    chromium from, 30;
    mercury in, 35
sheet piling, 159–60, *160*
short-circuiting EZVI, 375
silicate clays. *See* clays
Silicate Technology Corporation (STC), 211–12, *213*
Simon, F.-G., 251
slurry, for bioremediation, 284–89, *285*, *286*, 302–5, *303*, *305*
slurry walls:
    for isolation, 155–58, *155–58*;
    for SVE, 235–36
smectites, 83
smelting:
    arsenic from, 25;
    lead from, 32;
    zinc from, 40

smog, 3, *3*
SMP test, 281
SOD. *See* soil oxidant demand
soil amendments:
    biosolids as, *409*, 413;
    CAFOs for, 408, *409*;
    FGD as, 414–15;
    fly ash as, 414;
    fungi for, 415–16, *416*;
    manures as, *409*, 413–14;
    for revegetation, 408–19, *409–12*, *416*, *418*
soil flushing, 171–98, *172*, 389–90, *389–90*;
    by-products of, 185–86;
    contaminant forms and concentrations for, 180–82;
    effects on soil properties, *183*, 182–83;
    environmental impacts of, 185;
    good and bad of, 185–86;
    at King of Prussia Technical Corporation, 187–91, *189*, *190*;
    for metals, 173–77, *175*;
    for organics, 178–85, *179*, *180*;
    site preparation for, 184–85;
    subsurface drains for, 183–84;
    vehicle traffic and, 184;
    vertical and lateral contaminant distribution from, 183
soil oxidant demand (SOD), 350–51
soils:
    actinomycetes in, 92–93, *93*;
    arsenic in, 27;
    bacteria in, 88–90, *89*;
    barium in, 27–28;
    biological component of, 86–93;
    buffering capacity of, 81;
    bulk density of, 97;
    cadmium in, 29–30;
    for capping of isolation, 162–63;
    chromium in, 30–32;
    colloidal fraction of, 82–86;
    fungi in, 91–92, *91*;
    groundwater and, 98–104;

hydraulic conductivity of, 101–4, *102*, 138, 232;
lead in, 32–35, *33*;
mercury in, 35–38, *37*;
nickel in, 38–39;
PAHs in, 58;
particle density of, 96;
particle size distribution in, 79, *80*;
permeability of, 101, *101*, 232, 325;
phase I ESAs for, 120, *121*;
pH of, 80–81;
physical properties of, 93–97;
phytoremediation and, 325;
porosity of, 97, 232;
redox of, 81;
remedial action for, 79–106;
selenium in, 39–40;
SERP for, 351–54, *353*;
structure of, 96, *96*;
texture of, 94–95, *95*;
water-holding capacity of, 104;
zinc in, 40–41.
See also permeability; porosity; *specific applicable topics*
soil vapor extraction (SVE), 221–40, 222, 355–56, 391–94, *392*, *393*;
biofilters for, 231;
carbon for, 230;
catalytic oxidation units for, 230;
condensation and, 229–30;
diffuser stacks for, 231;
at Fairchild Superfund site, 235–38, *237*;
field design for, *225–29*, 225–31;
geomembranes for, 228–29;
at Hanford Site, 233–35;
oxidants for, 230–31;
site characterization for, 223–25, *224*;
surface seals for, *227*, 228;
technical considerations, 231–33;
vapor combustion units for, 230
soil washing, 172, *173*
solidification and stabilization (S/S), 199–219, *381*, 391, *392*;
cement for, 202–3, *204*;
ex situ treatment, 204–8, *205–8*;
in situ treatment, 206–10, *207–10*;
of organics, 202–3, 210;
at Selma Pressure Treating (SPT), 210–15, *213–15*;
at SPT, 210–15;
technical considerations for, 208–9;
technologies for, 209–10
Solid Waste Disposal Act of 1965, 5
solubility, of hydrocarbons, 71
solutions for, 172–73
SOMS. *See* swellable organically modified silica
sour crude, 61
South Korea, PCBs in, 17
specific gravity, of hydrocarbons, 71
sprinkler irrigation, for groundwater, 168
SPT. *See* Selma Pressure Treating
S/S. *See* solidification and stabilization
[S,S]ethylenediaminedisuccinic acid (EDDS), 185
STC. *See* Silicate Technology Corporation
Steam Enhanced Recovery Process (SERP), 351–54, *353*
Steele, M. C., 174
Steinbeck, John, 345
styrene, *70*
subsurface barriers, to isolation, 154–60, *155–60*
subsurface drains:
for groundwater, 152–54, *153*, *154*;
for soil flushing, 183–84
Subtitle C, of RCRA, 5–6, 66
sulfates:
anaerobic respiration and, 266–67;
for bioremediation, 281;
cadmium and, 29;
cement-based S/S for, 201;
lead and, 33;
soil pH and, 81
sulfides:
anaerobic respiration and, 266–67;

cement-based S/S for, 202, 202;
  lead and, 32, 33;
  metal precipitation from, 168;
  selenium from, 39;
  zinc and, 40–41
sulfur (S):
  in humus, 85;
  in hydrocarbons, 61;
  for landfarming, 296;
  for selenium toxicity, 39–40;
  soil pH and, 81;
  SVE for, 394
sulfur dioxide:
  FGD and, 415;
  lead and, 35;
  NAAQS and, 4
sulfuric acid, 174
Superfund. *See* National Priorities List
Superfund Amendments and Reauthorization Act (SARA), 7;
  ESAs and, 108;
  phase I ESAs and, *117*;
  on remedial actions, 16
surface seals, for SVE, 228
surfactants:
  for chromium, 31–32;
  for organic from soil flushing, 178–80, *179*;
  for soil flushing, 173;
  for soil flushing of metals, 176
SVE. *See* soil vapor extraction
SVOCs. *See* semivolatile organic compounds
sweet crude, 61
swellable organically modified silica (SOMS), 371–72
synthetic membrane caps, 161, *162*

TCA. *See* 1,1,2-trichloroethane
TCAA. *See* trichloroacetic acid
TCE. *See* trichloroethylene
TCLP. *See* toxicity characteristic leaching procedure
TCP. *See* tetrachlorophenol

technetium, 373;
  in groundwater, 249
Tessier, A., 181
tetrachloroethylene, 56
tetrachlorophenol (TCP), 212–13
tetraethyl lead, 64
tetramethyl lead, 33, 64
Thirumalai Nivas, B., 176
tillage:
  bacteria and, 90–91;
  for landfarming, 299, 307–8;
  oxygen and, 263
tin, 3
toluene, 60, *70*;
  in gasoline, 64;
  from soil flushing, 178;
  SVE for, 221.
  See also benzene, toluene, ethylbenzene, and xylene
Torres, L. G., 176
tortuosity, electrokinetic remediation and, 348
total petroleum hydrocarbons (TPH), 140;
  SVE for, 224
toxicity characteristic leaching procedure (TCLP), for S/S, 208, 209, 210, 211
Toxic Substances Control Act (TSCA), 7
TPH. *See* total petroleum hydrocarbons
trace metals, 24, *25*;
  for bioremediation, 265;
  EDTA and, 177;
  for plants, 319
transpiration:
  nickel and, 38–39;
  phytoremediation and, 325
treatment, storage, and disposal (TSD), 6, *117*
trees, revegetation of, *426*, 426–31, *428–30*
Triano, R. M., 370
trichloroacetic acid (TCAA), 336
1,1,2-trichloroethane (TCA):

phytoremediation of, *337*;
SVE for, 237
trichloroethylene (TCE), 56;
  BNPs for, 367;
  EZVI for, 368;
  metalloporphyrinogens for, 371;
  phytoremediation of, *335*, 336;
  PRBs for, 245–47;
  redox for, 351;
  from soil flushing, 178;
  SOMS for, 371;
  SVE for, 237
trickling filters, *294*, 294–95
trivalent chromium, 30
TSCA. *See* Toxic Substances Control Act
TSD. *See* treatment, storage, and disposal

UEL. *See* upper explosive limit
unconfined aquifers, 99, *99*
underground storage tanks (USTs), 6, 6–7;
  aerial photographs of, 114;
  bioremediation and, 277;
  cost of remediation for, 385;
  electrokinetic remediation for, 346;
  EM for, 130–31;
  GPR for, 129;
  phase I ESAs for, 114, 117, 119, *120*;
  PIDs for, 145;
  sheet piling for, 160;
  soil flushing for, 184;
  SVE for, 394
Union Carbide, 7
United States Department of Agriculture (USDA):
  PMCs of, 426;
  soils map of, phase I ESAs and, 112, *112*;
  on soil texture, 94–95
United States Geological Survey (USGS), topographic map of, 110–11, *111*, *117*, 126–27
unsaturated zone, 138

upper explosive limit (UEL), 74, *75*
uranium, PRBs for, 251
urethane, 161
USDA. *See* United States Department of Agriculture
Used Oil Management Standards, of EPA, 68
USGS. *See* United States Geological Survey
USTs. *See* underground storage tanks

V. *See* vanadium
vadose zone, *98*, 99, 138;
  SVE in, 221, 232, 354–56
Valley of the Drums, Kentucky, 8, *9*
vanadium (V), 319
vapor combustion units, for SVE, 228
vapor pressure, of hydrocarbons, 73, *74*
VC. *See* vinyl chloride
vehicle traffic, soil flushing and, 184
vermiculites, 83
vertical auger application, for in situ S/S, 206, *206–8*
very low-density polyethylene (VLDPE), 161
vinyl chloride (VC):
  BNPs for, 367;
  cement based-S/S for, 200;
  nanoremediation for, 375
viscosity:
  of hydrocarbons, 73;
  ISV and, 360;
  phase II ESAs and, 136
VLDPE. *See* very low-density polyethylene
volatile organic components (VOCs), 140–41;
  bioremediation for, 302;
  phytoremediation of, 333;
  pneumatic fracturing for, 355–56;
  RFH for, 356–58;
  SOMS for, 371;
  SVE for, 222, 227–29, 393

Wada, K., 34
water:

for bioremediation, 265–66;
in plants, *318*
water-holding capacity, of soils, 104
water table, *98*, 99;
phase I ESAs for, 120
weathering:
of gasoline, *135*;
phase II ESAs and, 132
West, Mae, 385
wetlands:
constructed, 166, *167*, 318;
phase I ESAs for, 120
Whitman, Walt, 365
witherite, 27

Xing, B., 378
Xu, Y., 369
xylene, 58, *70*;
in gasoline, 64;
from soil flushing, 178;
S/S for, 209;
SVE for, 221, 237.
See also benzene, toluene, ethylbenzene, and xylene

yeasts:
for bioremediation, *276*;
mercury and, 36

Zeng, Q., 177
zeolite, 251, *252*;
for PRBs, 396
zerovalent iron (ZVI), 242, 244, 246;
EZVI, 367–68, *368*, 347–75, *376*;
for nanoremediation, 367;
for PRBs, 395;
for SOMS, 372
zinc (Zn), 40–41;
cadmium and, 29;
cement-based S/S for, 200;
from incinerators, 3;
metal from soil flushing of, 174;
nanoremediation for, 378;
PBRs for, 252;
phytoextraction of, 327;
for plants, 319
zone of influence, 225–26
ZVI. *See* zerovalent iron

About the Author

John Pichtel is a professor of natural resources and environmental management at Ball State University in Muncie, Indiana. He is a Certified Hazardous Materials Manager. Dr. Pichtel holds memberships in the Institute of Hazardous Materials Managers, the Sigma Xi Scientific Society, the International Association of Arson Investigators, and the Indiana Academy of Science. Dr. Pichtel received a PhD in environmental quality/agronomy from Ohio State University in 1987. At Ball State University, he conducts research in remediation of contaminated sites, environmental chemistry, and hazardous waste management. Dr. Pichtel was twice awarded a Fulbright Fellowship. He was appointed docent in remediation science at Tampere University of Technology in Finland. He has served as a consultant in field remediation projects and has conducted environmental assessments and remediation research in the United States, the United Kingdom, Ireland, Finland, and Poland. Dr. Pichtel enjoys painting, sculpting, gardening, and study of world history.